Springer Series in Microbiology

Editor: Mortimer P. Starr

"*Actually, with me it's not so much the heat, it's that awful sulphur smell.*"

This book is dedicated to all those who suffered through a Yellowstone "hell" so that our research could be completed.

Thomas D. Brock

Thermophilic Microorganisms and Life at High Temperatures

Springer-Verlag New York Heidelberg Berlin

Thomas D. Brock
University of Wisconsin, Madison
Department of Bacteriology
1550 Linden Drive
Madison, Wisconsin 53706/USA

Series Editor:
Mortimer P. Starr
University of California
Department of Bacteriology
Davis, California 95616/USA

With 195 figures.

The figure on the frontcover is a photograph of the runoff channel of Grand Prismatic Spring, Yellowstone National Park, showing the extensive development of blue-green algae and photosynthetic bacteria.

Library of Congress Cataloging in Publication Data

Brock, Thomas D
   Thermophilic microorganisms and life at high temperatures.
   (Springer series in microbiology)
   1. Micro-organisms, Thermophilic.   2. Microbial ecology.   3. Hot
spring ecology—Yellowstone National Park.   I. Title.
QR107.B76   576   78-6110

Printed in the United States of America.
9 8 7 6 5 4 3 2 1

ISBN 0-387-90309-7   Springer-Verlag New York
ISBN 3-540-90309-7   Springer-Verlag Berlin Heidelberg

Preface

From 1965 through 1975, I conducted an extensive field and laboratory research project on thermophilic microorganisms. The field work was based primarily in Yellowstone National Park, using a field laboratory we set up in the city of W. Yellowstone, Montana. The laboratory work was carried out from 1965 through 1971 at Indiana University, Bloomington, and subsequently at the University of Wisconsin, Madison. Although this research project began small, it quickly ramified in a wide variety of directions. The major thrust was an attempt to understand the ecology and evolutionary relationships of thermophilic microorganisms, but research also was done on biochemical, physiologic, and taxonomic aspects of thermophiles. Four new genera of thermophilic microorganisms have been discovered during the course of this 10-year period, three in my laboratory. In addition, a large amount of new information has been obtained on some thermophilic microorganisms that previously had been known. In later years, a considerable amount of work was done on Yellowstone algal-bacterial mats as models for Precambrian stromatolites. In the broadest sense, the work could be considered geomicrobiological, or biogeochemical, and despite the extensive laboratory research carried out, the work was always firmly rooted in an attempt to understand thermophilic microorganisms in their natural environments. Indeed, one of the prime motivations for initiating this work was a view that extreme environments would provide useful models for studying the ecology of microorganisms.

As a result of this 10-year research project, I published over 100 papers. Although these were all published in reputable and widely available journals, the work has perforce been scattered. Because of a current widespread interest in thermophilic microorganisms, especially from the biochemical and evolutionary point of view, it seemed appropriate to summarize not only our own work, but that of others, with special emphasis on those organisms living under the most extreme environments. I hope that this book will not only be a useful reference to past work on these organisms, but will provide some insight into the directions future research might take, especially for field-oriented work. To this end, I have attempted to give detailed maps of well-studied Yellowstone thermal areas, and to provide good photographs of habitats that have been widely studied.

My own research work could not have been done without the collaboration of a large number of people. Because there may be some interest in the social history of a research effort such as this, I have provided a few pages of personal history in a terminating chapter, and credit is given there to the individuals who have been involved in this work. My own research has been supported financially by the National Science Foundation, the Atomic Energy Commission (subsequently the Energy Research and Development Agency and the Department of Energy), Indiana University, and the Wisconsin Alumni Research Foundation. Because of the faith they showed me at the inception of this work, I am especially grateful to the National Science Foundation, without whose initial support this research project would probably never have gotten off the ground.

May, 1978   THOMAS D. BROCK

Contents

Contents

CHAPTER 7

The Genus *Chloroflexus*

CHAPTER 8

The Thermophilic Blue-green Algae

CHAPTER 9

The Genus *Cyanidium*

Contents xi

CHAPTER 14

Some Personal History 441

Personnel Involved in Yellowstone Research Project 444
Bibliographic Note 445
Public Service 446
Movies and Television 446
The West Yellowstone Laboratory 448
The Decision to Quit 449

Subject Index 451

Chapter 1
Introduction

There are several ways in which the work to be discussed in this book could be characterized. Studies that attempt to relate the characteristics of organisms to their ecology are best called *physiological ecology* and much of the work to be discussed in this book fits under this heading. Another major theme of this book is *biogeochemistry,* the study of chemical processes taking place in nature which are being carried out by living organisms. At one time, the whole body of work might have been called *geobiology;* indeed, the eminent Dutch scientist Baas-Becking once wrote a book with this title, which included the kinds of things I am going to discuss here. And finally, since much of the work deals with the structure, taxonomy, and biochemistry of microorganisms, it could (perhaps most aptly) be called *general microbiology*.

I was fortunate when my work began that a wide variety of techniques were available for the sophisticated study of natural ecosystems. Many of these techniques were not available to earlier workers in the Yellowstone habitats, which accounts in part for the tremendous amount of new information we were able to obtain rapidly. The most important technique was radioactive tracing, using ^{14}C, ^{32}P, ^{35}S, and ^{3}H. Counting by gas flow or liquid scintillation counting, or by microscopic autoradiography, made possible a wide variety of experiments. A Nuclear-Chicago gas flow detector and scaler was used for most work, but in later years we had a Beckman Beta-Mate liquid scintillation counter. I also had available a very good microscope, a Carl Zeiss Universal, with phase, fluorescence, and Nomarski optics. Spectrophotometers are not especially new, but had not been used in any previous Yellowstone work. We had a B and L Spectronic 20 for routine chlorophyll and chemical assays, and a Beckman DB-G spectrophotometer, with recorder, for scans. Temperature measurements in the field were made with Yellow Springs Instrument Co. thermistor probes and bridges, and over the 10 years we wore out four YSI bridges and innumerable probes. pH Measurements were made with Corning or Orion pH meters with combination glass electrodes. For one summer, we had a Packard-Becker model 419 gas chromatograph with flame photometric detector, for use in our sulfur work. Aside from these pieces of equipment, most of the work was done with simple glassware, plastic ware, test tubes, tanks of gases, immersion heaters for water baths and aquaria, etc.

Extreme Environments

What is an extreme environment? It is not appropriate to define it anthropocentrically, as we should be the first to admit that human life is not everywhere possible. More appropriate is its definition as a condition under which some kinds of organisms can grow, whereas others cannot. If we accept this definition it means that an environmental extreme must be defined taxonomically. Instead of looking at single species, or groups of related species, we must examine the whole assemblage of species, microbial and multicellular, living in various environments. When we do this we find that there are environments with high species diversity and others with low species diversity. In some environments with low species diversity we find that whole taxonomic groups are missing. For instance, in saline and thermal lakes there are no vertebrates and no vascular plants, although they may be rich in microorganisms and very high in the numbers of organisms of the species that do live there. In many of the most extreme conditions we find conditions approaching the pure culture, with only a single species present.

Such environments are clearly of enormous interest to the microbiologist, especially one interested in the ecological and evolutionary relationships of organisms. In such biologically simple environments experimental microbial ecology is most easy to carry out, since microbial interactions are minimized or absent. Compare the problems of the soil microbiologist, faced with thousands of species in a mere handful of dirt, with those of the thermal microbiologist, who can often lift from his hot waters grams of essentially monotypic bacterial protoplasm!

Since by definition many organisms cannot grow in the extreme environment, we can ask two sets of questions: (1) how does the environmental extreme affect organisms that cannot tolerate it? (2) How is it possible for organisms which *are* adapted to overcome the effects of the extreme factor? One point which we must keep in mind, however, is the other environmental factors of the extreme environment, which must be adequate for the support of life. For instance, the bottom of the ocean floor not only has a high hydrostatic pressure but is also dark and low in organic nutrient concentration. The organisms that we find there will thus be those that are adapted to all the relevant factors, of which high hydrostatic pressure is only one. For this reason it is desirable to study habitats in which a gradient exists for the factor of interest, from extreme to normal, with all other factors remaining the same.

Environmental Extremes

The most commonly considered environmental extremes are for the factors temperature, redox potential (Eh), pH, salinity, hydrostatic pressure, water activity, nutrient concentration, and electromagnetic and ionizing radiations. A brief general review of these factors has been given by Vallentyne (1963). For virtually all of these factors, it is found that organisms with simple structures can grow under more extreme conditions than organisms with more complicated structures, hence the preeminence of microorganisms in extreme environments. There are three ways in which an organism can adapt to an environmental extreme: (1) it can develop a mechanism for excluding the factor from its structure; (2) it can develop a mechanism for detoxifying the factor; (3) it can learn to live with the factor. In some cases, it is the latter situation that obtains, and in fact it sometimes occurs that evolutionary adaptation is such that the organism actually becomes dependent on the factor which for other organisms is lethal. In this chapter we will take some of the environmental factors listed above in turn, and consider the natural habitats where the factor is relevant, and the biochemical mechanisms involved in adapting to it. It should be emphasized, however, that an environmental factor might not always affect an organism directly, but might do so indirectly through its effects on other environmental factors. High temperature, for instance, reduces greatly the solubility of O_2, reduces the viscosity of water, and increases its ionization. We must be certain, therefore, that the effects we are measuring are not due to such indirect effects of the environmental variable.

Temperature as an Environmental Extreme

Temperature is one of the most important environmental factors controlling the activities and evolution of organisms, and is one of the easiest variables to measure. Not all temperatures are equally suitable for the growth and reproduction of living organisms, and it is apt, therefore, to consider which thermal environments are most fit for living organisms. For such a study, high-temperature environments are of especial interest, in that they reveal the extremes to which evolution has been pushed. The high-temperature environments most useful for study are those associated with volcanic activity, such as hot springs, since these natural habitats have probably existed throughout most of the time in which organisms have been evolving on earth.

Although the words *high temperature* will often be used in this book without qualification, the viewpoint of the observer or the group of organisms under discussion will often determine if a given temperature is to be considered high. Thus, a temperature of 50°C would be considered critically high when referring to a multicellular animal or plant but would be considered moderate or even rather low if certain thermophilic bacteria were under discussion. Where necessary, I will qualify the word *high* by appending actual numbers. Another point which should be emphasized at the outset is that one must distinguish between temperatures an organism can tolerate and those at which it can carry out its whole life processes. Thus certain invertebrates can survive exposure to temperatures approaching boiling, although they cannot grow at temperatures above 50°C (Carlisle, 1968; Hinton, 1960). No animal has been found that can carry out its complete life cycle at a temperature over 50°C. In terms of evolutionary success in high temperature environments only organisms able to carry out their complete life cycles are of interest.

Although there may be some doubt that the cytoplasm of an organism in high salt or low pH is subject to the environmental extreme, there is no doubt that microorganisms are isothermal with their environments. A thermophile cannot be a biological submarine with an air-conditioning system, since there is no way for a heat pump to operate in an isothermal environment. (On the other hand, it is possible for organisms to be warmer than their environments if they contain their own internal heating systems, driven by metabolic energy; witness the whale and the submarine.) Thus thermophiles cannot survive at high temperatures because of a thick layer of insulation. Because all of this book deals with high temperature, further discussion of temperature extremes will be reserved for later chapters.

Low pH

"One man's poison is another man's meat" well characterizes the situation for those organisms that flourish in acid environments. In habitats where other organisms rapidly die, these organisms thrive, often multiplying to high density. And the diversity of organisms found in these acid environments is surprisingly high, including not only bacteria, but also algae, protozoa, and invertebrates. Upon understanding the environment, this biological diversity seems more reasonable, for environments made acidic with sulfuric acid are quite widespread on earth, providing many opportunities for evolutionary adaptation.

Acidity is more common in nature than alkalinity, because acidity develops in oxidizing environments, and the biosphere is predominantly aerobic. Sulfuric acid arises from the oxidation of sulfides such as hydrogen sulfide (H_2S) and pyrite (FeS_2) (Dost, 1973). Sulfides are very common in volcanic and other geothermal habitats, in bogs and swamps, and in the sea. In the absence of oxygen, sulfides are stable, but if conditions change and air enters, rapid oxidation, brought about either spontaneously or by action of special sulfur-oxidizing bacteria, quickly leads to sulfuric acid production. As one example, vast areas of low-lying soils along the southeastern coast of the United States (South Carolina, Georgia, North Carolina) were at one time flooded by the sea, and are rich in sulfides. When these soils are drained for agriculture, acidity rapidly develops, and the soils become completely unsuitable for plants (Dost, 1973). Other areas where such acid soils occur extensively are in The Netherlands, Africa, and Southeast Asia. In The Netherlands such soils are called "cat clays" (kattekleigronden, in Dutch), the word "katte" meaning in the Dutch vernacular any harmful, mysterious influence or quality. Such completely barren soils can be a blight on the landscape, and in underdeveloped countries where every bit of agricultural land is needed they present a real challenge for the agricultural scientist.

Perhaps the most well-known situation of sulfuric acid intrusion is "acid mine drainage." This is found primarily in coal-mining regions, but occurs to a small extent from copper, lead, zinc, and silver mines. In eastern U.S.A., over 10,000 miles of streams and rivers are affected by acid mine drainage, and are unsuitable for human use. Acid mine drainage develops in those coal-mining regions where the coal and overlying rock are high in sulfides, predominantly pyrite. As long as these sulfides are buried beneath the earth, everything is fine, but when mining occurs, air is introduced, and oxidation of the sulfides to sulfuric acid occurs. As rainwater percolates through the earth, it leaches this acid out of the rocks, and streams and lakes that develop are highly acidic. Frequently they are also rich in iron, rendered soluble by the acid waters, and this iron subsequently precipitates and forms a scum for miles around in areas affected by acid mine drainage.

Another form of acid pollution which has recently become known is the formation of acid rains as a result of air pollution. In many parts of the world, rains are falling that are essentially dilute solutions of sulfuric acid. As a result, lakes and rivers are becoming acidic, with important biological consequences. Acidification of rain occurs because of sulfur dioxide pollution of the atmosphere. When sulfur-rich coal or oil is burned to make electricity, the sulfur is oxidized to sulfur dioxide, which spews out into the atmosphere unless special controls are instituted. In clouds or moist air this sulfur dioxide oxidizes to sulfuric acid, which dissolves in the moisture and makes it acidic. Acid rains have been especially noted in Scandinavia and New England, but occur wherever excessive sulfur dioxide pollution occurs.

Seawater is slightly alkaline, with a pH of 8, and many lakes and river waters are just about neutral, with a pH of 7. Soft water lakes are usually slightly acidic, pH 5 or 6, and many useful agricultural soils also have pH values in this area. Tomato juice has a pH around 4, and vinegar around 3; lemon juice is even more acidic and has a pH around 2. Stomach fluids, which are quite acidic, have pH values between 1 and 2. (However, the acidity of fruit juices is not due to sulfuric acid, but to organic acids such as citric acid, acetic acid, and lactic acid, and the acid of stomach juices is hydrochloric acid.)

The most dramatic development of sulfuric acid environments occurs in geothermal areas. Volcanic gases, coming from deep within the earth, are often rich in hydrogen sulfide, and as this gas reaches the surface of the earth it meets with the oxygen of the air and oxidizes, first to elemental sulfur, and subsequently to sulfuric acid. The classic habitat of this type is Solfatara (the Forum Vulcani of the Romans), a small steaming volcanic crater along the Bay of Naples near Pozzuoli, familiar to tourists for hundreds of years and popular today as a camping site. In Italian, solfatara means "sulfur mine," so-called because sufficient crystalline elemental sulfur was present so that at one time it was mined, and the name solfatara has subsequently been given to the many other sulfur-rich acid geothermal areas around the world. There are many solfataras in Yellowstone National Park, most notably at Roaring Mountain and the Norris Geyser Basin. Solfataras are most easily recognized by the crumbling and bleached nature of the rocks. Sulfuric acid corrosively attacks the rock fabric and causes the rocks to disintegrate into crumbling bits. Elements such as iron, which give rocks their normal color, are leached out, leaving behind a whitish residue containing predominantly quartz, the only mineral stable in very acid conditions. Often the rocks are so destroyed by acid that walking in solfataras is unsafe; beneath a thin surface crust the earth is hollow and hot, and one can easily thrust a foot through into a steaming pool of hot sulfuric acid.

Yet in all these diverse kinds of acidic environments organisms not only live, but flourish. Some, in fact, will live nowhere else. (See Chapters 5, 6, 9 and 12)

Sulfuric acid environments may be as acidic as any environment on earth. Values in solfatara soils are usually pH 2 or below, and acid mine drainage usually has a pH between 2 and 3, but some volcanic crater lakes are even more acidic, with values less than 1. I have even had water samples from small steaming pools which have had pH values as low as 0.2. In an area where sulfuric acid leaches into a hot soil, the evaporation of water may lead to an unusual concentration of the sulfuric acid that is left behind, and we once measured a pH in a Yellowstone solfatara soil of 0.05; even at this low pH we found a living organism, the alga *Cyanidium caldarium*.

Saline Environments

Aqueous habitats exist with total ion concentration varying from essentially zero up to saturation. As ion concentration of the water goes up, some ions precipitate before others. The most soluble are the cations Na^+, K^+, and Mg^{2+} and the anions Cl^-, SO_4^{2+}, and HCO_3^- (or CO_3^{2-} depending on pH). It is thus not surprising that it is these ions that appear in the largest amounts in sea water and the saline lakes. Species diversity in sea water is high and the oceans can hardly be considered extreme environments. Hypersaline waters are devoid of fish and low in diversity of invertebrates; they are clearly extreme.

The precise ion composition of a saline lake is determined by the nature of the rocks surrounding the lake basin. Great Salt Lake, Utah, and many smaller lakes of the U.S. Great Basin, have essentially the ion proportions of sea water in more concentrated form (Livingstone, 1963); such lakes are sometimes called thalassohaline. The Dead Sea, Palestine, is quite different from Great Salt Lake in that it is higher in Mg^{2+} than Na^+, and is low in SO_4^{2-}. Lakes high in bicarbonate-carbonate are often called alkali lakes because they are of high pH, values over 10 being reported. One lake in British Columbia is almost pure $NaHCO_3$. Still other lakes are low in chloride and high in sulphate; one lake in British Columbia is composed mostly of magnesium sulfate (Livingstone, 1963). It should also be noted that most of these lakes differ greatly in concentrations of some of the minor but biologically important ions, such as F^-, Br^-, Fe^{3+}, Hg^+, etc.

It must be emphasized, however, that few workers have established that growth of the organisms was taking place at the salinities at which they were found. The mere presence of an organism in a water body does not indicate that it has been grown there; it may have drifted in from fresh water sources. Furthermore, even if the organism can be shown to be alive and growing, the salinity measurements must be made precisely at the site and time at which it is found. Saline lakes vary in salinity considerably throughout the year, due to variations in freshwater drainage. Great Salt Lake, for instance, has a lower salinity in winter than in summer, due to entry of freshwater from snow melt in the surrounding mountains. Also, freshwater is less dense than salt water, and hence will float on top. Organisms developing in this freshwater layer might be mistaken for halophiles.

Even if reproduction is taking place, this does not imply that the organism is optimally adapted to that environment. Only in the case of the halophilic bacteria is the situation reasonably clear; many of these isolates grow optimally in culture in media of about 25% (w/v) NaCl, and show a minimum requirement for growth of 12–20% NaCl (Larsen, 1967). They are thus true halophiles, and even though experiments have apparently never been performed to show that these organisms grow in nature best at high salt concentrations, we would be very surprised indeed if we did not find that to be so. In salt pans, where solar salt is being made, the water at the time its salinity reaches close to saturation often acquires a brilliant pink cast due to the growth of halobacteria, whereas in the more dilute waters, before evaporation is completed, these organisms are not seen. It seems very unlikely that the appearance of these organisms could be due to their mere concentration from diluter waters, and therefore they must be growing. Most halophilic bacteria in culture have apparently been isolated from solar salt, or from food products preserved with solar salt. Thus, it seems reasonable to conclude that the natural habitats of these extreme halophilic bacteria are salt pans and salt lakes.

Dry Environments

The work in this laboratory on water potential arose out of work David Smith did to explain the presence and activity of the acidophilic, thermophilic alga *Cyanidium caldarium* in Yellowstone soils. In some ways saline and dry environments are similar, since both have low water activities. However, there are some major differences (Griffin, 1972). In saline environments, ions are present that organisms might pump inside so that the osmotic pressure of the environment can be balanced. In dry environments, on the other hand, water activity is reduced by way of adsorption phenomena, the remaining water molecules being held tightly to clay or other particles (so-called matric phenomena). In dry environments, no possibility exists for organisms to pump in ions to balance the matrically controlled water activity, so that generally organisms have more difficulty coping with dry than with saline environments.

The organisms found in dry environments are predominantly fungi. Blue-green algae form crusts in dry deserts, but they don't really grow when it is dry, merely waiting for the occasional brief rains, at which time they quickly flair up (Brock, 1976).

Other Environmental Extremes

Space does not permit more than a brief mention of some other kinds of extreme environments that have been studied microbiologically. Environments of high hydrostatic pressure in the depths of the oceans yield not only microbes but higher animals, including vertebrates. Thus, by our definition such environments are not extreme, although they may be biologically interesting. Surprisingly, although bacteria capable of growing at high hydrostatic pressure have been isolated, it is not certain that true barophilic bacteria exist.

Anaerobic environments including environments high in H_2S concentration are undoubtedly extreme. Of the animals, only certain protozoa, and a few other invertebrates, can live anaerobically and of the plants probably only the blue-green algae. Many bacteria are of course obligately anaerobic, although the reasons for this are still obscure (Morris, 1975). Highly anaerobic environments are found in lake and estuarine sediments, animal intestinal tracts, the rumen, sewage plants, and a few other areas. Although H_2S is more toxic than cyanide to most aerobes, many microorganisms tolerate high amounts. Lackey et al. (1965) have discussed this problem in an interesting article that deserves wider attention. They provide an extensive list of procaryotic and eucaryotic algae and protozoa that can tolerate H₂S. A number of species remained alive (i.e., were still actively motile) after 48 hr in water through which a stream of H₂S gas was passed continuously. Although it is usually assumed that H₂S, like cyanide, affects the cytochrome system, many of the organisms Lackey et al. used had cytochromes, either in photosynthetic or respiratory systems. Sulfur bacteria such as *Thiothrix*, which only live in habitats where free H₂S is present, are strict aerobes, and must have a respiratory system unusually resistant to H₂S.

Another environmental extreme, which has been inadvertently explored by several generations of continuous culturists, is the essential nutrient in limiting quantities. The effect of limiting substrate on growth rate has been explored primarily as a laboratory phenomenon, but substrate-limited growth is a common natural phenomenon in the aquatic environments.

The effects of ultraviolet light on organisms are well known. Although we usually assume that ultraviolet light is not a natural environmental factor, this is not so, as near ultraviolet is present in significant amounts, especially at high altitudes. Findenegg (1966) has shown that algal photosynthesis in high alpine lakes is definitely inhibited in surface waters by near ultraviolet light from the sun, although not all lakes showed this phenomenon. Adaptation to near ultraviolet by the reduction of total pigment concentration or by the production of a photo-protective pigment which resides at the cell periphery seems possible. The role of carotenoids in the protection of bacteria that live in habitats where they are exposed to bright visible light is well established (Mathews and Sistrom, 1959). It is interesting to note that sensitivity to ultraviolet and visible light is more likely to occur in microorganisms than in macroorganisms, since the latter have only a small fraction of their cells exposed to the outside world.

Evolutionary Considerations

The study of the adaptations of microorganisms, which make it possible for them to live in extreme environments, provides us with some of the clearest insights into the phenomena of microbial evolution. This is because we can concentrate on the single extreme factor to the virtual exclusion of all others. Organisms living in such environments reveal for us the extremes to which evolution can be pushed. The limited species diversity in extreme environments suggests that many organisms are unable to adapt successfully. On the other hand (and I am trying not to reason in a circular fashion), it is the limited species diversity in a particular environment that tells us that it is extreme. In an illuminating article, Vallentyne (1963) has argued that the limited species diversity in a habitat does not necessarily indicate that it is extreme. "One can instructively reverse the point of view that has been taken here and ask why it is that most organisms live under 'common' conditions. The answer is, of course, because life as a whole is selectively adapted to growth in common environments. If the waters of the earth were predominantly acid, growth at neutral pH values would be regarded as an oddity. Thus, the fact that most living species conform physiologically and ecologically to average earth conditions should not be taken to indicate any inherent environmentally based physicochemical conservatism of living matter. Adaptation has taken place'' (Vallentyne, 1963).

Although I appreciate Vallentyne's argument fully, I do not agree with it. I believe that even if acid pH values were common, there would still be a restricted species composition due to the potential effects of the H^+ ion on the stability of macromolecules and cellular structures. It is a fact that DNA, ATP, and a variety of other essential cell constituents are acid labile, and no amount of biological evolution can alter the inherent properties of such biochemical compounds. Unless an organism were to evolve with an entirely new ground plan in which these acid labile compounds could be dispensed with, it seems unlikely that acid environments will ever be other than biological oddities. The same can be said for high-temperature environments. Since thermal environments, in particular, have probably existed as long as the earth has, there has been plenty of time for the evolution of organisms adapted to these environments. The fact that we do not find photosynthetic procaryotes living at temperatures above 73–75°C suggests that there is some inherent physiochemical limitation, perhaps of the photosynthetic apparatus, which is impossible for these organisms to overcome by further evolutionary changes.

We are just at the beginning of an exciting era in the study of evolution. Molecular biology, microbial ecology, and paleomicrobiology are converging to provide us with new insights into the origin, evolution, and nature of life itself. Microorganisms, with their great diversity, will prove the most useful objects of evolutionary study, and those living in extreme environments will be especially relevant.

References

Brock, T. D. 1976. Effects of water potential on a *Microcoleus* (Cyanophyceae) from a desert crust. *J. Phycol.* **11**, 316–320.

Carlisle, D. B. 1968. Triops (Entomostraca) eggs killed only by boiling. *Science* **161**, 279–280.

Dost, H. (editor). 1973. *Acid Sulphate Soils.* International Institute for Land Reclamation and Improvement, Wageningen, The Netherlands, Vol. I.

Findenegg, I. 1966. Die Bedeutung kurzwelliger Strahlung für die planktische Primarproduktion in den Seen. *Verhandl. Intern. Ver. Limnol.* **16**, 314–320.

Griffin, D. M. 1972. *Ecology of Soil Fungi.* London, Chapman and Hall.

Hinton, H. E. 1960. A fly larva that tolerates dehydration and temperatures from −270°C to +102°C. *Nature* **188**, 336–337.

Lackey, J. B., E. W. Lackey, and G. B. Morgan. 1965. Taxonomy and ecology of the sulfur bacteria. Bulletin of the Florida Engineering and Industrial Experiment Station, Gainesville, Fl., Ser. No. 119.

Larsen, H. 1967. Biochemical aspects of extreme halophilism. *Adv. Microbial Physiol.* **1**, 97–132.

Livingstone, D. A. 1963. Chemical composition of rivers and lakes. U.S. Geol. Survey Prof. Paper, No. 440-6.

Mathews, M. M. and W. R. Sistrom. 1959. Function of carotenoid pigments in nonphotosynthetic bacteria. *Nature* **184**, 1892–1893.

Morris, J. G. 1975. The physiology of obligate anaerobiosis. *Adv. Microbial Physiol.* **12**, 169–246.

Vallentyne, J. R. 1963. Environmental biophysics and microbial ubiquity. *Ann. N.Y. Acad. Sci.* **108**, 342–352.

Chapter 2
The Habitats

One of the attractions of the research on extreme environments is that it must, by necessity, involve study of natural environments. Thus, the work immediately becomes habitat-orientated or geographic. This appeals to many, as they imagine the glamour of traveling to exotic places. And it is a strange thing that good geothermal areas are, to a great extent, situated in exciting locales. This indeed adds a certain attraction to the work, but one soon becomes immersed in the microscope or the Teflon homogenizer, and the excitement palls. When I tried to publish a picture of a light-reduction setup in a Yellowstone hot spring (Figure 10.5), one of the reviewers for the *Journal of Bacteriology* demanded that the picture be removed, as all it would do would be to elicit envy in the reader. A strange reason! (The picture stayed in.)

Origins of Thermal Environments

There are four distinct causes of thermal environments: solar heating, combustion processes, radioactive decay, and geothermal activity. In man-influenced environments, most heating is due to combustion processes, but recently heat production as a result of radioactivity has become of significance and interest.

Solar Heating

Solar heating can lead to soil temperatures as high as 60°C. Schramm (1966) measured temperatures of this magnitude on black anthracite wastes in eastern Pennsylvania. Although such high temperatures occur only during the daylight hours, they are sufficient to prevent the successful establishment of higher plants (Schramm, 1966). Gates et al. (1968) measured a temperature of 60°C on the surface of a desert soil at midday; in the same location the temperatures of desert leaves were in general lower. They were usually no more than 3°C above air temperature. In aquatic environments, solar heating leads to marked temperature increases only in shallow waters or in those with unusual density characteristics. In shallow marine bays, temperatures up to 40°C can occur (Gunter, 1957), although again this is a transitory effect seen only at midday. In certain aquatic environments solar heating can apparently lead to stable increases in temperature. Thus, Lake Vanda, a permanently ice-covered, density-stratified lake in Antarctica, has a bottom temperature of 25°C (Hoare, 1966). In this case the possibility of geothermal heating has apparently been ruled out, and it is thought that the highly saline water on the bottom remains as a stratified layer that absorbs solar radiation and retains it. In a stratified pond near Eilath, Israel, temperatures as high as 50°C arise as a result of solar heating (Por, 1969).

Combustion Processes

Combustion processes can be either biological or nonbiological. The study of the self-heating of hay and other organic materials has a long history (Hildebrandt, 1927; Miehe, 1907; Norman et al., 1941), and it is well established that the temperatures, which reach up to 70°C or higher, are the result of microbial heat evolution. It is thought that the habitat of many terrestrial thermophilic microorganisms, both fungi and bacteria, may be in locations where organic waste materials become naturally piled in windrows by wind or water and undergo self-heating (Cooney and Emerson, 1964). Other habitats where natural self-heating exists, such as the nests of the intriguing incubator birds *(Megapodiidae),* are discussed by Cooney and Emerson (1964). Another combustion process, which is generally man-made but may also occur naturally, is the burning of coal and coal-refuse piles (Haslam and Russell, 1926; Myers et al., 1966; Stahl, 1964). In several burning coal-refuse piles we have measured temperatures ranging from 45 to 150°C. (One small acid stream draining such a pile had a temperature of 34–36°C even when the air temperatures were less than 0°C.)

Not all such combustion processes are man-made. In a fascinating account, Burns (1969) describes the smoking cliff of the Horton River in the western Arctic. This was reported by Johann August Miertsching in his diary on the H.M.S. Investigator, 1850–1854.

4 September [1850]. . . . In the night we had passed Horton River. As soon as it was day we perceived thick smoke rising. . . . The captain ordered a boat to be lowered. . . . After rowing for two hours we finally arrived at this great column of smoke and found neither tent nor man, but a thick smoke emerging from various vents in the ground, and a smell of sulphur so strong that we could not approach the smoke-pillar nearer than ten or fifteen feet. Flame there was

none, but the ground was so hot that it scorched the soles of our shoes. There was no rock or stone in site; the vertical coastline was composed of a fused mass like rubble, neither clay nor sand, but like a very soft pumice-stone, and in colour green, grey, brown, and mostly sulphur-coloured; where the smell was strongest the heated earth resembled a thick or tough dough; long oars were thrust into it, and each time the smell of sulphur was redoubled, but no flame could be seen; we counted thirty or forty smoke-pillars rising from the earth in different places: Alas, alas! that none of us had knowledge of chemistry; the whole place seemed to me like a huge chemical laboratory. We took samples of the particoloured dough or earth with us to the ship; but later to our sorrow we learned that our pocket-handkerchiefs in which we brought the specimens to the ship—and also the wood in which we laid them—were scorched. The captain, with whom we shared each type of earth, arranged them on the table in his cabin, and a little later was startled to find his splended mahogany table burnt full of holes as if vitriol had been spilt on it. We disposed of our samples in thick-covered glasses. On other parts of the coast there was an appearance as of a brick-kiln; whole layers of burnt strata, a half-inch thick and varying in extent; very similar to tiled roofs, in colour reddish and also grey. In such places we found small reservoirs full of water which tasted sour.

The existence of such a combustion process so far from human existence bears witness to natural heating. The reason is that the rock strata here are from the Bituminous Zone of the Upper Cretaceous series, and in addition to organic matter contain much iron sulfide, selenite, and elemental sulfur. Self-heating may be partly microbial in origin, and is continuing today, over 100 years after it was first reported.

Geothermal Activity

Geothermal activity is probably responsible for the creation of the most numerous high temperature environments. The temperatures in active volcanoes are much too hot for living organisms (molten lava can have a temperature of 1000°C or over), but hot springs and fumaroles associated with volcanic activity often have temperatures with a more reasonable range and are prime candidates for the development of thermophilic organisms. A recent survey (Waring, 1965) reveals that thermal waters are widely distributed over the face of the earth, although springs are often concentrated in restricted areas. The largest concentrations of hot springs and fumaroles are found in Yellowstone National Park (Keefer, 1971), Iceland (Barth, 1950), New Zealand, Japan, and the USSR. Temperatures of hot springs range from the 30s up to boiling (90–100°C depending on altitude). Fumaroles which consist only of stream vapor can have temperatures considerably higher than 100°C and seem to be devoid of living organisms.

It is sometimes not realized that geothermal heating is a normal event throughout the earth's surface, although in most areas the heat production is so small that it is unrecognized. Likens and Johnson (1969) calculated the heat production of Stewart's Dark Lake, a small meromictic lake of about 0.69 ha in northwestern Wisconsin. The bottom of this lake received essentially no heating from solar radiation, and the heat production from geothermal sources was estimated as 66 cal/cm^2/year or about 7×10^6 kcal/year for the lake as a whole. A comparison of this heat production to that of a hot spring may be of interest. For instance, a rather small Fijian hot spring produced 3×10^9 kcal/year (Healy, 1960), which is almost 3 orders of magnitude higher. Likens and Johnson (1969) note that on a normal land surface the flux of solar energy is 4–5 orders of magnitude larger than the flux of geothermal heat, but, clearly, within the localized region of a hot spring the geothermal flux is much higher.

An extensive survey of the thermal waters of the world (Waring, 1965) provides a broad view of their distribution. The antiquity of many individual springs is noteworthy. Historical records of springs in the Mediterranean go back to the ancient Greeks and Romans (Brock, 1967). Chemical studies on mineral springs were initiated by Robert Boyle in the seventeenth century, and were greatly systematized in the nineteenth century by Bunsen (1847), who was apparently the first to see the value of chemical studies in interpreting the origin of hot springs.

Man-made Sources of Heat

Man-made sources of heat are indeed common in this industrial age. Space does not permit a detailed survey of such thermal habitats, although I discuss the development of *Thermus aquaticus* in such habitats in Chapter 4. Some habitats which have received biological attention include hot water heaters, sugar refineries, power plants and their cooling towers, paper mills, and textile mills.

Constancy of Temperature

I must confess that in my early work on hot springs, I was somewhat sanguine about the constancy of the temperatures of geothermal habitats. It is true that the temperatures of hot spring *sources* are often remarkably constant, so that where organisms in the source are being studied (see Chapter 10), things are greatly simplified. But in most of the work on algal mats, it is not the source but the outflow channel that is important. And the temperature of the outflow channel is influenced by two factors, flow rate and weather. Some springs vary drastically in flow rate (Mosser and Brock, 1971), and this leads to wide fluctuations in temperature at specific sites along the drainway. Although organisms have learned to cope with wide temperature fluctuations (see Chapter 8), for many studies it is essential to have constant temperatures.

The important point is that the temperature fluctuations must be either very small, a few degrees Celsius, or they must have a long time constant, considerably longer than the generation times of the organisms. As the water flows away from the source, it cools gradually, and the rate of cooling is a function of initial water temperature, air temperature, channel dimensions, and volume of water. Because most of the cooling occurs at the air–water interface, variations in wind speed will cause variations in cooling, and this factor is probably more important than fluctuation in air temperature in affecting the cooling rate.

Change in temperature with time of a hot body follows Newton's law of cooling, which is an exponential function. Since distance from the source of a spring is approximately equal to time of cooling, this means that cooling is faster near the source than farther away. Thus, there is a shorter stretch of outflow channel with a high temperature than with a low temperature. I discovered this graphically about 1966 when I was attempting to locate a variety of thermal habitats along White Creek, a thermal creek in the Lower Geyser Basin of Yellowstone. The creek begins at a temperature of 55°C, and by walking down the creek it was relatively easy in a few hundred meters to locate temperatures from 55°C to 40°C, but after a kilometer or two of walking I still hadn't found a place where the temperature dropped to 35°C. When the creek spread out into a broad channel and filtered out of site through the porous sinter, I gave up.

In two locations where we did a lot of work we were careful to take frequent periodic measurements of temperature. The Octopus Spring (Pool A) mat, in a side area of the pool (see Figure 11.11), showed temperature variations of only a degree or two even though the flow rate of the source varied markedly. This was because the side area where the mat was present was so far from the source that the fluctuations in flow were considerably damped. Even here the temperature of the water surface fluctuated much more than the temperature of the mat surface. This was because hot water is less dense than cold water, and floats. This led to an evening out of the fluctuations at the mat surface.

In our work on Mushroom Spring, where we were trying to draw some specific conclusions about the adaptation of algae along an outflow channel to their thermal habitats, it was important for us to know the temperature of each station along the channel. During the period of virtually daily visits to Mushroom Spring, we measured the temperature at each station along the channel frequently. (A plot of these temperature measurements is shown in Figure 2.1.) As seen the temperature variation at a single station is within 6–10°C. Although this variation is larger than would be desirable, it is small in comparison with the range of 30–75°C for the whole thermal gradient. In addition, the temperature ranges of the various stations do not overlap.

Long-term Constancy

We tend to get misled by the example of Old Faithful Geyser. Springs and geysers are anything but faithful, and some of my favorite study springs have given up on me, for no obvious reason. When I began working in

Figure 2.1. Average temperature and variation of temperature at given locations along the thermal gradient of Mushroom Spring. (From Brock, 1967.)

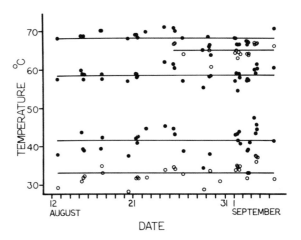

Yellowstone, Dr. Donald White of the U.S. Geological Survey told me: "The one thing you can be certain about is that springs will change." I discounted this statement, believing he was talking about geological time, and that over the duration of interest for biologists, change would not be a factor. However, in my 10 years in Yellowstone, I learned to appreciate what Dr. White was really talking about. The problem was greatest in acid thermal areas, where springs would dry up, reappear, be replaced by new ones, or change drastically in temperature or chemistry, for no obvious reasons. In some cases, large changes in acid thermal areas could be tied to earthquakes in the area, and earthquakes are frequent in volcanic areas where hot springs exist. But in other cases, the changes occurred willy-nilly. Fortunately, it was often possible to make use of such changes in research studies, and also, fortunately, some of our best-studied springs did not change, but I would make a warning to anyone beginning a long-term study of acid springs to count on change. In the case of the alkaline springs, changes are less common, but still occur. Many examples can be found in the report of Marler (1973). Silica deposition may clog a spring orifice, or violent geyser action may blow out the top, or there may be unobserved changes underground that lead to significant alterations.

The Death of Mushroom Spring

During the years that we had Mushroom Spring under study, we had the opportunity to watch it die and be reborn. At the same time, we observed the birth of a small geyser, which we called Toadstool Geyser, whose rise was clearly responsible for the fall of Mushroom Spring. And later Toadstool Geyser succumbed and Mushroom Spring returned.

Only one brief study of Mushroom Spring had been done before we began our studies in 1965. Schlundt and Moore (1909) had measured the radioactivity and temperature of the water in 1906. This temperature measurement, 75.0°C, is at the upper end of the range of measurements that we made over the 4-year period (1965–1968) before Mushroom Spring began to change. Thus, we concluded that Mushroom Spring had changed little between 1906 and the time we began our studies. During the 3 years 1965 through 1967, the flow rate and source temperature of Mushroom Spring showed no changes (Table 2.1).

During the last of this 3-year period, the forerunner of Toadstool Geyser first made its appearance about 2 meters from the lip of Mushroom Spring and about 20 meters from the source. This was a small superheated pool occupying the hole that previously had been cold and filled with plants. (The fact that it was a hole suggests that at some time in the past it had been a spring. It may have had a slight steam exudate, although we didn't notice it at that stage.) This superheated hole, which became known to us as Mushroom Annex, remained relatively constant in temperature throughout the summer of 1967, and we used it as an incubation site for some high-temperature studies. In February 1968 the water level in Mushroom Annex was considerably higher and the temperature considerably lower. At the time we attributed this change to effects of melting snow.

However, in late May of the same year, Mushroom Annex had become a flowing spring but was not superheated or eruptive. Throughout the summer of 1968, the temperature and flow rate of Mushroom Annex remained relatively constant. During the same period that Mushroom Annex began to flow, the average temperature of Mushroom Spring was about 4°C. lower than it had been over the previous three summers (Table 2.1). The flow rate of Mushroom Spring was visually unchanged, however, and although we did not measure flow rate quantitatively, several wooden channels and

Table 2.1. Temperatures of Mushroom Spring and Toadstool Geyser during 1965–1974

	Mushroom Spring			Toadstool Geyser		
Year	Average temperature (°C)	Flow	Average temperature (°C)	Flow	Eruptions
1965	72.5	Good	Not present	—	—
1966	72.1	Good	Not present	—	—
1967	72.6	Good	91.8	None	None
1968	68.2	Good	84.5	Good	None
1969	54.6	None	95.0	Good	2–3 meters
1970	54	None	95.0	Good	1 meter
1971	58	Fair	92.5	Good	Slight
1972	62	Fair	—	—	—
1973	68	Good	87–89	Less	None
1974	67	Good	72	Slight	None

other markers that we had installed previously provided visual clues of an unchanged flow rate.

During the early fall of 1968, the temperature of Mushroom Annex had risen from an average of 83.7–90.5°C. Unfortunately, observations had to be terminated on October 3, 1968, and the next observations were made 4 months later on January 30 and February 1, 1969. At this time, Mushroom Annex was flowing more extensively and was semi-eruptive whereas Mushroom Spring had ceased flowing completely and its temperature had dropped to 53.2°C, about 20°C lower than the normal average. Its water level had also dropped about 0.3 meters. Although the exact time at which Mushroom Spring ceased flowing is not known, it was probably sometime in January as the effluent channel of Mushroom Spring was still free of snow on January 30, 1969, although away from the spring about 2 meters of snow was on the ground.

By May 1969, the activity of Mushroom Annex had increased even further, and it was now a real geyser. Eruptions of 1–2 meters in height occurred irregularly, and eruptions of 0.5 meter were occurring about every minute, followed by quiescent periods when water overflowed the lip gently. During one series of temperature measurements with the thermistor at a depth of 2 meters, the temperature rose to 96.5°C just before an eruption, dropping to 93.5°C at the time of eruption. The flow rate had also increased considerably, the effluent extending considerably farther from the spring. A number of lodgepole pine trees were killed in early 1969 by the effects of the hot water. Because of its malevolent effects on Mushroom Spring, we named this new geyser Toadstool Geyser.

It seems clear that the 4°C drop in temperature of Mushroom Spring in 1968 was associated with the commencement of flow of Mushroom Annex. Some of the energy previously supplied to Mushroom Spring must have been transferred to the new spring, although the extent of this energy transfer was not sufficient to affect the flow rate of Mushroom Spring. During the latter part of 1968 and early 1969 a much greater transfer of energy must have occurred, since Mushroom Spring dropped about 20°C in temperature and ceased flowing. Since energy dissipation by Mushroom Spring must have been primarily through the water leaving the source along the effluent channel, at the time Mushroom Spring ceased to flow the bulk of the energy dissipated by the system must have been transferred to Toadstool Geyser. Although the flow rate of the Toadstool Geyser was less than that of the former Mushroom Spring, its higher temperature and eruptive nature probably account for the additional energy loss from Mushroom Spring.

The two thermal features clearly have some sort of underground connection. Such underground connections are not uncommon in Yellowstone, although many thermal features as close as Mushroom Spring and Toadstool Geyser show no signs of such connection.

The actual cause of the transfer of energy from Mushroom Spring to

Toadstool Geyser during the years 1967–1969 is unknown. No major earthquakes occurred in the vicinity during this time period, although conceivably only a minor tremor might have initiated the changes. Unfortunately, at the time we first observed the cold hole adjacent to Mushroom Spring we had no idea that within a few years it would be an eruptive geyser.

A good algal mat developed in the newly formed outflow channel of Toadstool Geyser. The complete community formed in less than 6 months (Brock and Brock, 1969). Making the best of adversity, we began studying this mat (Doemel and Brock, 1974). However, Toadstool Geyser was not fated to be a long-lived feature. By 1971 it had essentially stopped erupting and had cooled down a little, and Mushroom Spring was flowing again, although with much less flow than previously. By 1973, Mushroom Spring had warmed up to its 1968 temperature, and was flowing as well as previously, and Toadstool Geyser had cooled down more and was flowing noticeably less. In 1973 we had an excellent algal mat at Toadstool Geyser and did a lot of studies (Doemel and Brock, 1974), but by 1974 Toadstool was flowing so slightly that no algal mat was present. Another year or two, and it might be gone again, and its existence never suspected, except for the good color photograph taken in 1971 which we have published (Brock and Brock, 1971; Brock, 1974).

Locations of Hot Springs Studied

Giving directions for locating a single, isolated hot spring is no problem, but unfortunately most hot springs occur in groups with large numbers of other hot springs. Detailed maps have become available in the last few years for many of the Yellowstone thermal areas, a result of the large U.S. Geological Survey study begun around 1965. Portions of these maps have been abstracted for use here, to locate the springs that we have studied most intensively.

In other parts of the world, location of suitable springs was catch-as-catch-can, but in Yellowstone there was time to select springs carefully. We concentrated our work in six areas, two with predominantly neutral-alkaline pH springs and four with predominantly acid springs. The neutral-alkaline areas were in the Lower Geyser Basin (Figure 2.2), the largest single thermal area in the Park, and the one alkaline area that had the least amount of tourist development. Some of our studies were in the Fairy Meadow-Boulder Spring-Sentinal Meadow area but the main work was done in the vicinity of Great Fountain Geyser, along a road that has been variously called the Firehole Lake Loop Road and the Firehole Lake Drive.

Norris Geyser Basin provided a number of the acid springs that we studied (Figure 2.3). We had to select our sites carefully in this area

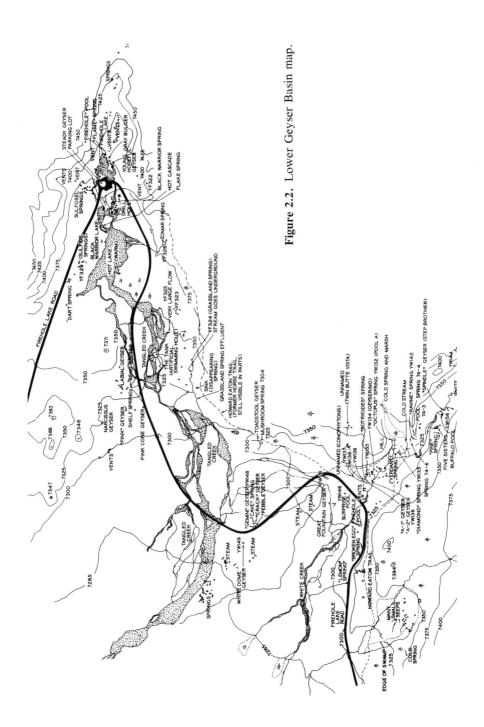

Figure 2.2. Lower Geyser Basin map.

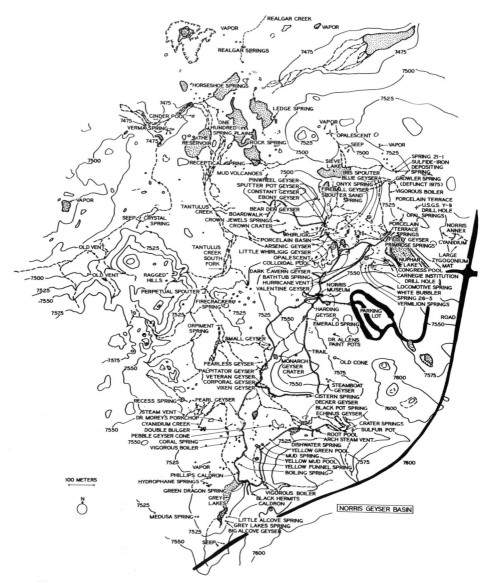

Figure 2.3. Norris Geyser Basin map.

because a museum, boardwalks, and ranger-naturalist activities encroached on many of the best sites. A lot of the work on *Cyanidium caldarium* was done at the outflow channel of a series of springs on the south side of Norris Geyser Basin, a site dubbed "Cyanidium Creek" by Doemel (Figure 9.6). Our work on *Sulfolobus* was concentrated in the Congress Pool-Locomotive Spring area and the Procelain Basin (Figure 2.3). It may be of interest that Cyanidium Creek could only be studied after

1966, when the road through the Norris Geyser Basin was moved. The old road ran almost abutting Cyanidium Creek, making this interesting habitat so public as to be impossible to study. When the road was moved, the creek suddenly became part of the back reaches of Norris, and was hardly ever visited by anyone. The moving of this road came at the same time that Doemel joined us as a graduate student. I decided that with this habitat available *Cyanidium* now became a good organism for study, and turned him loose on Cyanidium Creek. It was a wise, and fruitful, decision (Doemel, 1970).

One of our favorite study areas for acidophilic organisms was the Sylvan Springs area. This beautiful series of springs was about 0.6 km from the road and could only be reached after crossing the Gibbon River on a log, and slogging through a broad meadow. We never saw a tourist in this area, even though it is fully visible from the road. It has an excellent collection of acid hot springs (Figure 2.4), and we were able to do a lot of experimental modifications which would have been impossible elsewhere (Mosser et al., 1974; Brock and Mosser, 1975).

The third acid area we studied intensively was in the Mud Volcano area (Figure 2.5). The front part of this area is jammed with tourists, but the back reaches are relatively untrodden. One of our favorite pools, which Jerry and Ann Mosser dubbed Moose Pool (after a moose they saw there; later all we ever saw there were buffalo), was just invisible to tourists who were standing on the boardwalk at Black Dragon's Cauldron. The other major spring that we studied in this area, Sulfur Caldron, was much more public. A parking lot rested just above it, and it was possible for tourists to chuck beer cans into the spring while we were working on it. Since the Park Service was somewhat leery of us working in view of tourists, we carried out very few time-consuming studies there, restricting our work mostly to taking quick samples and leaving. For our turnover studies, though, we needed to add about 45 kg (100 lbs) sodium chloride to the spring. This massive amount of salt obviously could not be added in full view of tourists, so Ann and Jerry Mosser went out at 6:00 A.M., and by dint of careful organization had the salt in the pool by 6:30 A.M. (In case any one is interested, we did *not* ask the Park Service's permission to do this experiment. We knew that the salt would have no influence on the spring, and that it would eventually dilute out. It was gone by the next summer.)

The fourth acid area we studied was Roaring Mountain, a large solfatara area that had the most acidic flowing spring in the Park (Figure 2.6). This mountain hardly roars, but constitutes a massive area of fumaroles and bubbling pools. The northern part of the mountain is in full view of the road and is impossible to study, but by a quirk of geology the southern part of the mountain, where the best flowing springs are, is just out of view of the road and tourist area. We did a lot of studies on *Cyanidium* and *Sulfolobus* here (Doemel and Brock, 1971; Fliermans and Brock, 1972; Smith and Brock, 1973).

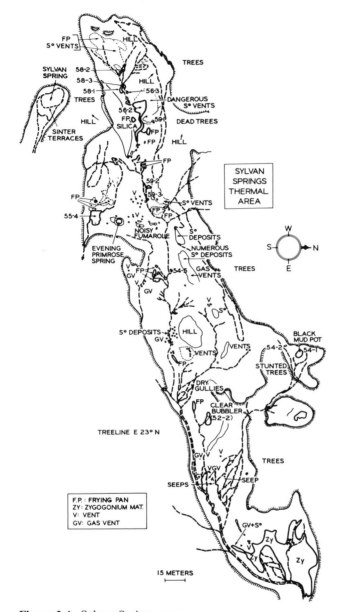

Figure 2.4. Sylvan Springs map.

Some other areas that we studied occasionally were the Mammoth Hot Springs area (travertine and sulfide-rich springs), Amphitheater Springs (hot acid springs and *Cyanidium* mats), Clearwater Springs (good pH gradient), Nymph Creek and Nymph Lake (the creek has the best *Cyanidium* mat in the Park; the lake is an excellent acid lake); Gibbon Hill Geyser Basin (Brierley, 1966); Terrace Spring (travertine-depositing; very public

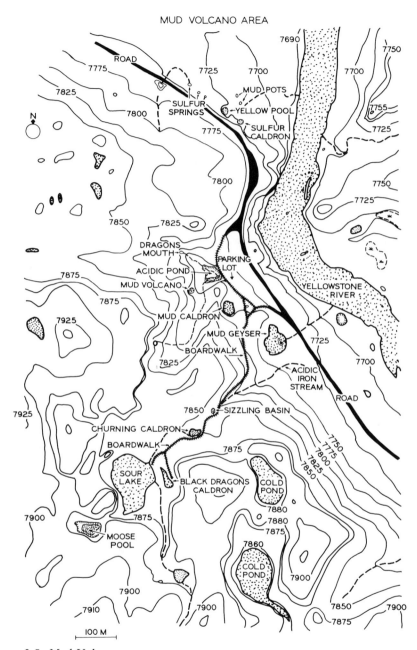

Figure 2.5. Mud Volcano map.

Figure 2.6. Roaring Mountain. The southern effluent drains the narrow valley seen. Most of the steam rises from fumaroles.

but the most accessible thermal area in the winter); Upper Geyser Basin (very public, with Old Faithful et al., but large number of superheated springs); West Thumb Geyser Basin (my first love, but too public for most studies); Shoshone Geyser Basin (one of the more interesting areas, but too far from the road for continuous study; a 15–18-km hike or an all-day canoe trip to reach it); Heart Lake Geyser Basin (large number of very high pH springs, greater than 10.5); Turbid Lake (a low-temperature acid lake, virtually a pure culture of sulfur bacteria); Great Sulfur Spring in the Crater Hills (a low pH–high chloride area, very interesting geochemically, but full of grizzly bears during most of the time we were in the Park); Canyon-Clear Lake area (cold, acid lake and a number of acid hot springs); Washburn Hot Springs (an unusual area geochemically; springs high in NH_4^+ and H_2; resemble acid springs but are of neutral pH; a fair cross-country hike).

We also studied several rivers and creeks. The Firehole River, which runs through the main Geyser Basins, was an excellent model of a thermally polluted river, and is the subject of a chapter later in this book. Obsidian Creek, which partly drains out of Roaring Mountain, begins as a cold acid creek and becomes progressively neutral. It provided an interesting pH gradient for some of the work on the lower pH limit of blue-green algae (Brock, 1973). Lemonade Creek (pH 2.5) drains out of the Amphitheater Springs area. It is more acidic than Obsidian Creek and spreads out in the Beaver Lake area into a large marsh that provided some insight into the lower pH limit for higher plants. Yellowstone also has a large number of interesting lakes. We were able to locate lakes with pH values ranging from 1.9 to over 9.0. Several have already been mentioned above: Clear Lake (pH 2), Turbid Lake, Nymph Lake. Others included Nuphar Lake (pH 4.2) in the Norris Geyser Basin, Sour Lake in the Mud Volcano area (the most acidic lake in the Park, pH around 2.0); North and South Twin Lakes (virtually touching, but North Twin has pH 3.5 and South Twin pH 5.9).

Choosing Springs for Study

The springs that one might carry out long-term research studies on are not necessarily those of most interest to geologists or others. Several criteria are important in choosing a spring for detailed study:

1. It should have chemical, physical, and biological characteristics of interest.
2. It should be relatively constant in characteristics.
3. If flowing habitats are under study, the flow rate should be relatively constant (or predictably variable, if effects of variation in flow rate are of interest) (see Mosser and Brock, 1971).
4. For Nos. 2 and 3 to obtain, the spring should be of relatively large size.
5. In a tourist area, the spring should be away from and roads and trails that might attract people to it. The spring should not be visible from the road, or be especially attractive, or be listed in tourist brochures.

If detailed studies are to be carried out, the spring should be carefully marked with station markers, so that repeated measurements and samplings at specific sites can be done. For neutral-alkaline springs, large nails with attached plastic tags make suitable markers, written on with indelible marking pens. The markers should be checked periodically to be certain that they remain legible. In acid areas, nails will disappear quickly, so that wooden stakes, pointed at one end must be used.

Effect of Temperature on Physical and Chemical Parameters

Temperature affects a wide variety of properties of water, and some of these effects are summarized in Table 2.2. Certain of these temperature effects are of more practical significance than others. For instance, the marked decrease in density of water at higher temperatures means that when hot water flows into cold, it usually floats on top and only mixes gradually. The decreased viscosity of water at high temperature means that it flows more easily than cold. (Hot water also sounds different from cold when flowing, probably due to the viscosity effect.)

The effect of temperature on oxygen solubility may be of considerable biological significance. For instance, at 90°C oxygen will dissolve in water at less than 2% of the amount at which it dissolves at 20°C. On the other hand, silica is present in solution in hot spring waters to concentrations considerably higher than in cool water (Fournier and Rowe, 1966). At 25°C the pH of pure water is 7.00, whereas at 0°C it is 7.47, and at 40°C, 6.77. However, of more interest is the effect of temperature on the pH of dilute

Table 2.2. Effects of Temperature on Physical and Chemical Properties of Water

Property	Effect of an increase in temperature
Physical	
Density	Decrease
Viscosity	Decrease
Surface tension	Decrease
Volume	Increase
Dielectric constant	Decrease
Vapor pressure	Increase
Heat capacity	Decrease
Compressibility	Decrease
Refractive index	Decrease
Diffusion	Increase
Chemical	
Ionization	Increase
pH	Decrease
Oxygen solubility	Decrease
Solubility of most organic and inorganic compounds	Increase

aqueous solutions of anions and cations such as are found in nature. For instance, a solution of potassium phosphate which has a pH of 6.86 at 25°C decreases to only pH 6.84 at 40°C.

Effect of Temperature on Biologically Active Substances

Of considerable evolutionary significance are the effects of temperature on the stability of certain biochemical and macromolecular constituents. Such key compounds as adenosine triphosphate (ATP) and nicotinamide adenine dinucleotide (NAD) are somewhat heat labile, although the extent of such lability and its practical significance have not been investigated. Macromolecules such as RNA and DNA are heat sensitive under certain conditions; divalent cations such as Mg^{2+} stabilize them greatly, and it seems that at ionic concentrations found in the cell these substances may be stable even to boiling. Although molecular biologists have studied some effects of heat on macromolecules, heat has more often been looked upon as a physical-chemical tool rather than an environmental factor.

Chemistry of Hot Springs

The early geochemical work is reviewed by Allen and Day (1935). Chemical analyses have also been made by balneologists seeking an explanation of the reputed curative properties of certain springs (Waring, 1965). Unfortunately for the biologist, many chemical elements of biological significance are not assayed by either the geochemist or balneologist, but the results of the analyses do show that there are many chemical types of hot springs. The pH of hot springs varies and values as low as 1.0 and greater than 11.0 have been recorded (Waring, 1965). Many, but by no means all, hot springs have significant amounts of hydrogen sulfide. The concentration of such interesting elements as fluoride, arsenic, rare earths, and gold varies very much from spring to spring. Many springs are highly radioactive (Schlundt and Moore, 1909), whereas others have no more radioactivity than normal ground waters. Some springs precipitate silica, others deposit travertine ($CaCO_3$), and still others form elemental sulfur. When one considers chemical, hydrologic, thermal, and geographical variation, it is clear that every hot spring can be considered as an individual, differing in minor or major ways from other springs. However, many springs are more similar than different. For instance, in the geyser basins of Yellowstone National Park, virtually all springs contain mildly alkaline waters which deposit silica (Allen and Day, 1935), although even these springs show differences in that they contain varying and significant amounts of dissolved gases, including the biologically important gas methane. Unfortunately, the geochemist has

different interests than the biologist, and therefore assays different substances. For the geochemist, substances such as silica, potassium, sodium,
and rare gases are of interest, whereas the biologist is interested in nitrogen, phosphorus, organic carbon, sulfide, etc. Thus, the vast series of
chemical analyses of Yellowstone springs presented by Allen and Day
(1935) provide little of value.

We have thus been forced to carry out our own chemical analyses,
which is probably just as well, since there are often marked temporal
variations in some chemical species, and certain of the biologically important substances are not stable and must be assayed immediately after
sampling. In 1971, we were fortunate to have working with us Prof. Kimio
Noguchi and his associates from Tokyo, Japan, and they carried out a

Table 2.3. Chemical Analysis on Yellowstone Hot Springs[a] (*continued on pp. 32 and 33*)

Name of spring	Temperature (°C)	pH	Cl mg/l	SO$_4$ mg/l	Na mg/l	K mg/l	Ca mg/l	Mg mg/l	Fe mg/l	Mn mg/l
I Upper Geyser Basin										
Old Faithful Geyser	—	9.3	453	21	364	22.8	2.1	0.1	0.01	0.006
Crested Pool	93.0	9.1	388	21	413	19.8	0.0	0.0	0.01	0.004
Tortoise Shell Spring	93.0	8.7	391	28	408	22.0	0.0	0.0	0.01	0.003
Giant Geyser	94.0	8.4	292	16	434	20.1	0.0	0.0	0.02	0.006
Morning Glory Pool	78.0	8.1	296	21	414	20.4	0.8	0.0	0.01	0.003
Three Sisters Springs	71.0	8.8	414	29	380	19.8	0.0	0.0	0.02	0.005
II. Lower Geyser Basin										
Firehole Pool	91.0	8.3	323	21	363	13.4	0.3	0.0	0.01	0.004
Geyserino	94.5	8.6	300	20	368	12.4	1.1	0.4	0.01	0.006
Octopus Spring	91.2	8.3	256	23	321	16.6	0.8	0.0	0.01	0.006
Stepbrother	93.0	8.6	257	33	320	16.0	1.6	0.2	0.24	0.010
Great Fountain Geyser	94.5	8.6	350	25	372	14.7	0.8	0.0	0.01	0.006
Bead Geyser	87.0	8.7	368	31	358	24.7	0.0	0.0	0.03	0.006
Ojo Caliente	94.5	7.9	325	27	332	11.5	1.4	0.6	0.02	0.003
Boulder Spring	95.0	8.8	306	24	333	10.2	1.6	0.0	0.02	0.006
Red Terrace Spring	95.0	9.0	263	20	322	11.3	1.3	0.2	0.02	0.004
Steep Cone	94.0	8.2	260	24	310	12.9	0.0	0.0	0.01	0.003
Queen's Laundry	89.0	8.2	237	25	327	14.2	0.0	0.0	0.02	0.003
III. Midway Geyser Basin										
Grand Prismatic Spring	67.5	8.0	285	18	423	18.2	1.0	0.2	0.01	0.027
Excelsior Geyser Crater	88.0	7.6	274	19	402	15.2	0.6	0.0	0.02	0.025
IV. Miscellaneous Acid Areas										
Roaring Mountain Upper Spring	92.0	2.4	43	435	48.2	42.0	3.8	0.0	3.70	0.050
Roaring Mountain Lower Spring	92.0	2.4	41	426	48.6	41.4	4.3	2.4	3.60	0.050
Nymph Creek Spring	59.5	2.6	30	291	59.5	40.0	8.8	0.2	4.22	0.087
Frying Pan	90.0	2.3	3	497	55.2	12.6	4.1	0.0	1.38	0.018
V. Norris Geyser Basin										
Amphitheater Spring 1	79.5	2.3	10	496	36.0	22.4	5.6	0.0	4.62	0.008

number of useful analyses on many of the springs we were working on. Chemical analyses for some of the springs we have worked on are given in Table 2.3.

Bimodal pH Distribution of Hot Springs of the World

As part of an extensive study of the distribution of bacteria in hot springs as a function of temperature and pH, my associates and I have measured pH values of thermal springs in many parts of the world. It became clear early in this study that there was a bimodal distribution curve for pH, with many

Cr mg/l	Hg μg/l	As mg/l	Cu mg/l	Zn mg/l	H_2S mg/l	SiO_2 mg/l	HBO_2 mg/l	NH_4 mg/l	NO_2^- mg/l	NO_3^- mg/l	PO_4 mg/l
0.017	0.04	1.60	0.003	0.010	0.5	338	17.7	0.79	0.01	0.14	0.53
0.012	0.04	1.70	0.003	0.003	1.6	322	16.2	1.32	0.00	0.00	1.07
0.005	0.03	1.78	0.001	0.001	1.0	285	15.7	1.48	0.00	0.07	0.51
0.015	0.04	1.36	0.001	0.002	0.4	240	12.3	0.62	0.00	0.11	0.68
0.010	0.18	1.36	0.003	0.005	0.1	245	9.3	0.03	0.02	0.11	1.57
0.012	0.02	1.20	0.009	0.014	0.3	293	16.7	0.19	0.01	0.07	1.37
0.012	0.06	2.26	—	—	0.6	302	12.3	0.02	0.00	0.14	3.28
0.012	0.03	2.26	0.001	0.003	0.8	244	11.3	0.96	0.00	0.35	0.62
0.010	0.15	1.28	0.002	0.003	0.2	254	10.3	0.00	0.01	0.28	1.70
0.011	0.03	1.38	0.003	0.006	0.6	294	8.8	0.40	0.00	0.00	0.79
0.010	0.04	2.42	0.002	0.004	0.9	254	14.7	1.24	0.00	0.09	0.74
0.016	0.00	1.98	0.001	0.004	0.8	280	15.2	0.32	0.00	0.15	1.56
0.010	0.02	1.41	0.005	0.006	2.0	219	16.2	0.84	0.00	0.19	0.19
0.005	0.04	1.47	—	—	3.2	139	16.7	1.74	0.00	0.22	0.61
0.012	0.02	0.89	0.001	0.003	0.6	319	12.3	1.34	0.00	0.19	0.60
0.012	0.00	0.92	0.000	0.002	0.3	345	16.2	0.54	0.00	0.08	0.28
0.015	0.01	0.89	0.001	0.002	0.8	290	11.3	0.26	0.00	0.07	0.77
0.014	0.07	1.38	—	—	0.6	329	12.3	0.03	0.00	0.14	1.91
0.012	0.13	1.57	0.001	0.001	0.4	261	10.5	0.02	0.00	0.19	2.27
0.012	0.03	0.10	0.001	0.011	0.5	335	7.3	2.70	0.00	0.59	0.02
0.015	0.07	0.10	—	—	0.0	337	6.9	3.08	0.00	0.70	0.11
0.010	0.01	0.05	—	—	0.3	216	3.4	1.81	0.00	0.58	0.09
0.007	0.39	0.19	—	—	1.8	215	3.0	2.66	0.02	0.31	0.07
0.004	0.04	0.025	—	—	3.3	292	0.0	1.52	0.01	0.55	0.05

Table 2.3. Chemical Analysis on Yellowstone Hot Springs[a] (*continued*)

Name of spring	Temperature (°C)	pH	Cl mg/l	SO₄ mg/l	Na mg/l	K mg/l	Ca mg/l	Mg mg/l	Fe mg/l	Mn mg/l
Amphitheater Spring 2	78.5	2.3	10	499	36.9	23.5	4.4	0.3	4.62	0.048
Congress Pool	79.0	2.4	735	404	457	76.4	3.7	1.1	1.68	0.043
Locomotive Spring	93.0	2.3	88	512	71.4	39.4	5.6	3.8	22.8	0.132
Small Triangular Pool	93.0	1.9	4	969	1.05	2.8	1.4	0.0	0.61	0.004
Spring near Locomotive Spring	80.0	2.3	79	505	—	—	—	—	—	—
Basin Geyser	94.0	5.9	571	47	337	63.5	2.9	2.1	0.33	0.187
Growler Spring	91.5	2.2	5	500	2.58	6.3	1.1	0.0	0.87	0.006
Emerald Spring	91.0	4.1	654	79	396	64.6	1.6	1.3	0.20	0.032
Cistern Spring	93.0	6.8	455	76	298	63.0	0.8	0.1	0.03	0.016
Cyanidium Creek Spring	75.5	2.8	387	103	229	45.2	4.0	0.4	0.91	0.094
Cyanidium Creek	52.0	2.8	386	120	—	—	—	—	—	—
Echinus Geyser	84.0	3.3	115	281	156	53.6	6.4	0.0	2.59	0.187
Black Hermit Caldron	82.0	2.4	66	499	88.5	49.5	5.4	1.8	5.64	0.910
Green Dragon Spring	89.0	3.0	449	244	281	60.3	4.1	0.3	3.71	0.146
Yellow Funnel Spring	78.5	3.0	699	95	405	69.8	3.2	1.7	0.22	0.097
Monarch Geyser Crater	88.0	4.6	433	120	—	—	—	—	—	—
Realgar Spring	29.5	2.3	429	391	—	—	—	—	—	—
Horseshoe Spring	90.0	2.9	384	169	227	57.0	4.9	0.5	1.82	0.043
Cinder Pool	86.5	3.9	661	153	404	70.7	9.3	0.4	1.91	0.033
VI. Sylvan Springs										
Sylvan Spring₁	80.0	2.2	365	1479	321	29.6	4.0	2.0	4.46	0.139
Sylvan Spring₂	88.0	1.9	5	608	12.7	3.6	8.0	1.1	2.60	0.014
Sylvan Spring₃	86.0	5.3	553	214	428	67.7	6.1	3.6	0.19	0.031
Sylvan Spring₄	91.0	3.9	462	283	—	—	—	—	—	—
Sylvan Spring₅	91.0	2.0	42	691	—	—	—	—	—	—
Sylvan Spring₆	—	3.5	14	—	—	—	—	—	—	—
Sylvan Spring₇	—	4.8	—	—	—	—	—	—	—	—
Sylvan Spring₈	85.0	5.3	366	210	298	27.2	7.2	1.9	0.96	0.029
Sylvan Spring₉	90.5	7.2	343	82	436	22.8	2.8	0.9	0.16	0.017
Sylvan Spring₁₀	45.0	8.3	584	77	—	—	—	—	—	—
Sylvan Spring₁₁	92.5	1.8	2	2039	4.50	10.8	5.6	4.2	33.4	
Sylvan Spring₁₂	90.5	1.8	4	1781	—	—	—	—	—	—
VII. Mud Volcano Area										
Sulphur Caldron	70.0	1.6	3	3595	—	—	—	—	—	—
Black Dragon Cauldron	79.0	2.0	5	3678	—	—	—	—	—	—
Moose Pool	75.0	1.7	6	3889	—	—	—	—	—	—
Sour Lake	28.7	2.0	12	1690	—	—	—	—	—	—
Mud Geyser	64.0	1.7	13	2274	—	—	—	—	—	—

[a]For locations, refer to the maps in this chapter or to Marler (1973). Analysis done by Prof. Kimio Noguchi and associates. Biological unstable elements were analyzed within hours of collection at a field laboratory in West Yellowstone, Montana. All samples were collected in the summer of 1971. It should be emphasized that there may be marked temporal variations in spring chemistry, so that the analyses ser mainly to give an indication of chemical differences and should not be assumed to still obtain.

Cr mg/l	Hg μg/l	As mg/l	Cu mg/l	Zn mg/l	H_2S mg/l	SiO_2 mg/l	HBO_2 mg/l	NH_4 mg/l	NO_2^- mg/l	NO_3^- mg/l	PO_4 mg/l
0.015	0.04	0.050	—	—	3.1	294	0.0	1.68	0.00	0.35	0.05
0.024	0.03	2.58	0.002	0.024	0.2	592	55.6	1.60	0.00	0.19	1.59
0.015	0.20	0.38	0.002	0.006	0.0	149	10.8	1.56	0.01	0.23	0.64
0.023	2.52	0.001	0.006	0.017	0.3	80	19.7	0.45	0.02	0.12	0.14
—	—	0.34	—	—	—	283	10.8	2.53	0.02	0.36	0.68
0.023	0.02	2.12	0.002	0.008	0.6	507	39.4	3.14	0.05	0.01	0.90
0.010	0.48	0.01	0.004	0.010	0.5	308	2.5	65.0	0.04	0.12	0.23
0.023	0.06	2.43	—	—	0.8	522	46.2	3.83	0.00	0.00	0.12
0.001	0.00	1.68	—	—	0.4	468	29.6	1.54	0.00	0.02	0.25
0.015	0.40	1.21	—	—	0.0	288	25.6	1.71	0.00	0.41	1.07
—	—	1.10	—	—	0.0	342	26.1	1.61	0.00	0.19	1.28
0.015	0.25	0.13	—	—	0.2	286	6.8	1.99	0.01	0.49	0.28
0.015	0.13	0.78	—	—	0.6	346	7.4	2.33	0.01	0.55	0.63
0.020	0.07	1.20	—	—	1.0	320	31.0	8.20	0.03	0.41	0.19
0.020	0.04	2.44	—	—	1.0	467	47.7	4.85	0.01	0.14	0.05
—	—	1.58	—	—	0.9	332	31.0	3.28	0.01	0.31	0.65
—	—	1.31	—	—	—	301	29.0	—	—	—	—
0.012	0.03	1.08	—	—	2.8	247	25.1	—	—	—	—
0.008	0.24	2.32	—	—	1.7	215	46.7	—	—	—	—
0.015	1.78	1.59	—	—	0.7	282	26.6	0.31	0.04	0.41	0.39
0.011	0.57	0.14	—	—	1.3	169	3.9	0.18	0.05	0.37	9.45
0.016	2.90	2.54	—	—	1.6	130	39.4	0.52	0.02	0.26	0.06
—	—	2.24	—	—	1.0	550	34.5	3.30	0.01	0.21	0.05
—	—	0.20	—	—	0.9	153	5.4	0.50	0.01	0.47	0.19
—	—	—	—	—	—	—	—	—	—	—	—
—	—	—	—	—	—	—	—	—	—	—	—
0.015	1.48	1.79	—	—	1.6	261	27.1	0.79	0.02	0.00	0.77
0.010	0.02	1.63	—	—	2.3	259	20.2	0.19	0.00	0.33	0.24
—	—	3.16	—	—	—	707	36.4	1.26	0.02	0.32	1.52
—	—	0.060	—	—	0.8	263	7.3	0.89	0.00	0.12	0.99
—	—	0.010	—	—	1.3	293	7.8	0.68	0.00	0.63	0.28
—	—	0.000	—	—	0.5	230	9.3	0.21	0.03	0.12	1.93
—	—	0.025	—	—	—	259	0.4	0.94	0.02	0.26	7.65
—	—	0.010	—	—	—	260	0.4	1.15	0.04	0.39	1.28
—	—	0.000	—	—	0.0	299	0.4	1.53	0.12	0.52	0.05
—	—	0.000	—	—	0.6	219	0.4	0.74	0.06	0.37	1.28

springs in the pH range of 2–4 and 7–9, but few between these two regions (Brock, 1971). The bulk of the data obtained is presented in Table 2.4. Data from some areas studied are not included, mainly the Azores, Italy, El Salvador, and Japan, either because the data were too few or because a thorough sampling program had not been undertaken. It is unlikely that addition of the data not included would alter the picture obtained. The data of Table 2.4 are presented graphically in Figure 2.7, and the bimodal distribution curve can readily be seen.

Table 2.4. pH Values for Thermal Springs in Various Parts of the World

pH Interval	Yellowstone[a]	Western U.S.[b]	New Zealand[c]	Iceland[d]	Summation
<1.5	1				2
1.5–2.0	7		3	0	10
2.1–2.5	14	1	14	0	29
2.6–3.0	33	2	14	1	50
3.1–3.5	16		2	4	22
3.6–4.0	7		4	3	14
4.1–4.5	3	1	2	3	9
4.6–5.0	3	2	0	3	8
5.1–5.5	3		1	2	6
5.6–6.0	5		3	5	13
6.1–6.5	4		7	1	12
6.6–7.0	12	3	6	0	21
7.1–7.5	16	8	2	0	26
7.6–8.0	16	3	2	0	21
8.1–8.5	25		7	1	33
8.6–9.0	24	1	1	2	28
9.1–9.5	8	2	0	4	14
9.6–10.0	6	1		8	15
10.1–10.5	1			2	3
Totals	204	24	68	40	336

Number of springs in each pH interval are given.
[a]Yellowstone data were obtained from the following main thermal areas: Upper Geyser Basin, Lower Geyser Basin, Shoshone Geyser Basin, Heart Lake Geyser Basin, Norris Geyser Basin, Sylvan Springs, Gibbon Creek, Mud Volcano, and minor springs around Fishing Bridge and Obsidian Creek.
[b]Western U.S. data were obtained from the following: Mt. Lassen Volcanic National Park, The Geysers, California, Steamboat Springs, Nevada, and Beowawe, Nevada.
[c]The New Zealand data were obtained from the following areas: Tikitere, Whakarewarewa, Ohinomutu, Waimangu, Waiotapu, Wairakei, Waikite Springs, Orakei Korako, Taupo, Lake Rotokawa, Ketetahi.
[d]The Iceland data were obtained from springs in the Myvatn, Husuvik, Krisuvik, Geysir, and Laugarvatn areas.
(From Brock, 1971.)

Figure 2.7. Graphic representation of data of Table 2.4.

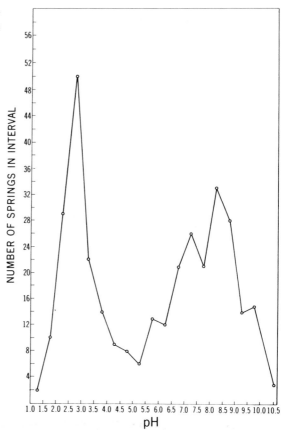

In general, the three main thermal areas of Yellowstone, New Zealand, and Iceland show similar bimodal distribution curves when plotted individually, but the Iceland curve is skewed more toward the alkaline side. Thus, there were no springs found in Iceland with pH values of 7–8, whereas a large number of springs had pH values greater than 9. In New Zealand, springs with pH values greater than 9 were not found. In Yellowstone, springs with pH values greater than 9 seem to be localized in certain areas, such as the Upper Geyser Basin and the Heart Lake Geyser Basin.

The bimodal distribution curve obtained provides a quantitative basis for the separation of springs by Allen and Day (1935) into acid and alkaline. (Another group of springs, travertine depositing, was not studied by us.) In chemical terms, these two kinds are characterized as sulfuric acid and bicarbonate springs. It might be noted that the pK for sulfuric acid is 1.92, and those for carbonic acid are 6.37 and 10.25 at 25°C. Silicate, another common anion in thermal springs, has one pK at 9.7. Presumably, the buffering effect of these anions plays a significant role in controlling the pH of hot springs.

Drilling and Subsurface Chemistry

The U.S. Geological Survey carried out an extensive drilling program in Yellowstone during the years 1967–1968 (White et al., 1975). The data obtained provided considerable insight into heat flow and source of heat, but also information on underground chemistry. Little of biological interest was obtained, but I was able to obtain one water sample from the drill hole in the Porcupine Hills, Lower Geyser Basin (Figure 2.8). This sample was obtained for me by Dr. Robert Fournier of the U.S. Geological Survey. The way the sample is obtained, there is no loss of gases or other constituents, but the temperature is lowered to normal ambient air temperature. The sample I obtained was from a depth of about 50 meters.

Chemically the water differed significantly from the nearby geyser water in two respects: high sulfide and low pH. The drill hole water had a pH of 5.5 and smelled heavily of sulfide, wheras the geyser nearby had a pH of around 8.5 and had almost no sulfide. The differences probably arise as a result of loss of CO_2 to the atmosphere when the water reaches the surface, and oxidation of sulfide upon contact with oxygen (see Chapter 12).

A 100-ml sample of this water was filtered through a membrane filter and examined microscopically for bacteria by epifluorescence. None were seen, although bacteria could be readily observed from a much smaller sample of geyser water. It is unlikely that deep subsurface microbial activity occurs, due to the lack of O_2 (needed as an electron acceptor by sulfide oxidizing bacteria).

Figure 2.8. Sampling from a drill hole in Yellowstone Park. Stainless steel tubing is lowered to the desired depth from the reel in the operator's hands.

References

Allen, E. T. and A. L. Day. 1935. *Hot Springs of the Yellowstone National Park.* Carnegie Institution of Washington Publication No. 466, Washington, D.C., 525 pp.

Barth, T. F. W. 1950. *Volcanic Geology, Hot Springs, and Geysers of Iceland.* Carnegie Institution of Washington Publication No. 587, Washington, D.C., 174 pp.

Brierley, J. A. 1966. Contribution of chemoautotrophic bacteria to the acid thermal waters of the Geyser Springs group in Yellowstone National Park. Ph.D. thesis, Montana State University, Bozeman.

Brock, T. D. 1967. Life at high temperatures. *Science* **158**, 1012–1019.

Brock, T. D. 1971. Bimodal distribution of pH values of thermal springs of the world. *Bull. Geol. Soc. Am.* **82**, 1393–1394.

Brock, T. D. 1973. Lower pH limit for the existence of blue-green algae: evolutionary and ecological implications. *Science* **179**, 480–483.

Brock, T. D. 1974. *Biology of Microorganisms,* 2nd ed. Prentice-Hall, Inc., Englewood Cliffs, N.J., 852 pp.

Brock, T. D. and M. L. Brock. 1969. Recovery of a hot spring community from a catastrophe. *J. Phycol.* **5**, 75–77.

Brock, T. D. and M. L. Brock. 1971. *Life in the Geyser Basins.* Yellowstone Library and Museum Association.

Brock, T. D. and J. L. Mosser. 1975. Rate of sulfuric-acid production in Yellowstone National Park. *Geol. Soc. Am. Bull.* **86**, 194–198.

Bunsen, R. 1847. Ueber den inneren Zusammenhang der pseudovulkanischen Erscheinungen Islands. *Liebig's Ann.* **62**, 1–59.

Burns, F. H. 1969. The smoking pillars of Horton River. *The Beaver,* Spring 1969, 40–43.

Cooney, D. G. and R. Emerson. 1964. Thermophilic fungi. W. H. Freeman, San Francisco, 188 pp.

Doemel, W. N. 1970. The physiological ecology of *Cyanidium caldarium.* Ph.D. thesis, Indiana University, Bloomington.

Doemel, W. N. and T. D. Brock. 1971. The physiological ecology of *Cyanidium caldarium. J. Gen. Microbiol.* **67**, 17–32.

Doemel, W. N. and T. D. Brock. 1974. Bacterial stromatolites: origin of laminations. *Science* **184**, 1083–1085.

Fliermans, C. B. and T. D. Brock. 1972. Ecology of sulfur-oxidizing bacteria in hot acid soils. *J. Bacteriol.* **111**, 343–350.

Fournier, R. O. and J. J. Rowe. 1966. Estimation of underground temperatures from the silica content of water from hot springs and wet-steam wells. *Am. J. Sci.* **264**, 685–697.

Gates, D. M., R. Alderfer, and E. Taylor. 1968. Leaf temperatures of desert plants. *Science* **159**, 994–995.

Gunter, G. 1957. Temperature. In: *Treatise on Marine Ecology and Paleoecology,* J. W. Hedgepeth, ed. Geol. Soc. Am. Mem. 67, *1*:159–184. Geological Society of America, Boulder, Colorado, 1296 pp.

Haslam, R. T. and R. P. Russell. 1926. *Fuels and their Combustion,* 1st ed. McGraw-Hill, New York, 809 pp.

Healy, J. 1960. The hot springs and geothermal resources of Fiji, N.Z. *Dept. Sci. Ind. Res. Bull.* **136**, 77 pp.

Hildebrandt, F. 1927. Beiträge zur Frage der Selbsterwärmung des Heues. *Zentralbl. Bakteriol.* 2 Abt. **71**, 440–490.

Hoare, R. A. 1966. Problems of heat transfer in Lake Vanda, a density stratified Antarctic lake. *Nature* **210**, 787–789.

Keefer, W. R. 1971. The geologic story of Yellowstone National Park. U.S. Geological Survey Bulletin No. 1347, 92 pp. Washington, D.C.

Likens, G. E. and N. M. Johnson. 1969. Measurement and analysis of the annual heat budget for the sediments in two Wisconsin lakes. *Limnol. Oceanogr.* **14**, 115–135.

Marler, G. D. 1973. *Inventory of Thermal Features of the Firehole River Geyser Basins and Other Selected Areas of Yellowstone National Park.* U.S. Department of Commerce, National Technical Information Service, PB-221 289, 639 pp. Springfield, Va.

Miehe, H. 1907. *Die Selbsterhitzung des Heus.* Fischer, Jena, 127 pp.

Mosser, J. L. and T. D. Brock. 1971. Effect of wide temperature fluctuation on the blue-green algae of Bead Geyser, Yellowstone National Park. *Limnol. Oceanogr.* **16**, 640–645.

Mosser, J. L., A. G. Mosser, and T. D. Brock. 1974. Population ecology of *Sulfolobus acidocaldarius.* I. Temperature strains. *Arch. Microbiol.* **97**, 169–179.

Myers, J. W., J. J. Pfeiffer, E. M. Murphy, and F. E. Griffith. 1966. Ignition and control of burning of coal mine refuse. U.S. Bur. Mines Rep. Invest. No. 6758, 24 pp.

Norman, A. G., L. A. Richards, and R. E. Carlyle. 1941. Microbial thermogenesis in the decomposition of plant materials, Part 1, An adiabatic fermentation apparatus. *J. Bacteriol.* **41**, 689–697.

Por, F. D. 1969. Limnology of the heliothermal solar lake on the coast of Sinai (Gulf of Eilat). *Verh. Int. Verein. Limnol.* **17**, 1031–1034.

Schlundt, H. and R. B. Moore. 1909. Radioactivity of the thermal waters of Yellowstone National Park. U.S. Geol. Survey Bull. No. 395.

Schramm, J. R. 1966. Plant colonization studies on black wastes from anthracite mining in Pennsylvania. *Trans. Am. Phil. Soc.* **56**, Part 1. Philadelphia, 194 pp.

Smith, D. W. and T. D. Brock. 1973. The water relations of the alga *Cyanidium caldarium* in soil. *J. Gen. Microbiol.* **79**, 219–231.

Stahl, R. W. 1964. Survey of burning coal-mine refuse banks. U.S. Bur. Mines Inform. Circ. No. 8209, 39 pp.

Waring, G. A. 1965. Thermal springs of the United States and other countries of the world. A summary. U.S. Geol. Surv. Prof. Paper No. 492, 383 pp.

White, D. E., R. O. Fournier, L. J. P. Muffler, and A. H. Truesdell. 1975. Physical results of research drilling in thermal areas of Yellowstone National Park, Wyoming. U.S. Geol. Surv. Prof. Paper No. 892, 70 pp.

Chapter 3
The Organisms: General Overview

In keeping with the definition of an extreme environment presented in Chapter 1, the organisms of most interest are those that live near the upper limits.

There are two ways in which we can obtain information on the upper temperature limits of different taxonomic groups. One is by looking at various thermal environments throughout the world, and the other is by looking at the species distribution along the thermal gradient in a single spring. In the latter case, we are dealing with organisms all living in water with the same chemical properties.

If we analyze the observations which have been made in all kinds of thermal environments (see Brock, 1967a, and Brock, 1970, for references to the early literature) we can tentatively construct the scheme shown in Table 3.1. Note that we are considering here not the ability of an organism to survive or endure a given high temperature, but its ability to carry out its complete life cycle. The limits given are not for all members of a group, but only for certain members, which would then be called thermophiles. For instance, there are many species of bacteria that are heat sensitive, even though certain bacteria are able to live and reproduce at the boiling point of water.

In evaluating these results, the problem of habitat suitability (other than temperature) and competition with other organisms merits attention. Thus, it is surprising that eucaryotic algae are not common at temperatures above 40°C, whereas eucaryotic fungi are found up to 60°C (Cooney and Emerson, 1964). This difference may merely be due to the fact that the niche into which the eucaryotic algae could fit is already occupied by blue-green algae. This idea is strengthened by the observation that the eucaryotic alga *Cyanidium caldarium* does live at temperatures near 60°C (Doemel and

Table 3.1. Upper Temperature Limits for Growth of Various Microbial Groups

Group	Approximate upper temperature (°C)
Animals	
Fish and other aquatic vertebrates	38
Insects	45–50
Ostracods (crustaceans)	49–50
Plants	
Vascular plants	45
Mosses	50
Eucaryotic microorganisms	
Protozoa	56
Algae	55–60
Fungi	60–62
Procaryotic microorganisms	
Blue-green algae (Cyanobacteria)	70–73
Photosynthetic bacteria	70–73
Chemolithotrophic bacteria	>90
Heterotrophic bacteria	>90

Brock, 1970), but is restricted to very acid hot springs (pH less than 4) where blue-green algae do not grow (Brock, 1973). Even among the blue-green algae, competition can be seen. For instance, the cosmopolitan thermophile *Mastigocladus laminosus* lives at higher temperatures in Iceland than in many Yellowstone hot springs probably because it does not meet competition in Iceland from *Synechococcus* (Castenholz, 1969a). In Yellowstone, many of the high-temperature environments in which *Mastigocladus* could grow are already well colonized with *Synechococcus*.

I discuss in more detail the upper temperature limit for eucaryotes later in this chapter.

Temperature and Species Diversity

The relationship between diversity and temperature for a single taxonomic group can best be illustrated by data on water beetles and blue-green algae (Table 3.2). The significance of such data in terms of ecosystem function are clear: as the temperature increases, the population structure becomes progressively simpler. From an evolutionary viewpoint, such correlations are less easy to explain. One explanation often advanced is that the species living at the upper limit are not optimally adapted to their environment, but have been able to extend their range into this region because they do not meet competition from other forms. Although this may be true for the animals and plants (see Mitchell, 1974), it is unlikely to be true for the microorganisms, since the members found at the highest temperatures seem to be optimally adapted to conditions close to if not identical with the extreme (Brock, 1967b; Brock et al., 1971).

Is There an Upper Temperature for Life?

This is a question of considerable fundamental importance.

In antiquity, the presence of organisms in hot springs was noted. Pliny the Elder, in his Natural History, noted: "Green plants grow in the hot springs of Padua, frogs in those of Pisa, fishes at Vetulonia in Etruria near the sea." The hot springs at Padua (Abano) still flow, and they are colonized by blue-green algae, perhaps the green plants of Pliny. The algae of hot springs were described by many early workers, but it was Ferdinand Cohn (1862) who first realized the general biological significance of organisms living in hot springs: "Even simple visual observations of the different colors show that different species exist at different temperatures in the water. Such observations are not merely of passing interest, since even if most aquatic plants and animals cannot live at temperatures above 37°C . . . it is important to know the highest temperature at which organic life, *no matter how organized,* can exist." (My translation and emphasis.) He did not find any algae growing in Karlsbad waters at temperatures above 55°C, a condition that generally agrees with Löwenstein's later observations (1903).

Table 3.2. Number of Species of Water Beetles and Blue-green Algae Collected in Hot Springs at Different Temperatures[a]

Water beetles		Blue-green algae	
Temperature (°C)	Number of species	Temperature (°C)	Number of species
30	60	10–15	42
31	58	15–20	54
32	55	20–25	76
33	52	25–30	86
34	47	30–35	90
35	46	35–40	86
36	45	40–45	76
37	35	45–50	60
38	33	50–55	25
39	33	55–60	24
40	22	60–65	2
41	15	65–70	1
42	10	70–75	1
43	6		
44	4		
45	2		

[a]From Brock (1975).

Hoppe-Seyler (1875) extended Cohn's work and pointed out two basic technical problems: (1) The necessity of showing that the organisms are indeed growing at the temperature in question and (2) temperature readings must be taken precisely where the organism is found, since the temperature even a few centimeters away from the organism may be quite different from that of the organism. From his observations on the hot springs of Padua (Abano), Sicily, and Ischia (Bay of Naples), he concluded that the upper temperature for algal growth was just above 60°C. He saw that his observations might have significance for an understanding of the origin of life and speculated that when the earth was cooling, chlorophyll-containing and hence CO_2-fixing and O_2-producing organisms could already live when the temperature was still over 60°C.

Before further discussion, it is necessary to examine what is meant by the question, what is the upper temperature of life? Temperature is only one of many variables influencing the growth of living organisms. It seems reasonable that environmental factors such as pH, nutrient quality or quantity, hydrostatic pressure, salinity, and light intensity (for photosynthetic organisms) will probably influence temperature responses, including the upper temperature at which an organism can grow. Heterotrophic bacteria cannot grow in hot-spring water in the absence of organic matter. We can thus imagine that the highest temperature at which some organism is now living somewhere on earth is not the upper temperature for life, but only the upper temperature at which all conditions for life are possible.

Probably the most careful early observations were those made by Setchell (1903); these were apparently never published in detail, but a lengthy unpublished manuscript exists in the University of California-Berkeley archives (personal communication from Prof. George F. Papenfuss). He made extensive observations in Yellowstone Park and reported that the upper temperature where algae were visible was 75–77°C, and that at which bacteria were found was 89°C. These upper limits occurred in siliceous waters having a slightly alkaline pH. In travertine-depositing springs, the upper temperature for algae was lower by about 10°C. Kempner (1963) reexamined this question, and cast doubt on Setchell's observations, suggesting that 73°C was the highest temperature at which organisms were found growing, but my own observations in Yellowstone confirm Setchell's conclusions.

Since the summer of 1966, I have made a large series of observations in Yellowstone and elsewhere. In the effluent channels of hot springs where the flow and temperature are usually constant, it is quite easy to determine the upper temperature for algal growth, the algal mats forming characteristic V-shaped patterns, which are due to the lower temperatures at the edge than those in the center of the channel (see Figure 3.1). Visible algal growth (of the unicellular blue-green *Synechococcus*) is found at temperatures up to 73–75°C, but not at higher temperatures (see also Castenholz, 1967). That these algae are growing and not merely existing can be easily shown by darkening the channel; within 5 to 7 days the algal cells have completely

Figure 3.1. The characteristic V-shaped pattern formed by the blue-green algae growing at the upper temperature limit for growth. This is a small spring in the West Thumb Geyser Basin, Yellowstone Lake in the background. The V develops because water cools more rapidly at the edges than in the center of the channel, so that the upper temperature limit is reached further upstream at the edges. Water flows toward the lake from the source, just out of the bottom of the picture.

disappeared (see Chapter 8). This is due to the fact that a steady-state exists between growth and washout of the algal cells. In the absence of light, growth no longer can occur (thus showing that in nature these algae are obligately phototrophic), and the existing algae are quickly washed away (Brock and Brock, 1968).

Bacteria are present in some, but by no means all, of the hot springs at temperatures much above 75°C. In the effluents of certain springs in the White Creek area I have found pinkish, yellowish, or whitish masses of filamentous bacteria at temperatures up to 88°C. These bacteria are present in such large amounts that they can be readily seen with the naked eye (Figure 3.2), and under the microscope are revealed as dense tangles of long filaments (Figure 3.3). Spectrophotometry of acetone extracts of these organisms revealed no chlorophyll. In other springs, with temperatures of over 90°C, filamentous and rod-shaped bacteria are visible only microscopically. These organisms are probably similar to those described by Setchell as "filamentous Schizomycetes." Bacteria at temperatures "near 90°C" were also seen by Van Niel (see Allen and Day, 1935). Water boils at about 92°C at Yellowstone.

That these bacteria are growing at these temperatures is shown by the fact that if an artificial substrate (a glass slide or piece of cotton string) is placed in the pool or effluent channel, it quickly becomes covered with

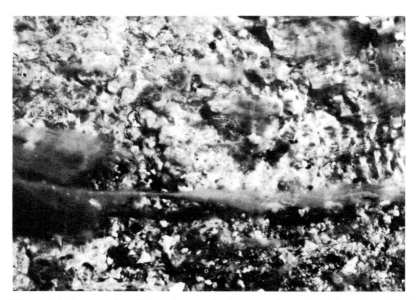

Figure 3.2. Large tufts of bacteria growing in an outflow channel at about 85°C. At this temperature, no photosynthetic organisms are seen. Only a relatively small percentage of Yellowstone springs show such massive accumulations, but virtually all springs of neutral to alkaline pH show microscopic evidence for extensive bacterial growth. Length of photo, about 10 cm.

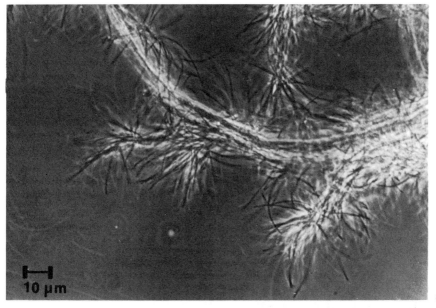

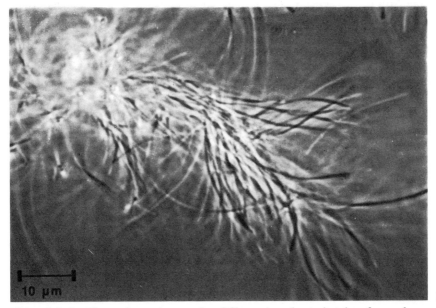

Figure 3.3a,b. Phase photomicrographs of filamentous bacteria taken from a large tuft in rapidly flowing water at 85°C in Octopus Spring (formerly called Pool A), Yellowstone Park.

bacterial cells. I have made a survey for bacterial growth in superheated pools of Yellowstone (at temperatures of 93.5–95.5°C) by immersing glass slides several feet down in these pools and retrieving them after a week to 10 days. In *every* pool, bacteria were present on the slides, and in over half the pools the bacteria so densely covered the slides as to form a film visible to the naked eye. In one pool, where the pink bacteria are present, string immersed in the pool (at temperatures that never go below 91°C) became covered with macroscopically visible pinkish masses. We have carried out several detailed studies on these high-temperature bacteria (Bott and Brock, 1969; Brock et al., 1971; Brock and Darland, 1970), and our results are clear-cut. In springs of neutral to alkaline pH, bacteria live at temperatures right up to the boiling point, and are optimally adapted to temperatures close to this. Since bacteria have not been observed in fumaroles at temperatures above boiling, where liquid water does not exist, I have concluded that life is possible at any temperature at which there is liquid water (Brock, 1967a).

The Work of Setchell

Because the pioneering work of this California phycologist has never been published, and because it provided some of the earliest and firmest evidence for the presence of life at high temperatures, it seems appropriate to present a brief excerpt here. As noted earlier, the lengthy unpublished manuscript of Prof. W. A. Setchell is in the archives of the University of California-Berkeley. (The copy I have was made for me by Dr. Michael Tansey.) Entitled "The Algae of the Yellowstone National Park," the manuscript was first written for the U.S. Geological Survey in 1899, and a brief note on the work was published in *Science* in 1903 (Setchell, 1903). The manuscript was later revised in 1915 but again not published. In 1932 Setchell wrote to Dr. David White, Head Geologist for the U.S. Geological Survey, requesting a small amount of funds to get the manuscript in final shape for publication. It is not known if this request for funds was denied, but the manuscript was not published. It is 215 pages long, and deals primarily with the taxonomy of the Yellowstone thermal algae. This portion of the manuscript has been rendered superfluous by the extensive work of Copeland in the 1930s (Copeland, 1936), but the observations on the high-temperature bacteria are still of some relevance and will be quoted here, in slightly abbreviated form.

Setchell's Manuscript

The Schizomycetes or bacteria are closely related to the algae, or at least to the Myxophyceae . . . and as several of the colorless filamentous forms of these plants were obtained by members of the U.S. Geological Survey, as well as the writer, it has seemed best to include here such an account of the forms met as shall be possible under present circumstances. The minute unicellular forms, in the strictest sense of the word, were not, of course, investigated, but two or three of the filamentous types occurred in such masses and in connection with such high temperatures as to make them noticeable, and also to make it desirable to incorporate into this monograph all that could be learned in regard to them. Undoubtedly the more minute forms occur, since several cases are mentioned in the literature. . . .

Chlamydothrix Calidissima (Setchell Manuscript)

The growth of this species occurs in the form of tufts which are very gelatinous. The color is whitish, or while in the water it seemed to be of a very light flesh color. The tufts are not high, being not over 3–5 millimeters. The filaments are long and slender, their diameter ranging from 0.3–0.5 μm. They are colorless and without distinct granules, the filaments breaking up and partially disintegrating into shorter or longer fragments at the tip, as is characteristic of the genus; the sheath is indefinite and gelatinous.

The name Chlamydothrix is given to this form as a provisional designation, by which this, one of the most interesting of the thermal organisms, may be known. It will be impossible to determine much about it without further investigation in its habitat. It seems to be either a Chlamydothrix or a Thiothrix, and in the absence of granules, it has seemed best to the writer to refer it for the present, to the former genus. It is impossible to compare it with the other species of this genus, however, and it is to be distinguished for the present, only by the high temperature which it undoubtedly prefers. It seems best to adopt Chlamydothrix as a generic name, following Migula. . . .

Chlamydothrix calidissima is found in the very hottest waters (89°C) in the Yellowstone National Park holding living organisms, so far as the writer can determine at the present time; the highest recorded temperature for any living organism being 93°C, recorded by Brewer (1866, p. 392), for unicellular algae at the so-called "geysers" in California [Brewer's observations are almost certainly wrong, TDB]. Brewer says that his specimens, growing at the highest temperatures, were of a bright green color, and consequently they must have been quite different from those of Yellowstone Park, which are filamentous and colorless. The writer, as has been stated, has not found any green species growing above the temperature of 77°C (*Gloeothece effusa*) and it is interesting to note that careful observations to determine this point throughout the Park demonstrated quite conclusively, that, as far as its hot springs are concerned, the inhabitants of the hottest waters are colorless and belong to the group of the Bacteriaceae or Schizomycetes.

The locality for this species is Black Sand Basin, in the Upper Geyser Basin (W.A.S., No. 1965). The specimens grew in the bowl of the spring, not far from the point where the overflow passes out. The species, although small and inconspicuous, yet grew in sufficient abundance to make itself plainly evident. It is certainly suggested by a study of this Chlamydothrix, that there may be species of Bacteria of inconspicuous size and habit, which do not easily catch the eye. It may be a significant fact that in the immediate vicinity of these tufts, associated with the filaments themselves, were found numerous tracheids of some coniferous wood. Although it cannot be demonstrated that the filaments of the Chlamydothrix are attached in any way to the tracheids, it is decidedly suggestive that they are present in intimate association.

Chlamydothrix Penicillata Sp. Nov.

The growth of this species is in the form of gelatinous salmon-colored penicillate masses in which the unbranched slender filaments are very numerous and more or less intertwined, yet with a general longitudinal course in the penicillate tufts of jelly. The filaments are of average length and composed of cells not much longer than their diameter. They are little more than 0.1 μm in thickness (without the gelatinous sheath-like covering) and are readily to be distinguished, particularly in stained preparations, from the filaments of the foregoing species (*Chl. calidissima*) with which they are associated. The greater mass of the filaments belong to the present species. The two species, however, are closely related, but seem distinct in their size and in habit, but may be simply forms of the same plants. They both break up and separate at the tips into short segments.

Chlamydothrix penicillata was collected on the writer's second visit to Yellowstone National Park in July 1905 from a small rill flowing out of "Firehole Pool" in Firehole Basin. It is No. 6131 of the writer's collections and was growing on pine tracheids. It is very satisfactory to have collected what seem to be living plants at a temperature closely approximating 89°C on two visits, in localities widely separated and thus to firmly fix this highest temperature at which organisms have been found living. No. 6131 is simply a small sample selected from an abundance of material existing at the place of collection.

Thiothrix carnea (Setchell Manuscript)

The filaments of this species form extended tufts of a flesh color attached to the bottom of quiet pools. The filaments are several centimeters long and about 0.3 μm thick. The sheaths are not readily demonstrated but the jelly which surrounds them and holds them together indicates the presence of structures which become diffluent into an amorphous gelatinous substance. The filaments do not show any septa and the cell contents contain granules presumably of sulphur, arranged in an irregular line.

The species described above under the provisional name of *Thiothrix carnea* resemble closely the succeeding species except for the absence of incrustation of crystals or lumps of sulphur which form so constant and so characteristic a feature of the masses of those species. The filaments of *Thio. carnea* are somewhat more slender than those of *Thio. luteola* and at the same time they are provided with granules which distinguish them from the plant described above under the name of *Chlamydothrix calidissima*. It is with considerable hesitation that the writer has appended any names at all to these several filamentous bacteria, but it has seemed best to do something of the kind with the understanding that the names thus applied are only provisional and merely for convenience in discussing the part played by these species in the thermal waters. The difficulty of determining just where the limits of such a form as *Phormidium tenuissimum* are situated as regards these forms, and where they begin to be distinct, is a matter to be mentioned here to emphasize the closeness of the genus *Phormidium* and the genera of the Schizomycetes just noted. It is quite possible that the writer has reckoned some of the filamentous species of the Schizomycetes under the head of *Phormidium tenuissimum*, since it is impossible to determine with certainty in all cases whether a filament less than one micromillimeter in thickness, contains phycochrome or not.

The temperature of the waters in which *Thiothrix carnea* was growing was determined with special care, since it was the highest that the writer had found in connection with living organisms up to that time and remains the highest save two, viz. that of *Chlamydothrix calidissima* and *Chl. penicillata* mentioned above.

Thiothrix carnea was found by the writer in only one locality in the Yellowstone National Park, viz. the pool incorrectly called Morning Glory Pool, Norris Geyser Basin, W.A.S., No. 1930.

There was a large quantity of the material lining the bottom of the pool and extending out beyond reach of the writer's arm. The plants within reach were submerged from 3 to 10 centimeters. The temperature of the tufts, taken in several different parts of the pool, was uniformly 82°C. This temperature is the second highest at which the writer has found living plants, as has already been stated.

Taxonomic Confusion between Blue-green Algae and Nonchlorophyllous Procaryotes (Bacteria)

In his discussion of high-temperature bacteria, Setchell (see above) alludes to the difficulty of distinguishing tiny filamentous blue-green algae from filamentous bacteria. Setchell himself seems to have accomplished this readily, and with microscopes we would consider crude. Unfortunately, later workers were not blessed with Setchell's powers of observations, and several erroneous conclusions have been drawn. In his extensive monograph, which has been widely quoted, Copeland (1936) described a number of high-temperature forms as blue-green algae which are almost certainly bacteria (Brock, 1968). When we began our work in Yellowstone, we were fortunate to have modern microscopes and spectrophotometers, and were able easily to separate true blue-green algae from the filamentous bacteria living at the very highest temperatures.

Entering Yellowstone primarily as a bacteriologist, it was easy for me to see that the whitish, yellowish, and pinkish masses living in certain hot spring effluents at temperatures close to the boiling point were bacteria and not algae. A simple solvent extraction of the pigments, followed by spectrophotometry, confirmed the absence of chlorophyll. Indeed, the high-temperature forms observed in the 1890s by Setchell in Firehole Pool were still there when I first saw this pool in 1965.

In his monograph, Copeland (1936) describes four taxa from thermal waters which resemble considerably the filamentous forms that I have observed. His descriptions of these forms follow:

> 1. *Oscillatoria filiformis,* sp. nov. Plant mass salmon-brown to tannish. Trichomes long, flexuous, entangled, 0.4–0.5 μ in diameter. . . . In general it grows on and among other Myxophyceae or at the higher temperatures alone. It is one of the most typical limital species of the basic springs. The temperature of 85.0°C which it tolerates may be a fraction of a degree above the limit for growth.
>
> 2. *Phormidium geysericola,* sp. nov. Stratum tough, fibrous, not lamellate, . . . up to 3 cm in thickness with corded streamerlike projections up to 10 cm in length and up to 1 cm in diameter, pale salmon to yellowish, whitish below. Filaments very long, straight . . . 0.4–0.6 μ in diameter, pale blue-green, often yellowish and almost colorless. . . . The species seems to occur only in very specialized habitats, where the water is hot and agitated; it is a well marked thermal limit species.
>
> 3. *Phormodium subterraneum,* sp. nov. Stratum extended, soft leathery, up to 1.5 mm thick . . . yellow in diameter . . . yellowish green. . . . This species resembles *P. bijahensis* . . . but is easily separated from it by the nonpigmented matrix. . . .
>
> 4. *Phormidium bijahensis,* sp. nov. Stratum salmon-pink, flesh-colored or red . . . Trichomes . . . 0.3–5 μm in diameter . . . extraordinarily high thermal tolerance . . . up to 85.2°C.

Note that all four of the above species are characterized by filaments less than 1 μm in diameter, by strata that are not obviously green or blue-green in color, and that they are thermal limit species. From my observations in Yellowstone, including many springs studied by Copeland, I would postulate that none of these forms is a blue-green alga. The predominant color of the strata in which these forms occur is salmon, yellow, or orange. Furthermore I have observed the form *P. bijahensis,* which is pink or flesh-colored, in a number of Yellowstone springs. The filaments do not fluoresce red, and spectrophotometry of acetone extracts shows no evidence of chlorophyll. Mann and Schlichting (1967), apparently following Copeland, presented *P. bijahensis* as a thermal limit species of a blue-green alga. I have studied some of the same springs that these workers have studied, found the same form, but can show that it lacks chlorophyll.

The taxonomic status of the thermophilic filamentous bacteria is uncertain. Although they may be related to the flexibacteria, I have not as yet found any evidence of gliding motility. Thus there is no reason to believe that these forms are apochlorotic Cyanophyceae, although further work is needed to clarify this point.

Having said all this, I must now confess that I myself erred when making observations on filamentous organisms living in association with blue-green algae at temperatures of 55–73°C. Having found that the high-temperature forms lacked chlorophyll, I was prepared to find that the thin-diameter filamentous forms associated with blue-green algae at temperatures of 55–73°C were also nonchlorophyllous bacteria. In my paper discussing the taxonomic confusion (Brock, 1968) I described the use of fluorescence microscopy to determine whether organisms in these mats were algae or bacteria. Chlorophyll a-containing organisms show a bright red fluorescence when observed with blue-light illumination, making it easy to spot them in mixtures. When samples from a variety of mats taken from temperatures of 55–70°C were observed, the only red-fluorescing organism was the unicellular blue-green alga *Synechococcus lividus* (see later in this chapter). Since the associated filamentous forms did not fluoresce, I deduced that they were nonphotosynthetic. Some autoradiographic studies using $^{14}CO_2$ seemed to confirm this; only the *Synechococcus* cells became labeled.

We now know that the predominant filamentous organism in these mats is the interesting photosynthetic bacterium *Chloroflexus aurantiacus* (Pierson and Castenholz, 1971, 1974a; Bauld and Brock, 1973; Madigan et al., 1974; Madigan and Brock, 1975). This organism contains bacteriochlorophyll c, which does not show red fluorescence. We missed the photosynthetic activity of this organism in our autoradiographic studies because it does not incorporate much $^{14}CO_2$; it grows primarily as a photoheterotroph (Doemel and Brock, 1977).

These problems point up some of the difficulties in working with organisms in extreme environments; there is no precedence for deciding the kinds of organisms that might be there.

Well-characterized Thermophilic Procaryotes

The diversity of procaryotes which have been cultivated at high or moderately high temperature is fairly large, although as the temperature is raised, the number of species able to grow drops markedly. I list the well-characterized genera and species of thermophilic procaryotes in Tables 3.3 and 3.4. Several precautionary notes about these tables follow: (1) I do not vouch for the validity of all the taxa; (2) the upper temperature limits for certain species may be found after more work to be either higher or lower than those listed; (3) it is likely that not all naturally occurring thermophilic bacteria are listed, because a concerted search for diverse types of thermophilic bacteria has not been made.

As seen in Table 3.3, the blue-green algae able to live at the highest temperatures are unicellular, non-nitrogen-fixing forms, referable to the genus *Synechococcus* (see discussion in Chapter 8). The upper temperature limit for the filamentous forms is somewhat lower, the nitrogen-fixing species *Mastigocladus laminosus* growing at temperatures up to 64°C, although only showing extensive development at temperatures under 60°C

Table 3.3. Procaryotic Microorganisms Growing at High Temperatures

Group, genus, species	Optimum temperature	Maximum temperature	Remarks	Reference[a]
Blue-green algae (Cyanobacteria)				
Chroococcales				
Synechococcus lividus	63–67	74	One strain (others, lower optimum)	40
S. elongatus		66–70		10
S. minervae		60		10
Synechocystis aquatilus		45–50		10
Aphanocapsa thermalis		>55		10
Chamaesiphonales				
Pleurocapsa sp.		52–54		10
Oscillatoriales				
Oscillatoria terebriformis		53		10
O. amphibia		57		10
O. germinata		55		10
O. okenii		>60		10
Spirulina sp.		55–60		10
Phormidium laminosum		57–60		10
P. purpurasiens		46–47		10
Symploca thermalis		45–47		10
Nostocales				
Calothrix sp.		52–54		10
Mastigocladus laminosus		63–64		10
Photosynthetic bacteria				
Chlorobiaceae				
Chloroflexus aurantiacus	55	70–73		2, 44
Chromatiaceae				
Chromatium sp.		57–60	Natural observations only	10
Nonphotosynthetic				
Spore formers[b]				
Bacillus acidocaldarius	60–65	70	Acidophile	12, 51
B. coagulans	37–45	60	Acidophile	4, 16, 19
Bacillus sp.	55–60	70	Hydrocarbon oxidizer	38
Bacillus YT-P	72	82	Proteolytic	23
Bacillus YT-G	80	85	Needs confirmation	22

(*continued on the following pages*)

Group, genus, species	Optimum temperature	Maximum temperature	Remarks	Reference[a]
B. thermocatenulatus		78		20
B. stearothermophilus	50–65	70–75	Versatile, widespread	16, 19, 21
B. licheniformis		50–55		19
B. pumilus		45–50		19
B. macerans		40–50		19
B. circulans		35–50		19
B. laterosporus		35–50		19
B. brevis		40–60		19
B. subtilis		55–70	Aromatic, heterocyclic, alcohol utilizers (many strains)	1
B. sphaericus		65–70	Carboxylic acid utilizers (several strains)	1
Clostridium thermosaccharolyticum	55	67		28, 50
C. thermohydrosulfuricum		74–76	Reduces sulfite	28
C. tartarivorum		67		28
C. thermocellum	60	68	Cellulose digester	5
C. thermoaceticum	55–60	65		5
C. thermocellulaseum	55–60	65	Cellulose digester	5
Clostridium sp.	60	75	Cellulose digester	34
Clostridium sp.	50–75		Butyric formation	35
Clostridium (many species)	35–45		Some pathogens	50
Desulfotomaculum nigrificans	55	70	Sulfate reducer	9, 27
Lactic acid bacteria				
Streptococcus thermophilus	40–45	50		14
Lactobacillus thermophilus	50–63	65		18
Lactobacillus (Thermobacterium) bulgaricus	40	52.5		42
Lactobacillus (various species)	30–40	53		47
Bifidobacterium thermophilum	46.5			47

Table 3.3. Procaryotic Microorganisms Growing at High Temperatures (*continued*)

Group, genus, species	Optimum temperature	Maximum temperature	Remarks	Reference[a]
Actinomycetes				
Streptomyces fragmentosporus	50–60			24
S. thermonitrificans	45–50			45
S. thermoviolaceus	50	60		45
S. thermovulgaris		60		45
Pseudonocardia thermophila	40–50	60		25
Thermoactinomyces vulgaris	60	70		31
T. sacchari	55–60	65		31
T. candidus		60		30
Thermomonospora curvata	50	65		36
T. viridis	50	60		31, 36
T. citrina	55–60	70–75		36
Microbispora thermodiastatica		55		11
M. aerata		55		11
M. bispora		60		11
Actinobifida dichotomica	50–58			11
A. chromogena	55–58			11
Micropolyspora caesia	28–45	55		11
M. faeni	50	60		11
M. rectivirgula	45–55	65		11
M. rubrobrunea	45–55	65		11
M. thermovirida	40–50	57		11
M. viridinigra	45–55	65		11
Actinoplanes				
Streptosporangium album var. *thermophilum*	50–55	70		37
Methane-producing bacteria				
Methanobacterium thermoautotrophicum	65–70	75		53
Sulfur-oxidizing bacteria				
Thiobacillus thiooxidans		55	Acidophile	17
Thiobacillus sp.	50	55–60		52
T. thermophilica	55–60	80	Spore former (valid species?)	15

Group, genus, species	Optimum temperature	Maximum temperature	Remarks	Reference [a]
Sulfolobus acidocaldarius	70–75	85–90	Acidophile	7, 41
Thermothrix thioparus	70–73	77–80	Facultative, filamentous, denitrifying	8
Sulfate-reducing bacteria				
Desulfovibrio thermophilus	65	85		48
Mycoplasma				
Thermoplasma acidophilum	59	65	Acidophile	3, 13
Spirochete				
Leptospira biflexa var. *thermophila*		54		26
Methane-oxidizing				
Methylococcus capsulatus	30–50	55		32
Pseudomonads				
Hydrogenomonas thermophilus	50	60		39
Gram-negative aerobes (uncertain affiliation)				
Thermomicrobium roseum	70–75	85		29
Thermus aquaticus	70	79		6
T. flavus	70–75	80		48
T. (Flavobacterium) thermophilus	70	85		43
T. ruber	60	80		33
Thermus X-1	69–71			46

[a]1, Allen, 1953; 2, Bauld and Brock, 1973; 3, Belly et al., 1973a; 4, Belly and Brock, 1974; 5, Breed et al., 1957; 6, Brock and Freeze, 1969; 7, Brock et al., 1972; 8, Caldwell et al., 1976; 9, Campbell, 1974; 10, Castenholz, 1969b; 11, Cross, 1974; 12, Darland and Brock, 1971; 13, Darland et al., 1970; 14, Deibel and Seeley, 1974; 15, Egorova and Deryugina, 1963; 16, Fields, 1970; 17, Fliermans and Brock, 1972; 18, Gaughran, 1947; 19, Gibson and Gordon, 1974; 20, Golovacheva et al., 1975; 21, Gordon et al., 1973; 22, Heinen, 1971; 23, Heinen and Heinen, 1972; 24, Henssen, 1969; 25, Henssen, 1974; 26, Hindle, 1932; 27, Hollaus and Klaushofer, 1973; 28, Hollaus and Sleytr, 1972; 29, Jackson et al., 1973; 30, Kurup et al., 1975; 31, Küster, 1974; 32, Leadbetter, 1974; 33, Loginova and Egorova, 1975; 34, Loginova et al., 1966; 35, Loginova et al., 1962; 36, Manachini et al., 1966; 37, Manachini et al., 1965; 38, Mateles et al., 1967; 39, McGee et al., 1967; 40, Meeks and Castenholz, 1971; 41, Mosser et al., 1973; 42, Orla-Jensen, 1942; 43, Oshima and Imahori, 1974; 44, Pierson and Castenholz, 1974a,b; 45, Pridham and Tresner, 1974; 46, Ramaley and Hixson, 1970; 47, Rogosa, 1974; 48, Rozanova and Khudyakova, 1974; 49, Saiki et al., 1972; 50, Smith and Hobbs, 1974; 51, Uchino and Doi, 1967; 52, Williams and Hoare, 1972; 53, Zeikus and Wolfe, 1972.

[b]The diversity of thermophilic bacilli almost certainly reflects the ease of isolation and the practical importance of this group, rather than any unusual adaptation to thermophily.

Table 3.4. Upper Temperature Limits for Genera
of Nonphotosynthetic Bacteria[a]

Neutral pH	Maximum
Thermomicrobium	85
Thermus	85
Bacillus sp.	82
Thermothrix thioparus	77–80
Bacillus stearothermophilus	75
Methanobacterium	75
Thermomonospora	70–75
Clostridium	74–76
Desulfotomaculum	70
Thermoactinomyces	70
Streptosporangium	70
Micropolyspora	65
Streptomyces	60
Pseudonocardia	60
Microbispora	60
Hydrogenomonas	60
Leptospira	54
Lactobacillus	53

Acid pH	Maximum
Sulfolobus	85–90
Bacillus acidocaldarius	70
Thiobacillus	55
Thermoplasma	59

[a] See Table 3.3 for details and references.

(Castenholz, 1969a). It is of some interest that the thermal limit species, *Synechococcus,* is missing from both Iceland (Castenholz, 1969a) and New Zealand (Brock and Brock, 1971a), two countries with extensive thermal areas and springs of types similar to those found in Yellowstone. *Synechococcus* is present in most North American and Central American springs of suitable character. (I have seen it throughout the Western United States and in El Salvador.) Why this organism is missing in Iceland and New Zealand is an interesting question in biogeography.

The only photosynthetic bacterium which shows extensive development in hot springs is *Chloroflexus,* which will be discussed in detail in a later chapter.

Chemolithotrophic Bacteria

Reduced sulfur compounds are common constituents of geothermal habitats; it should not be surprising to find thermophilic sulfur-oxidizing microorganisms. Two organisms, *Thiobacillus thiooxidans* and *Sulfolobus aci-*

docaldarius, have been found in acidic habitats (Fliermans and Brock, 1972; Brock et al., 1972). *Sulfolobus* is the most thermophilic autotroph available in pure culture, being able to grow at temperatures up to 85–90°C (see Chapter 6).

Very few sulfur-oxidizing bacteria able to grow at neutral pH under thermophilic conditions have been described. A spore-forming sulfur-oxidizing thermophile, *Thiobacillus thermophilica*, was isolated by Egorova and Deryugina (1963). The *Thiobacillus* isolated by Williams and Hoare (1972) grew only up to 60°C. Almost certainly, more thermophilic neutral pH thiobacilli (or other sulfur-oxidizers) exist, as sulfur crystals deposited in springs at temperatures close to boiling often are colonized by large numbers of bacteria (Brock, unpublished observations), but cultures have not been isolated. Attempts to isolate the sulfur bacteria living at temperatures over 90°C in Boulder Spring (Brock et al., 1971) were unsuccessful, although the ecological studies showed clearly that these bacteria were able to functioh at high temperatures (see Chapter 10). Specifically, CO_2 fixation by the Boulder Spring bacteria required the presence of sulfide, and a temperature optimum around 90°C was found. Further work on the Boulder Spring bacteria was abandoned when *Sulfolobus* was isolated, since this latter organism was considerably more amenable to ecological and cultural studies (Mosser et al., 1973; Brock et al., 1972; Shivvers and Brock, 1973; Mosser et al., 1974a; Mosser et al., 1974b; Bohlool and Brock, 1974; Brock and Mosser, 1975). Natural populations of *Sulfolobus* also oxidize ferrous iron, with a temperature optimum similar to that for the oxidation of elemental sulfur (Brock et al., 1976).

Methanogenic Bacteria

Methanogenic thermophiles can be readily isolated, even from sewage sludge (Zeikus and Wolfe, 1972), and also exist as a component of hot spring algal mats (Zeikus, personal communication). These bacteria can grow completely chemolithotrophically, and should provide interesting material for evolutionary studies.

Heterotrophic Bacteria

It is among the heterotrophic bacteria that most culture work has been done, and these bacteria are also of potential industrial importance. Although it is well established that certain bacteria thrive even in boiling water (Bott and Brock, 1969; Brock et al., 1971; Brock and Brock, 1971b; Brock, 1967a), cultures have been obtained only of organisms able to grow at somewhat lower temperatures. The highest temperature at which it has been possible to grow a bacterial culture continuously and reproducibly is about 80–85°C (Heinen, 1971).

Although much work on enzymes of thermophilic heterotrophic bacteria has been published, little ecological work has been done. The genus

Thermus, described by Brock and Freeze (1969) as the first nonsporulating extreme thermophile (see Chapter 4), has been isolated by a number of workers, and several new species have been described (see Table 3.3). Most hot spring waters probably have small amounts of organic matter, so that populations of bacteria can be maintained even in the absence of reduced sulfur compounds. Attempts to culture heterotrophic bacteria at temperatures of 90°C have so far failed, although in at least one spring (Octopus Pool, in Yellowstone Park), the temperature optimum of the resident population is about 90°C (Brock and Brock, 1971b).

At temperatures of 70°C or below, the blue-green algae which are able to grow and form mats excrete sufficient organic matter to support the growth of some heterotrophic bacteria (see Bauld and Brock, 1974, for documentation of excretion), and a diverse heterotrophic flora exists, although it has been little studied. Microscopic examination of blue-green algal mats reveals, in addition to the blue-green algae, primarily gram-negative rods similar to *Thermus aquaticus,* and this organism can be readily enriched from algal mats by incubation at 70°C (Brock and Freeze, 1969). Interestingly, the ubiquitous thermophile *Bacillus stearothermophilus* almost never appears in 70°C enrichments, although if an enrichment temperature of 55°C is used, it can be routinely isolated from the same mats (Brock, unpublished observations).

A variety of other thermophilic heterotrophs, both aerobic and anaerobic, are listed in Table 3.3. The involvement of thermophilic actinomycetes in self-heating of hay and compost is well established. I discuss the thermophilic, acidophilic *Thermoplasma,* present in self-heating coal refuse piles, in detail in Chapter 5.

Thermophilic Eucaryotes

Very early in my studies in Yellowstone, I perceived that eucaryotic microorganisms were unable to live at temperatures as high as those of procaryotes (Brock, 1967a). This conclusion was based on observations of the distribution of the alga *Cyanidium caldarium,* which occupies solely the niche of photosynthetic organisms able to grow in warm, acid waters. When Michael Tansey joined my laboratory as a post-doctorate, ripe with knowledge of thermophilic fungi, it seemed an opportune time to document more fully the upper temperature limit for eucaryotic microorganisms. This subject is described in some detail in a recent review (Tansey and Brock, 1978) and in an earlier article (Tansey and Brock, 1972).

Eucaryotic Algae

As noted, one species of eucaryotic algae stands out as being able to grow at high temperatures, *Cyanidium caldarium.* It has an upper temperature limit of 55–60°C, based on field observations, $^{14}CO_2$ incorporation in nature, and cultural studies in the laboratory (Doemel and Brock, 1970).

This alga has been fairly extensively studied in my laboratory, and will be discussed in some detail in Chapter 9. There is a large gap in upper limits between *C. caldarium* and other eucaryotic algae. High-temperature strains of *Chlorella* are known which grow at about 42°C (Sorokin, 1967) and some diatoms grow in hot springs at temperatures of 30–40°C. A number of reports (reviewed by Tansey and Brock, 1978) exist of diatoms able to grow at much higher temperatures. It is true that diatom *frustules* can be found even in boiling water, but this merely reflects the resistance of these siliceous structures to chemical attack. To prove that diatoms live at temperatures above 30–40°C, growth and viability data from culture studies would be required. The maximum temperature for growth for any diatom in culture is between 43°C and 44°C. This is for the hot spring diatom *Achanthes exigua,* which has a temperature optimum of 40°C (Fairchild and Sheridan, 1974).

As noted earlier in this chapter, it is likely that the reason eucaryotic algae (except *Cyanidium caldarium*) do not grow at temperatures much above 30–40°C is that at neutral pH they must compete with the much more successfully adapted blue-green algae. It is only in very acid waters, where blue-green algae cannot live, that the thermal limit species, *C. caldarium,* is able to grow.

Protozoa

Some discussion on protozoa in hot springs can be found in Tansey and Brock (1978). In a study to develop heterotrophic microcosms at a variety of temperatures, Allen and Brock (1968) were able to develop systems containing protozoa at temperatures up to 45°C, but not at 50°C or higher. Kahan (1969) has found four species of protozoa in hot springs in reasonable numbers at temperatures of 57–58°C. In culture, protozoa were able to reproduce at 56°C *(Cercosulcifer hamathensis)* and 55°C *(Vahlkampfia reichi)* and to survive for half an hour at 60°C *(V. reichi)* and 58°C *(C. hamathensis)*. Over a period of several weeks, *C. hamathensis* was adapted to live at temperatures of 58–59°C. In a later study, Kahan (1972) collected *Cyclidium citrullus* from hot springs at 50–58°C. In monoxenic culture the highest temperature at which some of the cells of this species would multiply was 47°C, although cells survived for a few days at 49°C. Some of the older work reporting growth of protozoa at much higher temperatures is discussed briefly by Tansey and Brock (1978) and the difficulty of accepting the validity of these reports is mentioned.

Thus, the upper temperature limit for protozoa seems to be about 55–58°C, in the same range as that for the eucaryotic algae.

Fungi

Mycologists have adapted a specific terminology to describe the temperature relationships of fungi living at high temperatures (Cooney and Emerson, 1964). *Thermophilic fungi* are defined as those that have a maximum temperature for growth at or above 50°C and a minimum temperature for growth at or above 20°C. *Thermotolerant fungi,* on the other hand, are those with maxima near 50°C but minima well below 20°C. In another microbial group, the terms obligate and facultative thermophile might be used to describe organisms with temperature relationships such as thermophilic and thermotolerant.

A wide variety of thermophilic and thermotolerant fungi are known, as outlined in Table 3.5. Although thermophilic and thermotolerant fungi can be isolated from a wide variety of habitats (see Tansey and Brock, 1978 for a review), there have been very few habitats where it has been possible to study the ecology of a thermophilic fungus *in situ.* In this respect, the mats of *Cyanidium caldarium* living in hot acid springs are unique (Belly et al., 1973b), and will be discussed in detail below.

Dactylaria gallopava, a Widespread Thermophilic Fungus

One thermophilic fungus of specific interest is *Dactylaria gallopava,* a causal agent of avian encephalitis, which has also been isolated from hot spring effluents, thermal soils, and self-heated coal refuse piles (Tansey and Brock, 1973). This hyphomycetous fungus was originally described as a causal agent of encephalitis of turkey poults. When Tansey began his isolations of thermophilic fungi from geothermal habitats and other thermal sources, he found this fungus to be widespread in thermal habitats with acid pH (pH 2.1 to 5.9). Because of the distinctive morphology of this fungus, it could be recognized microscopically directly in nature, either in natural material or on microscope slides incubated in the habitat. Attachment of the distinctive 1-septate, apiculate conidia of *D. gallopava* on the slides could be observed, followed by germination and the development of radial microcolonies which could be traced to germinated conidia. A branched, septate mycelium subsequently developed. Further evidence for the growth of *D. gallopava* was the occurrence in foam, which forms in turbulent regions of hot spring effluents, of hyphal networks traceable to spores and spores attached to conidiophores. One foam sample contained about 2×10^6 conidia per milliliter of liquid.

The self-heated coal refuse pile will be discussed in some detail in Chapter 5. Tansey isolated *D. gallopava* from two samples of coal refuse taken from piles in Pennsylvania and Indiana.

The pathogenicity of isolate 141-2 of *D. gallopava* from Yellowstone Park was established by H. G. Blalock and W. T. Derieux by intratracheal inoculation of day-old turkey poults; the fungus was reisolated from the brains of birds that died.

It thus seems reasonable to conclude that this fungus is widespread in acid thermal habitats. As temperatures in the range of mammalian and avian body temperatures exist in many geothermal habitats, it is conceiv-

Table 3.5. Thermotolerant and Thermophilic Fungi and Their Cardinal Temperatures[a] (continued on pp. 62–63)

	Cardinal temperatures (°C)		
	Minimum	Optimum	Maximum
Zygomycetes			
Absidia corymbifera (Cohn) Sacc. et Trott.	14	40	50
Mortierella turficola	25	47	55
Mucor miehei Cooney et Emerson	24–25	35–45	55–57
M. pusillus Lindt	20–27	35–55	55–60
Mucor sp. I	25	45	56
Mucor sp. II	25	42–45	55
Rhizomucor sp.	25–30	45–53	60–61
Rhizopus arrhizus	>15	40–45	55
Rhizopus cohnii Berl. et de Toni	10	42	55
R. microsporus v. Tiegh.	12	40	50
Rhizopus sp. I	16	40–45	55
Rhizopus sp. II	16	40	50
Rhizopus sp. III	28	48–50	60
Rhizopus sp. A	25	45–52	60
Ascomycetes			
Allescheria terrestris Apinis stat. conid. *Cephalosporium* sp.	22	42–45	55
Byssochlamys verrucosa Samson et Tansey stat. conid. *Paecilomyces*	15	40	55
Chaetomium britannicum Ames	b	b	b
C. thermophile La Touche	25–27	50	58–61
C. thermophile var. *coprophile* Cooney et Emerson	25–28	45–55	58–60
C. thermophile var. *dissitum* Cooney et Emerson	25–28	45–50	58–60
C. virginicum Ames	b	b	b
Chaetomium sp. A	14	40	50
Chaetomium sp.	>20	40–43	≥50
Emericella nidulans stat. conid. *Aspergillus*	10	35–37	51
Hansenula polymorpha	—	37–42	50
Myriococcum albomyces Cooney et Emerson [=*Melanocarpus albomyces* (Cooney et Emerson) v. Arx]	25–26	37–45	55–57
Sphaerospora saccata Evans	12	35–42	50
Talaromyces byssochlamydoides Stolk et Samson stat conid. *Paecilomyces*	ca. 25	40–45	>50
T. emersonii Stolk stat. conid. *Penicillium*	25–30	40–45	55–60
T. leycettanus Evans et Stolk stat. conid. *Penicillium*	18	42	55
T. thermophilus Stolk stat. conid. *Penicillium*	25–30	45–50	57–60
Thermoascus aurantiacus Miehe	20–35	40–46	55–62

Table 3.5. Thermotolerant and Thermophilic Fungi and Their Cardinal Temperatures[a] (*continued*)

	Cardinal temperatures (°C)		
	Minimum	Optimum	Maximum
T. crustaceus (Apinis et Chesters) Stolk stat. conid. *Paecilomyces*	20	37	55
Thielavia australiensis Tansey et Jack stat. conid. *Chrysosporium*	20	—	≥50
T. thermophila Fergus et Sinden stat. conid. *Sporotrichum* (=*Chrysosporium fergusii*)	20	45	56
Thielavia sp. A	16	40	52
Basidiomycetes			
Coprinus sp.	ca. 20	ca. 45	55
Phanerochaete chrysosporium Burds. [c]	≤12	36–40	50
Deuteromycetes			
Acremonium alabamensis Morgan-Jones	≤25	—	>50
Acrophialophora fusispora (Saksena) Samson	14	40	50
Aspergillus candidus	10–15	45–50	50–55
A. fumigatus Fres.	12–20	37–43	52–55
Calcarisporium thermophile Evans [=*Calcarisporiella thermophila* (Fergus) deHoog]	16	40	50
Cephalosporium sp.	—	—	≥50
Chrysosporium sp. A [See also Basidiomycetes, above]	25	40–45	55
Humicola grisea var. *thermoidea* Cooney et Emerson	20–24	38–46	55–56
H. insolens Cooney et Emerson	20–23	35–45	55
H. lanuginosa (Griff. et Maubl.) Bunce	28–30	45–55	60
H. stellata Bunce	<24	40	50
Malbranchea pulchella var. *sulfurea* (Miehe) Cooney et Emerson	25–30	45–46	53–57
Paecilomyces puntonii (Vuillemin) Nannfeldt	>30	—	>50
P. variotii Bainier	ca. 5	35–40	ca. 50
Paecilomyces spp. Group b, 7 str.	<30	45–50	55–60
Paecilomyces spp. Group c, 6 str.	<30	45–50	55–60
Paecilomyces spp. Group d, 4 str.	<30	45–50	55–60
Penicillium argillaceum Stolk, Evans, et Nilsson	ca. 15	ca. 35	ca. 50
Penicillium sp. A	20	42	55
Penicillium sp.	>20	ca. 45	>50
Scolecobasidium sp. A (=*Diplorhinotrichum gallopavum* W. B. Cooke) [=*Dactylaria gallopava* (W. B. Cooke) Bhatt et Kendrick]	14–16	40	50–52

	Cardinal temperatures (°C)		
	Minimum	Optimum	Maximum
Sporotrichum thermophile Apinis [=*Chrysosporium thermophilum* (Apinis)]	18–24	40–50	55
Stilbella thermophila Fergus	ca. 25	35–50	ca. 55
Thermomyces ibadanensis Apinis et Eggins	31–35	42–47	60–61
Torula thermophila Cooney et Emerson	23	35–45	58
Torulopsis candida	—	≤36–37	≥50
Tritirachium sp. A [=*Nodulisporium cylindroconium* de Hoog]	16	40	55
Mycelia sterilia			
Burgoa—Papulaspora sp.	≤20	—	53
Papulaspora thermophila Fergus	29–30	ca. 45	52

[a]Identity presented as published. Temperatures indicated as a range of temperatures are derived from values reported by several investigators or were originally reported in this manner. [For references, see Tansey and Brock (1978).]

[b]Not determined but other data indicate probable ability to grow at 50°C or above.

[c]It is unclear how many valid taxa are encompassed by organisms reported variously as *Sporotrichum pulverulentum*, *S. pruinosum*, *Sporotrichum* sp., *Chrysosporium lignorum*, *C. pruinosum*, *Chrysosporium* sp., *Ptychogaster*, and most recently *Phanerochaete chrysosporium*.

able that these habitats might be natural reservoirs for pathogenic microorganisms. The temperature optimum of *D. gallopava* was around 37°C, and growth occurred at both 20°C and 50°C, although only slight growth was obtained at the last temperature.

Microscopic examination of mats of the alga *Cyanidium caldarium* revealed that *D. gallopava* was virtually the only fungus present. It was of interest to examine the relationship between the fungus and the alga, since it seemed likely that the alga was providing the nutrients for fungus growth. In a combined laboratory and field study, Belly et al. (1973b) showed that the alga *C. caldarium* excreted a portion of the organic carbon which it synthesized during photosynthesis. From 2% to 6% of the $NaH^{14}CO_3$ taken up by natural or laboratory populations of the alga was excreted as ^{14}C-labeled materials. The maximum excretion occurred at temperature, light, and pH values that were optimum for photosynthesis.

It was shown that *D. gallopava* could not utilize cellulose (Tansey and Brock, 1973), the main constituent of the *C. caldarium* cell wall. However, because of the ability of *D. gallopava* to use low molecular weight organic compounds, it was likely that it could use some of the excreted materials as a source of carbon and energy. Therefore, a study was made of the quantitative relationships between the fungus and alga (Belly et al., 1973b). Mats of various thickness were obtained by sampling at different temperatures and in areas of various flow rates. Most of the samples were taken from the thick *C. caldarium* mat at Nymph Creek. As shown in Figure 3.4, there is a good correlation between the number of algal cells and the fungal biomass. On the other hand, there seems to be no correlation between algal numbers and bacterial (primarily *Bacillus coagulans*) biomass in the same mats. Since fungal biomass is much greater than bacterial biomass, it is likely that the bacteria present in these mats are not getting their organic nutrients directly from the alga, but from some other source.

To investigate the ability of *D. gallopava* to grow on products released by the alga, a thick suspension of algal cells was heated in a boiling water bath for 5 minutes. When a portion of the heated suspension was placed in a basal salts medium, the fungus grew quite well, although it was unable to grow in the basal salts medium alone. In another series of experiments, a pure culture of *C. caldarium* in a mineral salts medium was inoculated with *D. gallopava* and incubated under light with constant bubbling of 5% CO_2 in air. *C. caldarium* grew quite well both in the presence and absence of the fungus, while in the tubes containing the fungus, germination and visible growth had occurred.

Thus it seems clear that *D. gallopava* is able to grow on the excretion products of the alga *C. caldarium,* and that in the acid hot springs in which it was studied, its major source of nutrients is derived from algal photosynthesis.

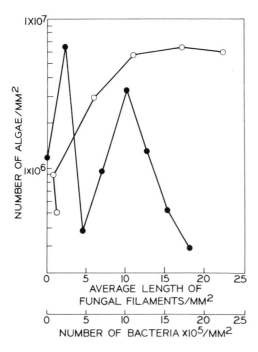

Figure 3.4. Quantification of algal, bacterial, and fungal components of *C. caldarium* mats. Samples were removed from various locations of a thick *C. caldarium* mat at Nymph Creek, Yellowstone Park, homogenized, and the microbial components quantified microscopically. Open circles: fungi; closed circles: bacteria. [For details, see Belly et al. (1973b).]

The Upper Temperature Limit for Eucaryotes

Because of the ease of isolation of thermophilic fungi, and the likelihood that such fungi would be able to live at higher temperatures than the other eucaryotic microorganisms, a concerted effort was made to determine the upper temperature limit for thermophilic fungi. It was reasoned that this upper limit would also likely be the upper limit for eucaryotic organisms as a group (Tansey and Brock, 1972). When making initial studies of the presence of thermophilic fungi in geothermal habitats, Tansey made the interesting observation that these fungi were much more common in acid thermal habitats than in those of neutral to alkaline pH. After this discovery, more detailed isolations were attempted in thermal acid soils and springs. Other habitats which were studied in detail were self-heating composts, hay, wood chip piles, stored grains, self-heated coal refuse piles, tobacco products (which undergo a heating process during curing), and sun-heated soil and mud. Another thermal source was alligator nesting material from Florida, since an abundance of thermophilic fungi has been isolated from this naturally heated plant material. Steam-line discharge sites were also studied, because they had also yielded a variety of thermophilic fungi.

In all, a total of 336 samples were tested, and temperature and pH were measured at the time each sample was collected. Samples were taken from habitats with pH values as low as 1 and as high as 11, and with temperatures as low as 10°C and as high as 90°C (Tansey and Brock, 1972). Details of enrichment procedures can be found in Tansey and Brock (1972). For enrichment of conventional species of thermophilic fungi, an incubation temperature of 50°C was used, since this is a temperature that is selective for thermophiles and that selects against normal fungi. To determine if fungi might exist that could grow at higher temperatures than previously known, an enrichment temperature of 62°C was used, since no known thermophilic fungus is able to grow at a temperature above 60°C. Despite occasional reports of growth of certain pure cultures of thermophilic fungi at 62°C and 63°C, it has been Tansey's experience that these same species grow extremely poorly above 61.5°C.

No fungi grew in any of the 62°C enrichments, although enrichments of the same samples at 50°C often yielded thermophilic fungi in profusion. Since reproducible and continuous growth of some thermophilic fungi is possible at 60–61.5°C, it seems reasonable to conclude that there is a definite upper temperature limit for eucaryotic organisms near this temperature. The negative data from the large number of samples collected in Yellowstone seem to be the most conclusive, since natural geothermal habitats provide relatively stable habitats of long duration.

Since many procaryotes grow optimally at temperatures higher than 60–61.5°C, and some procaryotes have a *minimum* temperature for growth near this value, the inability of any eucaryotes to grow at temperatures above 60–61.5°C is even more interesting.

When we consider a molecular explanation for the inability of eucaryotes to evolve members able to grow at high temperatures, we must first consider how eucaryotes differ from procaryotes. Functionally, the thermophilic fungi are quite similar to those procaryotic heterotrophs able to grow at high temperatures. Both groups consist of aerobes with simple nutritional requirements, able to grow on media with single carbon compounds as sole sources of carbon and energy. Presumably, they thus contain enzymes and other macromolecules of similar function. The enzymes and other macromolecules of high-temperature procaryotes are unusually heat stable. Thus, normal aerobic heterotrophic metabolism can be performed with thermostable macromolecules, and presumably there has been a sufficiently long period for the fungi to have evolved such thermostable components. Cellular functions such as active transport also occur in high-temperature procaryotes, and the plasma membrane of these organisms is unusually thermostable.

However, the key differences between procaryotes and eucaryotes are not at the macromolecular level, but at the level of the intracellular organelles. All eucaryotes have membranous intracellular organelles, such as nuclei and mitochondria, and thermophilic fungi that have been examined by electron microscopy also have these structures (S. F. Conti and W. Samsonoff, personal communication). It seems likely that the inability of eucaryotes to grow at high temperatures may thus reside in their inability to form organellar membranes that are both thermostable and functional. These intracellular organellar membranes differ from the plasma membrane in that they must allow the selective passage of large macromolecular components. Because of this, they must have large pores and be relatively leaky; conceivably, this leakiness is not compatible with thermostability. If so, it would be impossible for a eucaryote to construct intracellular organellar membranes that are both functional and thermostable, hence explaining the absence of eucaryotes at high temperatures.

References

Allen, E. T. and A. L. Day. 1935. *Hot Springs of the Yellowstone National Park.* Carnegie Institution of Washington, Publication No. 466. 525 pp.

Allen, M. B. 1953. The thermophilic aerobic sporeforming bacteria. *Bacteriol. Rev.* **17**, 125–173.

Allen, S. D. and T. D. Brock. 1968. The adaptation of heterotrophic microcosms to different temperatures. *Ecology* **49**, 343–346.

Bauld, J. and T. D. Brock. 1973. Ecological studies of *Chloroflexis,* a gliding, photosynthetic bacterium. *Arch. Mikrobiol.* **92**, 267–284.

Bauld, J. and T. D. Brock. 1974. Algal excretion and bacterial assimilation in hot spring algal mats. *J. Phycol.* **10**, 101–106.

Belly, R. T., B. B. Bohlool, and T. D. Brock. 1973a. The genus *Thermoplasma. Ann. N.Y. Acad. Sci.* **225**, 94–107.

Belly, R. T. and T. D. Brock. 1974. Widespread occurrence of acidophilic strains of *Bacillus coagulans* in hot springs. *J. Appl. Bacteriol.* **37**, 175–177.

Belly, R. T., M. R. Tansey, and T. D. Brock. 1973b. Algal excretion of ^{14}C-labeled compounds and microbial interactions in *Cyanidium caldarium* mats. *J. Phycol.* **9**, 123–127.

Bohlool, B. B. and T. D. Brock. 1974. Population ecology of *Sulfolobus acidocaldarius*. II. Immunoecological studies. *Arch. Microbiol.* **97**, 181–194.

Bott, T. L. and T. D. Brock. 1969. Bacterial growth rates above 90°C in Yellowstone hot springs. *Science* **164**, 1411–1412.

Breed, R. S., E. G. D. Murray, and N. R. Smith. 1957. *Bergey's Manual of Determinative Bacteriology,* 7th ed. The Williams & Wilkins Company, Baltimore.

Brewer, W. H. 1866. On the presence of living species in hot and saline waters in California. *Am. J. Sci. (Sillman's J.)* **41**, 391–394.

Brock, T. D. 1967a. Life at high temperatures. *Science* **158**, 1012–1019.

Brock, T. D. 1967b. Microorganisms adapted to high temperatures. *Nature* **214**, 882–885.

Brock, T. D. 1968. Taxonomic confusion concerning certain filamentous blue-green algae. *J. Phycol.* **4**, 178–179.

Brock, T. D. 1970. High temperature systems. *Ann. Rev. Ecol. Systemat.* **1**, 191–220.

Brock, T. D. 1973. Lower pH limit for the existence of blue-green algae: evolutionary and ecological implications. *Science* **179**, 480–483.

Brock, T. D. 1975. *Predicting the Ecological Consequences of Thermal Pollution from Observations on Geothermal Habitats. In: Environmental Effects of Cooling Systems at Nuclear Power Plants.* International Atomic Energy Agency, Vienna.

Brock, T. D., K. M. Brock, R. T. Belly, and R. L. Weiss. 1972. *Sulfolobus:* a new genus of sulfur-oxidizing bacteria living at low pH and high temperature. *Arch. Mikrobiol.* **84**, 54–68.

Brock, T. D. and M. L. Brock. 1968. Measurement of steady-state growth rates of a thermophilic alga directly in nature. *J. Bacteriol* **95**, 811–815.

Brock, T. D. and M. L. Brock. 1971a. Microbiological studies of thermal habitats of the central volcanic region, North Island, New Zealand. *N.Z. J. Mar. Freshwater Res.* **5**, 233–257.

Brock, T. D. and M. L. Brock. 1971b. Temperature optimum of non-sulphur bacteria from a spring at 90°C. *Nature* **233**, 494–495.

Brock, T. D., M. L. Brock, T. L. Bott, and M. R. Edwards. 1971. Microbial life at 90°C: the sulfur bacteria of Boulder Spring. *J. Bacteriol.* **107**, 303–314.

Brock, T. D., S. Cook, S. Petersen, and J. L. Mosser. 1976. Biogeochemistry and bacteriology of ferrous iron oxidation in geothermal habitats. *Geochim. Cosmochim. Acta* **40**, 493–500.

Brock, T. D. and G. K. Darland. 1970. Limits of microbial existence: temperature and pH. *Science* **169**, 1316–1318.

Brock, T. D. and H. Freeze. 1969. *Thermus aquaticus* gen. n. and sp. n., a nonsporulating extreme thermophile. *J. Bacteriol.* **98**, 289–297.

Brock, T. D. and J. L. Mosser. 1975. Rate of sulfuric-acid production in Yellowstone National Park. *Geol. Soc. Am. Bull.* **86**, 194–198.

Caldwell, D. E., S. J. Caldwell, and J. P. Laycock. 1976. *Thermothrix thioparus* gen. et sp. nov. a facultatively anaerobic facultative chemolithotroph living at neutral pH and high temperature. *Can. J. Microbiol.* **22**, 1509–1517.

Campbell, L. L. 1974. In *Bergey's Manual of Determinative Bacteriology,* 8th ed., R. E. Buchanan and N. E. Gibbons, eds. The Williams & Wilkins Company, Baltimore, pp. 572–573.

Castenholz, R. W. 1967. Environmental requirements of thermophilic blue-green algae. In Environmental Requirements of Blue-Green Algae. A. F. Bartsch. ed. Federal Water Pollution Control Administration, Corvallis, Oregon, pp. 55–79.

Castenholz, R. W. 1969a. The thermophilic cyanophytes of Iceland and the upper temperature limit. *J. Phycol.* **5**, 360–368.

Castenholz, R. W. 1969b. Thermophilic blue-green algae and the thermal environment. *Bacteriol Rev.* **33**, 476–504.

Cohn, F. 1862. Ueber die Algen des Karlsbader Sprudels, mit Rücksicht auf die Bildung des Sprudelsinters. *Abhandlungen der Schlesischen Gesellschaft für vaterländische Cultur. Abt. Naturwiss. Med.* **II**, 35–55.

Cooney, D. G. and Emerson, R. 1964. *Thermophilic Fungi. An Account of Their Biology, Activities, and Classification.* W. H. Freeman and Company, San Francisco.

Copeland, J. J. 1936. Yellowstone thermal Myxophyceae. *Ann. N.Y. Acad. Sci.* **36**, 1–229.

Cross, T. 1974. In *Bergey's Manual of Determinative Bacteriology,* 8th ed., R. E. Buchanan and N. E. Gibbons, eds. The Williams & Wilkins Company, Baltimore, pp. 861–863.

Darland, G. and T. D. Brock. 1971. *Bacillus acidocaldarius* sp. nov., an acidophilic thermophilic spore-forming bacterium. *J. Gen. Microbiol.* **67**, 9–15.

Darland, G., T. D. Brock, W. Samsonoff, and S. F. Conti. 1970. A thermophilic, acidophilic mycoplasma isolated from a coal refuse pile. *Science* **170**, 1416–1418.

Deibel, R. H. and Seeley, H. W., Jr. 1974. In *Bergey's Manual of Determinative Bacteriology,* 8th ed., R. E. Buchanan and N. E. Gibbons, eds. The Williams & Wilkins Company, Baltimore, pp. 490–509.

Doemel, W. N. and T. D. Brock. 1970. The upper temperature limit of *Cyanidium caldarium. Arch. Mikrobiol.* **72**, 326–332.

Doemel, W. N. and T. D. Brock. 1977. Structure, growth, and decomposition of laminated algal-bacterial mats in alkaline hot springs. *Appl. Environ. Microbiol.* **34**, 433–452.

Egorova, A. A. and Z. Deryugina. 1963. The spore-forming thermophilic thiobacterium *Thiobacillus thermophilica* Imschenetskii nov. sp. *Mikrobiologiya* **32**, 437–446.

Fairchild, E. and R. P. Sheridan. 1974. A physiological investigation of the hot spring diatom, *Achnanthes exigua* Grün. *J. Phycol.* **10**, 1–4.

Fields, M. L. 1970. The flat sour bacteria. *Adv. Food Res.* **18**, 163–217.

Fliermans, C. B. and T. D. Brock. 1972. Ecology of sulfur-oxidizing bacteria in hot acid soils. *J. Bacteriol.* **111**, 343–350.

Gaughran, E. R. L. 1947. The thermophilic microörganisms. *Bacteriol. Rev.* **11**, 189–225.

Gibson, T. and R. E. Gordon. 1974. In *Bergey's Manual of Determinative Bacteri-*

ology, 8th ed., R. E. Buchanan and N. E. Gibbons, eds. The Williams & Wilkins Company, Baltimore, pp. 529–550.

Golovacheva, R. S., L. G. Loginova, T. A. Salikhov, A. A. Kolesnikov, and G. N. Zaïtseva. 1975. Novyï vid termofil'nykh batsill—*Bacillus thermocatenulatus* nov. sp. *Mikrobiologiya* **44**, 265–268.

Gordon, R., W. Haynes, and C. Pang. 1973. *The Genus Bacillus.* U.S. Department of Agriculture Handbook No. 427, 283 pp.

Heinen, W. 1971. Growth conditions and temperature-dependent substrate specificity of two extremely thermophilic bacteria. *Arch. Mikrobiol.* **76**, 2–17.

Heinen, U. J. and W. Heinen. 1972. Characteristics and properties of a caldo-active bacterium producing extracellular enzymes and two related strains. *Arch. Mikrobiol.* **82**, 1–23.

Henssen, A. 1969. *Streptomyces fragmentosporus,* ein neuer thermophiler Actinomycet. *Arch. Mikrobiol.* **67**, 21–27.

Henssen, A. 1974. In *Bergey's Manual of Determinative Bacteriology,* 8th ed., R. E. Buchanan and N. E. Gibbons, eds. The Williams & Wilkins Company, Baltimore, pp. 746–747.

Hindle, E. 1932. Some new thermophilic organisms. *J.R. Microsc. Soc.* **52**, 123–133.

Hollaus, F. and H. Klaushofer. 1973. Identification of hyperthermophilic obligate anaerobic bacteria from extraction juices of beet sugar factories. *Int. Sug. J.* **75**, 237–241, 271–275.

Hollaus, F. and U. Sleytr. 1972. On the taxonomy and fine structure of some hyperthermophilic saccharolytic clostridia. *Arch. Microbiol.* **86**, 129–146.

Hoppe-Seyler, F. 1875. Ueber die obere Temperaturgrenze des Lebens. *Pflügers Arch. f.d.ges. Physiol.* **11**, 113–121.

Jackson, T. J., R. F. Ramaley, and W. G. Meinschein. 1973. *Thermomicrobium,* a new genus of extremely thermophilic bacteria. *Int. J. Syst. Bacteriol.* **23**, 28–36.

Kahan, D. 1969. The fauna of hot springs. *Verh. Int. Verein. Theor. Angew. Limnol.* **17**, 811–816.

Kahan, D. 1972. *Cyclidium citrullus* Cohn, a ciliate from the hot springs of Tiberias (Israel). *J. Protozool.* **19**, 593–597.

Kempner, E. S. 1963. Upper temperature limit of life. *Science* **142**, 1318–1319.

Kurup, V. P., J. J. Barboriak, J. N. Fink, and M. P. Lechevalier. 1975. *Thermoactinomyces candidus,* a new species of thermophilic actinomycetes. *Int. J. Syst. Bacteriol.* **25**, 150–154.

Küster, E. 1974. In *Bergey's Manual of Determinative Bacteriology,* 8th ed., R. E. Buchanan and N. E. Gibbons, eds. The Williams & Wilkins Company, Baltimore, pp. 858–859.

Leadbetter, E. R. 1974. In *Bergey's Manual of Determinative Bacteriology,* 8th ed., R. E. Buchanan and N. E. Gibbons, eds. The Williams & Wilkins Company, Baltimore, pp. 267–269.

Loginova, L. G. and L. A. Egorova. 1975. Obligatno-termofil'nye bacterii *Thermus ruber* v gidrotermakh kamchatki. *Mikrobiologiya* **44**, 661–665.

Loginova, L. G., R. S. Golovacheva, and M. A. Shcherbakov. 1966. Thermophilic bacteria forming active cellulolytic enzymes. *Mikrobiologiya* **35**, 796–804.

Loginova, L. G., A. E. Kosmachev, R. S. Golovacheva, and L. M. Seregina. 1962. Issledovanie termofil'noï mikroflory gory Yangan-Tau na yuzhnom Urale. *Mikrobiologiya* **31**, 1082–1086.

Löwenstein, A. 1903. Ueber die Temperaturgrenzen des Lebens bei der Thermalalge Mastigocladus laminosus Cohn. *Ber. Deutsch. Bot. Ges.* **21**, 317–323.

Madigan, M. T. and T. D. Brock. 1975. Photosynthetic sulfide oxidation by *Chloroflexus aurantiacus*, a filamentous, photosynthetic, gliding bacterium. *J. Bacteriol.* **122**, 782–784.

Madigan, M. T., S. R. Petersen, and T. D. Brock. 1974. Nutritional studies on *Chloroflexus*, a filamentous photosynthetic gliding bacterium. *Arch. Microbiol.* **100**, 97–103.

Manachini, P., A. Craveri, and R. Craveri. 1966. *Thermomonospora citrina*, una nuova specie di Attinomicete Termofilo isolato dal suolo. *Ann. Microbiol.* **16**, 83–90.

Manachini, P., A. Ferrari, and R. Craveri. 1965. Forme termofile di Actinoplanaceae. Isolamento e ceratteristiche di *Streptosporangium album* var. *thermophilum*. *Ann. Microbiol.* **15**, 129–144.

Mann, J. E. and H. E. Schlichting, Jr. 1967. Benthic algae of selected thermal springs in Yellowstone National Park. *Trans. Am. Microsc. Soc.* **86**, 2–9.

Mateles, R. I., J. N. Baruah, and S. R. Tannenbaum. 1967. Growth of a thermophilic bacterium on hydrocarbons: a new source of single-cell protein. *Science* **157**, 1322–1323.

McGee, J. M., L. R. Brown, and R. G. Tischer. 1967. A high-temperature, hydrogen-oxidizing bacterium—*Hydrogenomonas thermophilus*, n. sp. *Nature* **214**, 715–716.

Meeks, J. C. and R. W. Castenholz. 1971. Growth and photosynthesis in an extreme thermophile, *Synechococcus lividus* (Cyanophyta). *Arch. Mikrobiol.* **78**, 25–41.

Mitchell, R. 1974. The evolution of thermophily in hot springs. *Q. Rev. Biol.* **49**, 229–242.

Mosser, J. L., B. B. Bohlool, and T. D. Brock. 1974a. Growth rates of *Sulfolobus acidocaldarius* in nature. *J. Bacteriol.* **118**, 1075–1081.

Mosser, J. L., A. G. Mosser, and T. D. Brock. 1973. Bacterial origin of sulfuric acid in geothermal habitats. *Science* **179**, 1323–1324.

Mosser, J. L., A. G. Mosser, and T. D. Brock. 1974b. Population ecology of *Sulfolobus acidocaldarius*. I. Temperature strains. *Arch. Microbiol.* **97**, 169–179.

Orla-Jensen, S. 1942. *The Lactic Acid Bacteria*, 2nd ed. (D. Kgl. Danske Vidensk. Selskab, Skrifter, Naturvidensk. Og Mathem. Afd., 8 Raekke), 197 pp., Kobenhavn.

Oshima, T. and K. Imahori. 1974. Description of *Thermus thermophilus* (Yoshida and Oshima) comb. nov., a nonsporulating thermophilic bacterium from a Japanese thermal spa. *Int. J. Syst. Bacteriol.* **24**, 102–112.

Pierson, B. K. and R. W. Castenholz. 1971. Bacteriochlorophylls in gliding filamentous prokaryotes from hot springs. *Nature New Biol.* **233**, 25–27.

Pierson, B. K. and R. W. Castenholz. 1974a. A phototrophic gliding filamentous bacterium of hot springs, *Chloroflexus aurantiacus*, gen. and sp. nov. *Arch. Microbiol.* **100**, 5–24.

Pierson, B. K. and R. W. Castenholz. 1974b. Studies of pigments and growth in *Chloroflexus aurantiacus*, a phototrophic filamentous gliding bacterium. *Arch. Microbiol.* **100**, 283–305.

Pridham, T. G. and H. D. Tresner. 1974. In *Bergey's Manual of Determinative*

Bacteriology, 8th ed., R. E. Buchanan and N. E. Gibbons, eds. The Williams & Wilkins Company, Baltimore, pp. 748–829.

Ramaley, R. F. and J. Hixson. 1970. Isolation of a nonpigmented, thermophilic bacterium similar to *Thermus aquaticus. J. Bacteriol.* **103**, 527–528.

Rogosa, M. 1974. In *Bergey's Manual of Determinative Bacteriology,* 8th ed., R. E. Buchanan and N. E. Gibbons, eds. The Williams & Wilkins Company, Baltimore, pp. 669–676.

Rozanova, E. P. and A. I. Khudyakova. 1974. Novji bessporovji termofil'nyi organizm, vostablivayushchii sut'faty, *Desulfovibrio thermophilus* nov. sp. *Mikrobiologiya* **43**, 1069–1075.

Saiki, T., R. Kimura, and K. Arima. 1972. Isolation and characterization of extremely thermophilic bacteria from hot springs. *Agr. Biol. Chem.* **36**, 2357–2366.

Setchell, W. A. 1903. The upper temperature limits of life. *Science* **17**, 934–937.

Shivvers, D. W. and T. D. Brock. 1973. Oxidation of elemental sulfur by *Sulfolobus acidocaldarius. J. Bacteriol.* **114**, 706–710.

Smith, L. D. S. and G. Hobbs. 1974. In *Bergey's Manual of Determinative Bacteriology,* 8th ed., R. E. Buchanan and N. E. Gibbons, eds. The Williams & Wilkins Company, Baltimore, pp. 551–572.

Sorokin, C. 1967. New high-temperature *Chlorella. Science* **158**, 1204–1205.

Tansey, M. R. and T. D. Brock. 1972. The upper temperature limit for eukaryotic organisms. *Proc. Natl. Acad. Sci. U.S.A.* **69**, 2426–2428.

Tansey, M. R. and T. D. Brock. 1973. *Dactylaria gallopava,* a cause of avian encephalitis, in hot spring effluents, thermal soils and self-heated coal waste piles. *Nature* **242**, 202–203.

Tansey, M. R. and T. D. Brock. 1978. Microbial life at high temperatures: ecological aspects. p. 159–215. In: Microbial life in extreme environments (D. Kushner, editor). Academic Press, London.

Uchino, F. and S. Doi. 1967. Acido-thermophilic bacteria from thermal waters. *Agr. Biol. Chem.* **31**, 817–822.

Williams, R. A. D. and D. S. Hoare. 1972. Physiology of a new facultatively autotrophic thermophilic *Thiobacillus. J. Gen. Microbiol.* **70**, 555–566.

Zeikus, J. G. and R. S. Wolfe. 1972. *Methanobacterium thermoautotrophicus* sp. n., an anaerobic, autotrophic, extreme thermophile. *J. Bacteriol.* **109**, 707–713.

Chapter 4
The Genus *Thermus*

The discovery of the genus *Thermus* (Brock and Freeze, 1969) provided our first impetus for detailed studies on the physiology of thermophilic bacteria. Its discovery was one of the simpler things we did, because enrichment and culture were so easy. We used a synthetic salts medium which had been shown to be suitable for growth of hot spring algae, added relatively low concentrations of organic compounds, and incubated at 70°C. The latter is of most importance. Before we began our work, the standard temperature for isolating and growing thermophilic bacteria had been 55°C, a temperature that the food microbiologists had used since the time of Cameron and Esty (Farrell and Campbell, 1969). At this temperature, and with the usual media, all one generally gets in culture is *Bacillus stearothermophilus,* a widespread, soil-borne sporeformer. Although this organism is of considerable importance in spoilage of canned foods, it is hardly of much importance in natural thermal environments, despite the fact that it can be routinely isolated (Fields, 1970). Microscopically, one never sees organisms resembling *B. stearothermophilus.* This organism is gram-positive; if one does gram stains of natural material, one never sees anything gram-positive, although gram-negative bacteria abound.

Isolation Procedures and Habitat

The salts medium used for the culture of *Thermus* was Medium D (Castenholz, 1969), which had the following composition (in milligrams per liter of water): nitrilotriacetic acid, 100; $CaSO_4 \cdot 2H_2O$, 60; $MgSO_4 \cdot 7H_2O$, 100; NaCl, 8; KNO_3, 103; $NaNO_3$, 689; Na_2HPO_4, 111; $FeCl_3$, 0.28; $MnSO_4 \cdot H_2O$, 2.2; $ZnSO_4 \cdot 7H_2O$, 0.5; H_3BO_3, 0.5; $CuSO_4$, 0.016;

$Na_2MoO_4 \cdot 2H_2O$, 0.025; $CoCl_2 \cdot 6H_2O$, 0.046; pH adjusted to 8.2 with NaOH. This salts medium can be made in two components, the NTA and the salts on the list through Na_2HPO_4 being dissolved at a 10× stock, the $FeCl_3$ being dissolved in water at 100×, and the rest of the trace salts dissolved at 100× in water containing 0.5 ml H_2SO_4 per liter. To make the final medium, 1/10 volume of the macroelements, 1/100 volume of the iron solution, and 1/100 volume of the trace salts are diluted with water, the pH adjusted, and the medium autoclaved. To make the final medium for *Thermus*, one merely adds 0.1% tryptone and 0.1% yeast extract. This medium is fairly dilute in the organic constituents, an important point. *Thermus aquaticus* will grow at up to 0.3% of tryptone and yeast extract, but not higher, although another species, *T. thermophilus* (Oshima and Imahori, 1974) grows in concentrations of organic constituents up to 1%.

The original isolations of *Thermus aquaticus* were from hot spring algal mats. Incubation in aerobic liquid medium at 70–75°C for 1–2 days led to the formation of visible turbidity, often with clumps or a surface pellicle. Turbid cultures were then streaked onto a medium of the same composition containing 3% agar, the plates sealed with plastic film to prevent evaporation, and incubated at 70–75°C. After 1–2 days, compact spreading yellow colonies were seen, which could be readily purified by restreaking. Most attempts at isolation of *Thermus* from neutral to alkaline pH hot spring sources were successful (Brock and Freeze, 1969) provided incubations were at 70–75°C. No isolations were successful with temperatures greater than 80°C, although *T. thermophilus* can grow at temperatures up to 85°C (Oshima and Imahori, 1974).

In the early work (Brock and Freeze, 1969) one strain of *Thermus aquaticus* was isolated from a hot water tap in Indiana and other strains were isolated from a creek receiving thermal pollution (Brock and Yoder, 1971). At the same time, it was shown that cold springs and some other nonthermal habitats did not yield *T. aquaticus* isolates. Ramaley and Hixson (1970) also isolated a strain of *Thermus* (strain X-1) which was unpigmented, but which otherwise resembled *T. aquaticus* in characteristics. This strain was isolated from a small spring-fed creek receiving thermal pollution, where a pigmented strain had been isolated earlier (Brock and Freeze, 1969).

It thus seemed likely that *Thermus* was capable of growing in man-made habitats of high temperature. The most common such habitat is the hot water heater. Domestic hot water heaters generally run about 55°C, but commercial hot water heaters, especially in laundries, often have considerably higher temperatures. In a detailed study done in Madison, Wisconsin, we found hot water heaters in laundries with temperatures as high as 81–82°C, and many were in the range of 70–75°C (Brock and Boylen, 1973). Sixteen of the 28 hot tap waters sampled yielded cultures of thermophilic bacteria resembling *T. aquaticus*. Samples of cold water taken from each location never yielded thermophilic cultures. Most-probable-number determinations showed thermophile densities ranging from less than 40 to greater than 2400 bacteria per 100 ml of water. In general, the hotter sources more frequently yielded thermophiles, although some of the cooler tanks also provided thermophilic isolates.

In contrast to the hot spring isolates, the isolates from hot water tanks were rarely pigmented. Ramaley and Hixson (1970), commenting on their nonpigmented isolate, postulated that pigmentation may serve a photoprotective function, and thermophiles growing in dark habitats such as hot water heaters would not need such pigments. It is significant that the nonpigmented isolate of Ramaley and Hixson (1970) grew faster than *T. aquaticus* YT-1, and thus could have had a selective advantage in an environment where pigment was of no value.

It can be concluded that the thermophilic bacteria isolated from hot water tanks were actually growing in the tanks. Cold water rarely yields thermophilic isolates, so that it is unlikely that the organisms are merely being carried to and through the tanks. Growth probably does not take place in the free water, but on the walls of the hot-water heater. The bacteria found free in the water thus probably represent organisms that have sloughed off the liner of the tank. The source of nutrients for growth of these heterotrophic bacteria in the hot water tanks is probably the water itself. Surface water supplies generally have some organic matter, and ground water supplies, even if free of organic matter at the source, generally acquire organic matter as a result of water softening procedures.

In hot springs, *Thermus aquaticus* and related organisms can obtain organic materials necessary for growth from photosynthetic organisms, if they are living in the temperature range where such organisms grow (less than 72–73°C). However, we have frequently seen mass accumulations of *Thermus aquaticus*-like organisms in hot spring effluents at temperatures above those at which photosynthetic organisms grow. The effluent of the spring labeled Twin Butte Vista on the map of the Lower Geyser Basin (Figure 2.2) has large yellow masses of *T. aquaticus*-like filaments at temperatures of 75–80°C. Although I have not proved that all of this mass of material is *T. aquaticus* biomass, it is very easy to isolate *T. aquaticus* cultures from such accumulations. Morphologically, the natural material is more filamentous than *T. aquaticus* in culture, but the cell size and general appearance are similar. Another spring, Firehole Pool (see map, Figure 2.2), has large grayish-yellow masses. There is no chlorophyll in these natural accumulations, and electron donors for chemolithotrophic bacteria, such as sulfide, are low or absent in these particular springs. It seems most likely that *Thermus aquaticus* is growing in such hot spring effluents on the small amounts of organic matter dissolved in the water. In Firehole Pool, the source water assayed about 2 ppm of total organic matter (unpublished analysis done by Beckman infra-red carbon analyzer, by associates of Dr. John Wright, Bozeman, Montana). In a flowing water system, such a concentration of organic matter, although low, should be sufficient to support a reasonable standing crop, especially since at these temperatures there are no grazers to consume biomass. Unfortunately, little real ecology was ever done on *Thermus aquaticus*. It just didn't prove to be a favorable organism for detailed field work.

Morphology of *Thermus aquaticus*

Initial isolates of *Thermus* are usually filamentous, but during continued transfer the filaments become shorter. However, the morphology of the organism is greatly influenced by temperature of growth and by growth stage of the culture. At temperatures of 75°C and above, the organism is usually filamentous, and even at 65–70°C filaments are common in station-ary-phase cultures. A comparison of rod and filamentous forms is shown in Figure 4.1. Electron microscopy reveals that *T. aquaticus* has the cell wall typical of a gram-negative organism (Figure 4.2). Cell division resembles that of typical gram-negative bacteria, such as *Escherichia coli* (Brock and Edwards, 1970). A characteristic structure seen in many *T. aquaticus* cultures was called the "rotund body." This is a large round body which was first interpreted to be a spheroplast (Brock and Freeze, 1969) but was shown by electron microscopy to be a much more complex structure. As seen in Figure 4.3, a rotund body consists of several separate cells sur-

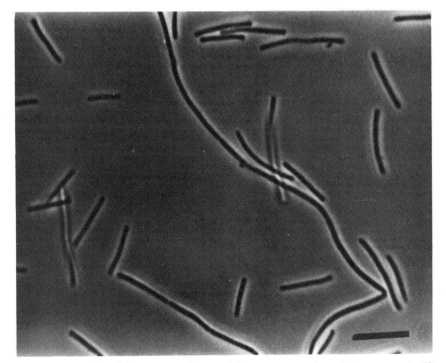

Figure 4.1. Rod and filamentous forms of *Thermus aquaticus*. Bar represents 10 μm. (From Brock and Freeze, 1969.)

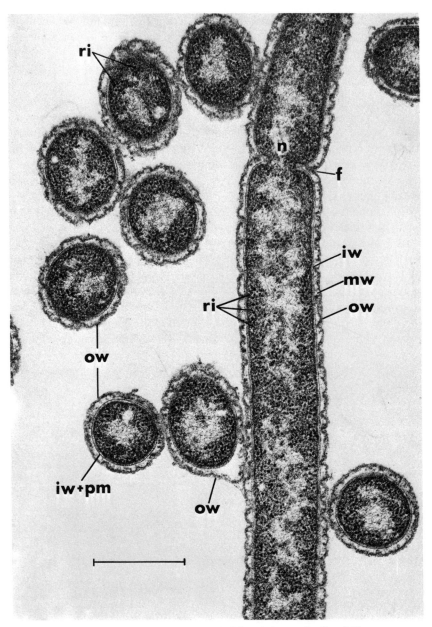

Figure 4.2. Longitudinal, transversal, and oblique views of cells of *Thermus aquaticus,* revealing nucleoplasm (n) with dense thin DNA fibrils, surrounded by cytoplasm containing numerous ribosomes (ri). Cell envelope comprises plasma membrane (pm) and wall exhibiting outer dense layers (ow), middle light zone (mw), and inner dense layer (iw). Note cell division by furrowing (f). Where two cells are in contact, the external wall layer (ow) has separated from the inner wall, which remains adherent to the plasma membrane. Bar represents 0.5 μm. (From Brock and Edwards, 1970.)

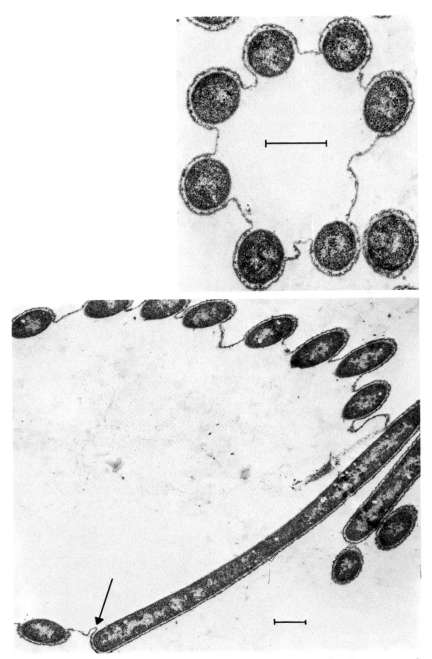

Figure 4.3. Upper: Cross section through a "rotund body," showing seven rods, which are connected by their outer wall to form a ring-like structure. Lower: A larger rotund body. Bar represents 0.5 μm. (From Brock and Edwards, 1970.)

rounded by a membrane. Detailed study showed that this membrane is composed of the outer cell envelope layer which has partly peeled away from the cell surface. It seems likely that formation of rotund bodies involves an association and interaction by way of the outer cell envelope of a number of separate cells. It is assumed that there is a partial separation of the outer wall layer, followed by a fusion of layers of adjacent cells. It is not known whether there is any function to the rotund body but it does seem to be a frequent structure formed by *Thermus aquaticus*. About half of the isolates from Madison, Wisconsin, hot water heaters also formed these structures.

Physiological and Nutritional Characteristics

The relationship between growth rate and temperature for strain YT-1 is shown in Figure 4.4. The optimal temperature for growth was 70°C, the maximum was 79°C, and the minimum was 40°C. The generation time at the optimum temperature was about 50 minutes. Several other species or strains of the genus have higher maximum temperatures, 85°C, although about the same temperature optimum as *T. aquaticus* (Oshima and Imahori, 1974). Although isolated on a complex medium, *T. aquaticus* was found to have no growth factor requirements. It would grow on the mineral salts medium with NH_4^+ as nitrogen source and acetate, sucrose, citrate, succinate, or glucose as carbon source, although growth was slower than in complex medium. Glutamate would also serve as a nitrogen source or as a combined source of carbon and nitrogen. Nitrate could not be used as a nitrogen source and no growth occurred if 0.1% tryptone and 0.1% yeast extract were made up in deionized water instead of basal salts. No explanation for this last finding has been made, although it may be related to the divalent cation content of the basal salts. The organism is an obligate aerobe.

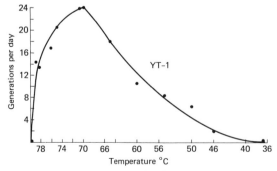

Figure 4.4. Growth rate of *Thermus aquaticus* YT-1 at different temperatures. (From Brock and Freeze, 1969.)

The pH optimum for growth was 7.5 to 7.8 and no growth occurred below pH 6 or above pH 9.5. The organism is sensitive to several antibiotics not normally active against gram-negative bacteria, including penicillin, novobiocin, and actinomycin. All of the laundry isolates showed similar high sensitivity to these antibiotics.

The DNA base composition of four strains of *T. aquaticus* was determined and found to be 67.5% GC (moles percent guanine plus cytosine); 65.4% GC; 65.4% GC; and 65.4% GC. Species of *Thermus* isolated by others have also had fairly high GC values.

I have never observed any motility in *Thermus aquaticus,* either at room temperature or upon heating the microscope slide to about 70°C. In agar slide cultures incubated at 70°C, slime tracks such as are formed by the gliding bacteria were not seen. Colonies on agar plates spread slowly, and the cells at the periphery of the colonies lie parallel to the circumference, so that the edge of the colony is not ragged but quite even, in contrast to the colonies formed by gliding bacteria. Flagella have not been detected either in flagella stains or in shadowed preparations observed with the electron microscope (Brock and Freeze, 1969). Endospores have never been observed, even on starch agar (Brock and Boylen, 1973).

The taxonomic position of *Thermus* is uncertain. As a yellow-pigmented nonmotile, gram-negative rod, the organism could have been considered a *Flavobacterium,* except that the genus *Flavobacterium* is itself ill-defined, and contains species with both low and high GC values. The high-GC strains of *Flavobacterium* are mostly motile by peritrichous flagella, and are facultatively anaerobic (Weeks, 1974). They are mesophiles, and never show a filamentous morphology. I preferred not to confuse things by classifying the present organisms in this ill-defined genus, and therefore created the new genus *Thermus.* Among other things, the establishment of this genus called attention to the unusual thermal characteristics of this group of organisms, and has probably motivated workers to study the organisms biochemically. In the 8th edition of Bergey's manual, *Thermus* is classified with the gram-negative aerobic rods and cocci as a genus of uncertain affiliation (Brock, 1974). A recent taxonomic comparison of a number of isolates from Europe and North America (Degryse et al., 1978) has shown how uniform strains of the genus *Thermus* really are.

Thermostable Enzymes of *Thermus aquaticus*

Whenever I discuss organisms such as *T. aquaticus* with biochemists, I am always asked how an organism such as this can survive, since enzymes are thermolabile. This is a typical example of a generalization from insufficient data: the enzymes of *mesophiles* are thermolabile, and since most of the organisms studied by biochemists are mesophiles, it has been easy to conclude that enzymes in general are thermolabile. At the time of Setchell's work (see Chapter 3), the existence of enzymes was hardly suspected, and thermolability had not been attributed to them, so that Setchell had no difficulty accepting the existence of bacteria even in water near the boiling point. By the 1920s and 1930s, enzymology had arisen as a discipline, and opinions had been formed on such matters. During the past decade, the pendulum has swung the other way, and everyone accepts the existence of thermostability.

I do not propose to review the history of the developments over the past generation on the molecular basis of thermophily. Much of this story can be found in reviews by Ingraham (1962), Brock (1967), and Singleton and Amelunxen (1973). A bit of confusion was introduced by the concept of the "dynamic nature of thermophily" (Allen, 1953), which suggested that thermophilic enzymes were not heat stable but were resynthesized as fast as they were destroyed. However, at about this same time, a group at the University of Nebraska under Militzer, using *Bacillus stearothermophilus*, showed clearly that some enzymes of thermophiles were thermostable. Henry Koffler made a major contribution by showing that flagellar proteins of thermophiles, easily purified to homogeneity, were more thermostable than their mesophilic counterparts (Koffler, 1957).

Thus, when we isolated *T. aquaticus*, the only important thing that seemed to be necessary was to show that its enzymes were even more thermostable than those of *B. stearothermophilus* (since it grew at considerably higher temperatures). This turned out to be easy to do. We chose the enzyme aldolase, because of its widespread distribution in bacteria, and because of the evolutionary studies that had been done by Rutter (1964). Hudson Freeze, then an undergraduate honors student, was quickly able to characterize and purify the enzyme, and demonstrate its dramatic thermostability (Freeze and Brock, 1970). The enzyme showed little activity at temperatures below 60°C and showed optimal activity at about 95°C. Even 43-fold purified enzyme was stable; at low protein concentrations it could be heated for 30 minutes at 97°C without significant inactivation, although it was rapidly inactivated at 105°C. (This last study required the use of a salt bath to get the temperature up and sealed tubes to prevent evaporation.) The heat inactivation curve of the purified enzyme is shown in Figure 4.5.

Except for thermostability, the aldolase of *T. aquaticus* resembled many other bacterial aldolases. The enzyme was activated by high concentrations of NH_4^+ and low concentrations of Fe^{2+} and Co^{2+}, it was strongly inhibited by ethylenediamine tetraacetic acid, a potent metal chelator, and it was activated by cysteine and other sulfhydryl compounds. In cysteine, the enzyme was more heat labile, but heat inactivation in the presence of cysteine was prevented by substrate. In the absence of cysteine, substrate partially labilized the enzyme to heat. The *T. aquaticus* enzyme resembled in most properties the class II aldolase of Rutter (1964), which is the class found in bacteria. The most significant difference with other class II aldolases was in the molecular weight, which in most class II enzymes is

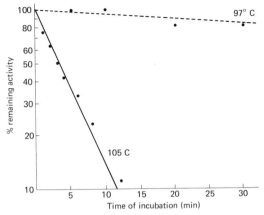

Figure 4.5. Heat inactivation of *Thermus aquaticus* aldolase. The 43-fold purified enzyme was heated at 7 μg of protein per ml at 97°C in 0.05 M Tris buffer (pH 7.3), and the enzyme was diluted 20-fold for assay after heating. Assay at 80°C. At 105°C, the enzyme was used at a protein concentration of 14 μg/ml and was diluted fourfold before assay. (From Freeze and Brock. 1970.)

about 70,000. The purified *T. aquaticus* enzyme had a molecular weight as determined by Sephadex G-200 gel filtration of 140,000. However, at least two other class II aldolases have molecular weights of 140,000 (see Freeze and Brock, 1970, for references). Furthermore, when treated with cysteine, the *T. aquaticus* enzyme is converted into a form with a molecular weight of 70,000. It is likely that both forms of the *T. aquaticus* enzyme are thermostable, since both could be assayed at high temperature.

A comparison of the *T. aquaticus* enzyme with that of *B. stearothermophilus* is of interest. The latter enzyme was rapidly inactivated at 75°C, whereas the *T. aquaticus* enzyme was stable even to 97°C. However, in catalytic properties and cofactor requirements, the two enzymes were similar. It was thus concluded that in all significant respects except thermostability the *T. aquaticus* aldolase resembles other class II aldolases of mesophiles or moderate thermophiles. Rutter (1964) had postulated that all class II aldolases were related and represented enzymes that, at least in the active site, were conserved during evolution. It seems likely that in the *T. aquaticus* enzyme, some portions of the molecule other than the active site must have been markedly altered, in order to confer thermostability.

There has been a large amount of work on enzymes of *Thermus* during the years since this first work on aldolase was published. Much of this work is summarized in a book edited by Zuber (1976), which presents the proceedings of a conference on thermophilic enzymes held in Switzerland. A number of thermophilic enzymes have now been purified to homogeneity and amino acid sequences have been determined. By comparison of the amino acid sequences with those of the same enzymes from mesophiles, it is possible to draw some conclusions about the molecular requirements for thermostability. It now seems clear (Amelunxen and Singleton, 1976; Hocking and Harris, 1976) that at least in some cases, the molecular differences between thermophilic and mesophilic enzymes are very tiny, representing changes in only a few amino acids. Such changes must have a significant effect on the folding of the polypeptide chains, so that three-dimensional configurations will have to be determined in order to draw final conclusions. But it seems clear that molecular interactions within polypeptide chains are sufficient to cause thermostability. Such interactions may involve increased numbers of ionic or hydrophobic bonds, or sulfide cross bridges. At least in the case of hydrophobic bonds, it is known that they are actually more stable at high temperatures than at low temperatures.

The industrial uses of thermophilic enzymes have been perceived but barely exploited. Thermostable enzymes almost certainly will have longer shelf lives in commercial preparations, although they can only be used in high temperature processes. There has been some interest in their use in laundry operations, where high temperatures are the rule. Hopefully, less mundane uses will be found. One can visualize the use of thermostable enzymes for chemical catalysis, to carry out specific reactions of interest. For such applications, the requirement of high temperature for the process is of no difficulty, and the increased reaction rates at high temperature (due at least in part to increased diffusion rates) may make the process economically attractive. The thermostability of the enzyme may make it easier to link it covalently in active form to an immobilizing substrate. The advantage of an organism such as *Thermus aquaticus* is that its enzymes are considerably more thermostable than those of *Bacillus stearothermophilus*.

No concerted effort has apparently been made to isolate organisms that might possess particular kinds of enzymes of interest. We isolated *T. aquaticus* not as a source of enzymes, but because it was a common organism in hot springs. By the use of different substrates or enrichment procedures, a wide variety of new organisms might be isolated, with enzymatic machineries of specific interest. When the use of enzymes in soap chips was a hot topic, a number of industrial companies isolated thermophiles which produced heat-stable proteases. In the past few years, this type of work has been increasingly concentrated in Japan, where both the industrial interest and the natural thermal environments exist.

Protein Synthesis

It was clear from the beginning that if a bacterium grew at high temperature, then it must be able to synthesize protein at high temperature. Either the protein-synthesizing machinery would be inherently thermostable, or it was somehow stabilized within the cell. The evidence now suggests that in the case of *Thermus*, both mechanisms are operative. The book edited by Zuber (1976) contains several articles related to this whole question.

The early work on thermostability of ribosomes and tRNA, using the organism *Bacillus stearothermophilus,* has been reviewed by Friedman (1968). In my laboratory, Zeikus carried out a study on the thermostability of ribosomes and tRNA of *Thermus aquaticus* (Zeikus et al., 1970), and on the aminoacylation of tRNA (Zeikus and Brock, 1971). The data showed that the tRNA and aminoacyl synthetases of *T. aquaticus* were biologically active at high temperatures, and that the structure and function of these macromolecules was considerably more thermostable than those of mesophilic bacteria. Of some interest was the fact that heating purified tRNA leads to its inactivation, probably by a depurination reaction, and that *T. aquaticus* tRNA was less inactivated by heating than tRNA of moderate thermophiles or mesophiles. Apparently, the tRNA of *T. aquaticus* is so structured that it is inherently thermostable.

Oshima and coworkers (1976) have carried out much more extensive work on the protein-synthesizing machinery of *Thermus thermophilus,* a species that grows at temperatures up to 80–85°C. They determined the nucleotide sequence of one tRNA from this organism, and compared it with that of *E. coli.* Three factors controlling thermostability of the tRNA were identified: (1) intrinsic stability due to the increased content of G-C base pairs and the tight stacking of base pairs in one region of the molecule; (2) protection from heat denaturation by magnesium ions, and (3) temperature-induced stability by thiolation of a ribothymidine (T) base to a 5-methyl-2-thiouridine base (m^5s^2U) at position 55 of the tRNA. As discussed by Quigley and Rich (1976): "The higher melting temperature is probably due to the fact that substitution of the sulfur atom in the 2 position . . . places more polarizable electrons at that site, thereby, increasing the strength of the stacking interactions." The thiolation mechanism is especially interesting because the thiolation reaction depends on the growth temperature of the organism. At a growth temperature of 50°C, the base is in the T configuration, and as the growth temperature is progressively raised, the proportion of m^5s^2U increases. There is a direct correlation between the fraction of m^5s^2U and the thermostability of the tRNA.

Ohno-Iwashita et al. (1976) have studied a cell-free protein synthesizing system from *Thermus thermophilus* and have shown that a polyamine was necessary in order for the system to be active at temperatures above 50°C. Spermine was active, and the action of spermine was to stabilize the ribosomes. This polyamine was not required for stabilization of either the tRNA or the aminoacylation reaction. Ohno-Iwashita et al. (1976) were able to purify spermine from *T. thermophilus,* as well as another polyamine, which they called thermine, which is similar in structure to spermine but has one less methyl group. Interestingly, they found thermine in all extreme thermophiles they examined, and spermine was found in moderate and extreme thermophiles, although scarcely found in mesophiles. (Spermine is known to be formed in mammalian systems, however, such as sperm, its first site of discovery.)

Because the technology for studying cell-free protein synthesis is so well worked out, it is likely that considerable detail on the mechanism of protein synthesis at high temperature will soon be known. The overall picture seems to be that the new requirements necessary for a thermostable protein-synthesizing system are not large, and that it requires a relatively small number of evolutionary steps to convert from a mesophilic to a thermophilic system. A thermostable restriction endonuclease from *Thermus aquaticus* with a unique specificity has been characterized by Sato et al. (1977).

Lipids and Membranes of *Thermus aquaticus*

Theoretical considerations on the molecular nature of thermophily (Brock, 1967) suggested that the plasma membrane would be a critical cellular component. Structures within the cell, such as proteins and nucleic acids, could, at least in part, be stablized by cations or other molecules (see for instance the work on the new polyamine thermine from *T. thermophilus,* Oshima, 1975), but the plasma membrane must perforce be exposed to the outside world, so that it must have inherent stability. Also, a single hole in the plasma membrane should lead to leakage and death. Therefore, when Paul Ray came to me to do a Ph.D. problem, I suggested the cell membrane of *T. aquaticus* as a likely problem of interest. When he got deeper into lipids than I had envisaged, I arranged that he work with David C. White (at that time at the University of Kentucky), and this fruitful collaboration led to some interesting studies in *T. aquaticus* biochemistry. In a sense, these studies had their forerunner in the work of Henry Koffler, who had the insight to realize that flagella were also exposed to the outside world and hence must be thermostable. The membrane, of course, is of much more critical importance in the life of the cell, although at the same time being much more difficult to study.

First it should be noted that the plasma membranes of mesophilic organisms are thermolabile (Ray and Brock, 1971). Protoplasts of *Sarcina lutea* and *Streptococcus faecalis* (Figure 4.6) undergo thermal lysis when heated to temperatures of 60°C or above. As an even more sensitive indicator of membrane damage, we took advantage of an amino acid incorporating system that I had studied some years ago, in which ^{14}C-glycine is allowed to accumulate in the cell pool. Under appropriate conditions all of the radioactivity is present as free glycine, and in whole cells this radioactivity is retained when the cells are suspended in glycine-free medium. As seen in Figure 4.7 when preloaded cells were heated at various temperatures, they lost ^{14}C-glycine rapidly at temperatures of 50°C and above. We interpreted these results in conjunction with those of Figure 4.6 as an indication that temperature had caused a destruction of the plasma membrane. The thermal lysis observed could be prevented by the addition

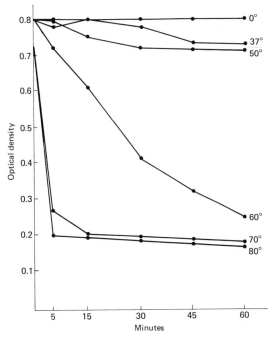

Figure 4.6. Effect of temperature on protoplasts of *Streptococcus faecalis* in isotonic medium. The protoplasts were stabilized in 1 M sucrose. The suspensions were incubated at various temperatures for the time periods given, and after cooling the optical density at 640 nm was measured. (From Ray and Brock, 1971.)

of the polyamines spermine and spermidine, and partially by cadaverine and streptomycin, two other basic organic compounds. Magnesium ions were also able to partially prevent thermal lysis, although calcium ions, lysine, ornithine, and putrescine were only weakly active or ineffective. The effective compounds were those that had the highest cation charge density. However, in the presence of high concentrations of NaCl or phosphate, the polyamines were no longer effective in preventing thermal lysis.

The spheroplasts prepared from *Thermus aquaticus* did not undergo thermal lysis even when boiled in distilled water (Figure 4.8). Since these spheroplasts were rapidly lysed by sodium lauryl sulfate, whereas normal rod-shaped *T. aquaticus* is not, the spheroplasts clearly depended on membrane and not wall for stability, and the stability at high temperature suggested that the plasma membrane of *T. aquaticus* was inherently thermostable. Further work by Paul Ray concentrated on the lipids and fatty acids of *T. aquaticus,* to attempt to understand the thermostability of the membrane. It might be pointed out that the relationship of lipids to thermophily and to temperature relationships of organisms in general has had a long history (reviewed in part by Ingraham, 1962). Our work was done

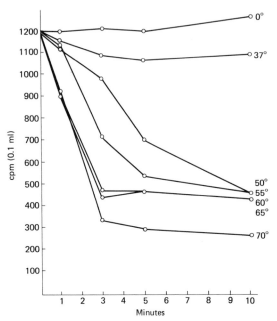

Figure 4.7. Effect of various temperatures on the release of ^{14}C-glycine from the pool of suspensions of whole cells of *Streptococcus faecalis*. (From Ray and Brock, 1971.)

before current tools for studying membrane fluidity (spin labeling, etc.) were available, but we had clearly in mind from the beginning a two-sided picture of the realtionship of membrane function to temperature: the membrane must be sufficiently rigid to resist the action of high temperature, yet it must be sufficiently fluid to function properly. The rigidity or fluidity of a membrane can in a crude way be related to the melting point of the predominant fatty acids. Rigidity requires the presence of saturated fatty acids; fluidity requires the presence of unsaturated fatty acids.

Although the fatty acids of *T. aquaticus* were fairly simple, the lipids turned out to be quite complex (Ray et al., 1971a). Phospholipids accounted for 30% of the total lipids and the minor components were identified as phosphatidylethanolamine (4%), phosphatidylglycerol (3%), phosphatidylinositol (10%), cardiolipin (3%), and phosphatidic acid (1%). The major phospholipid, which accounted for 80% of the lipid phosphate, was new and was not completely characterized. It contained an unsaturated undecylamine containing an hydroxyl group. This phospholipid contained three fatty acids per molecule and one glycerol per phosphate. It was periodate positive and ninhydrin negative. It had a minimum molecular weight of 1800. Further work on this phospholipid would be desirable, since it appears to be unique and may be related in some way to the thermostability of the membrane. In this respect, the novel glycolipid of *Thermus thermophilus* is of interest (Oshima and Yamakawa, 1974). This compound contains galactose, glucose, glucosamine, glycerol, fatty acid esters, and fatty acid amide in the ratio of 2:1:1:1:2:1. The presence of an amide-bound fatty acid in the glucosamine residue is unusual.

Carotenoids, responsible for the yellow pigment of the organism, accounted for 60% of the membrane lipid of *Thermus aquaticus* YT-1. The majority of the carotenoids were very polar, and proved difficult to purify and characterize. Mono- and diglucosyldiglyceride and the 35-, 40-, and 45-carbon vitamin K_2 isoprenologues were also found. All of the lipids were localized in the membrane of *T. aquaticus,* as shown by subjecting spheroplasts of the organism to repeated cycles of freeze and thaw and measuring lipid release.

The effect of growth temperature on the lipid composition of *T. aquaticus* was of interest. The organism can be grown conveniently over a temperature range from 50°C to 75°C. As growth temperature increased, there was a progressive increase in the total lipid content. The phospholipids increased two-fold, the carotenoids increased 1.8-fold, and the glucolipids increased four-fold between cells grown at 50°C and 75°C. The vitamin K_2 level did not change. The proportions of the individual phospho-

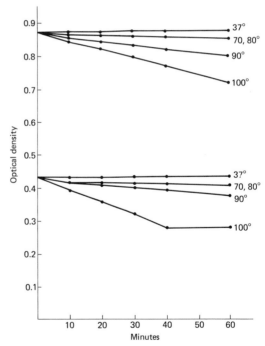

Figure 4.8. Effect of temperature on spheroplasts of *Thermus aquaticus.* The lysozyme-induced spheroplasts were suspended in distilled water and heated. Optical densities were measured at various times on portions of the suspension that had been cooled. (From Paul Ray, unpublished Ph.D. dissertation, Indiana University, Bloomington, 1969.)

lipids did not change as the growth temperature was raised. Studies were also done by means of ^{32}P pulse-chase experiments to determine whether there was any unusually rapid turnover of phospholipids. The turnover of the diacyl phospholipids during pulse-chase experiments was at rates comparable with mesophilic bacteria and the major phospholipid did not turn over significantly. Thus, there was no evidence for unusual thermal stress on the plasma membrane.

When grown at 70°C, *Thermus aquaticus* contained four major fatty acids, iso-C_{15} (28%), iso-C_{16} (9%), normal-C_{16} (13%), and iso-C_{17} (48%) (Ray et al., 1971b). Small amounts of iso-C_{12}, normal-$C_{12:1}$, iso-C_{13}, normal-C_{14}, iso-C_{14}, and normal-$C_{15:1}$ were also detected. Thus, *T. aquaticus* contained fatty acids with very few double bonds, as would be predicted, since unsaturated fatty acids have lower melting temperatures than saturated fatty acids. A change in growth temperature led to a striking change in the proportions of some of the fatty acids, as shown in Table 4.1. The proportions of the monoenoic and branched-C_{17} fatty acids decreased and the proportions of the higher-melting iso-C_{16} and normal-C_{16} fatty acids increased. The anteiso-C_{17}, which was undetectable at the higher growth temperatures, became one of the predominant fatty acids in cells grown at 50°C (Table 3.1), whereas the proportion of iso-C_{16} and normal-C_{16} decreased markedly. The iso- and normal fatty acids have considerably higher melting points, and hence thermostabilities, than the anteiso fatty acid. Thus, it appears that *T. aquaticus* uses the anteiso fatty acid to

Table 4.1. Effect of Growth Temperature on the Proportions of Fatty Acids of Exponentially Growing *Thermus aquaticus*[a]

Fatty acid methyl esters	Growth temperature (°C)					
	50	55	60	65	70	75
i-12	0.5	0.5	0.3	0.4	0.4	0.4
12:1	1.5	1.4	1.0	0.5	0.5	0.6
i-13	0.5	0.5	0.4	0.7	0.5	0.5
n-14	0.4	0.3	0.5	0.5	0.7	1.4
i-14	0.6	0.4	0.4	0.4	0.4	0.4
i-15	33.0	31.4	27.6	29.5	28.2	30.5
15:1	2.1	2.0	1.6	1.4	1.0	0.5
i-16	3.1	3.8	6.5	7.3	8.8	14.0
n-16	1.6	2.0	7.8	10.6	12.7	16.7
i-17	37.8	45.0	55.8	52.4	48.1	35.8
a-17	21.6	14.8	0.0	0.0	0.0	0.0

[a]Fatty acids were extracted, methylated, and analyzed by gas chromatography. The numbers are expressed as the percentage of the total at the temperature. The fatty acids are abbreviated as the number of carbon atoms prefaced with i for isobranching, a for anteiso-branching, n for unbranched, or followed by :1 for monoenoic fatty acids. (From Ray et al., 1971b.)

increase membrane fluidity at lower temperatures, rather than the unsaturated fatty acids used by other organisms.

The fatty acid composition of *T. aquaticus* is radically different from that of most gram-negative bacteria. Indeed, *T. aquaticus* resembles the gram-positive bacteria in that the majority of its fatty acids are branched. Also, no cyclopropane fatty acids are found, whereas these are common in some gram-negative bacteria.

However, although *T. aquaticus* lipids appear somewhat unusual, they are much more normal than those of the two acidophilic thermophiles we have isolated, *Thermoplasma* and *Sulfolobus,* which will be discussed in later chapters.

Final Words

Thermus aquaticus is an interesting organism, although in retrospect is not nearly as extreme a thermophile as we thought. At the time it was isolated, in 1967, it was the most thermophilic thermophile known, and because of this it elicited considerable interest. Its complement of thermostable enzymes, membranes, etc., is sufficient to explain its ability to grow at high temperatures. In most cases, the enzymes of *Thermus* are completely inactive at normal temperatures, thus explaining the obligately thermophilic nature of the organism. Because it is easy to grow at high temperatures, and large cell yields can be obtained, it has proved of considerable interest to enzymologists and others concerned with thermostable enzymes. Not necessarily related to thermostability, the organism has a restriction nuclease which is of some use in determining the sequence of DNA's (Sato et al., 1977).

Almost certainly, the primary habitat of *Thermus* is the hot spring, although it has extensively colonized man-made thermal habitats such as hot water heaters. The latter habitats are a useful source of new isolates of the organism. The ability of *Thermus* to become dispersed from place to place is obvious. There is no reason why this could not occur along surface waterways and channels. Zeikus and Brock (1972) and Brock and Yoder (1971) showed that *Thermus* was liberated from its growth habitat into surrounding streams and would survive for considerable distances in non-thermal water. In the case of the Firehole River study of Zeikus and Brock (1972) discussed in detail later in this book, it was possible to follow the die-off of *Thermus* downstream from the hot spring effluents. The organism did drop in numbers with time, but could still be picked up some kilometers below the final thermal effluents. When such surface waters are used as sources of domestic water, even with purification, it is likely that some viable cells will find their way into the water systems. Once such cells are established in a water heater, they provide a source of inoculum for other heaters, making possible a continual spread of the organism. In this way, the hot spring need only be a source of inoculation occasionally.

References

Allen, M. B. 1953. The thermophilic aerobic sporeforming bacteria. *Bacteriol. Rev.* **17**, 125–173.

Amelunxen, R. E. and R. Singleton. 1976. Thermophilic glyceraldehyde-3-P dehydrogenase. In *Enzymes and Proteins from Thermophilic Microorganisms,* H. Zuber, ed. Birkhäuser Verlag, Basel, pp. 107–120.

Brock, T. D. 1967. Life at high temperatures. *Science* **158**, 1012–1019.

Brock, T. D. 1974. Genus Thermus. In *Bergey's Manual of Determinative Bacteriology,* 8th ed., R. E. Buchanan and N. E. Gibbons, ed. Williams and Wilkins, Baltimore, p. 285.

Brock, T. D. and K. L. Boylen. 1973. Presence of thermophilic bacteria in laundry and domestic hot-water heaters. *Appl. Microbiol.* **25**, 72–76.

Brock, T. D. and M. R. Edwards. 1970. Fine structure of *Thermus aquaticus,* an extreme thermophile. *J. Bacteriol.* **104**, 509–517.

Brock, T. D. and H. Freeze. 1969. *Thermus aquaticus* gen. n. and sp. n., a nonsporulating extreme thermophile. *J. Bacteriol.* **98**, 289–297.

Brock, T. D. and I. Yoder. 1971. Thermal pollution of a small river by a large university: bacteriology studies. *Proc. Ind. Acad. Sci.* **80**, 183–188.

Castenholz, R. W. 1969. Thermophilic blue-green algae and the thermal environment. *Bacteriol. Rev.* **33**, 476–504.

Degryse, E., N. Glansdorff, and A. Pierard. 1978. A comparative analysis of extreme thermophilic bacteria belonging to the genus *Thermus. Archives of Microbiology,* in press.

Farrell, J. and L. L. Campbell. 1969. Thermophilic bacteria and bacteriophages. *Adv. Microbiol. Physiol.* **3**, 83–110.

Fields, M. L. 1970. The flat sour bacteria. *Adv. Fd. Res.* **18**, 163–217.

Friedman, S. M. 1968. The protein-synthesizing machinery of thermophilic bacteria. *Bacteriol. Rev.* **32**, 27–38.

Freeze, H. and T. D. Brock. 1970. Thermostable aldolase from *Thermus aquaticus. J. Bacteriol.* **101**, 541–550.

Hocking, J. D. and J. I. Harris. 1976. Glyceraldehyde-3-phosphate dehydrogenase from an extreme thermophile, Thermus aquaticus. In *Enzymes and Proteins from Thermophilic Microorganisms,* H. Zuber, ed. Birkhäuser Verlag, Basel, pp. 121–134.

Koffler, H. 1957. Protoplasmic differences between mesophiles and thermophiles. *Bacteriol. Rev.* **21**, 227–240.

Ohno-Iwashita, Y., T. Oshima, and K. Imahori. 1976. The effect of polyamines on the thermostability of a cell free protein synthesizing system of an extreme thermophile. In *Enzymes and Proteins from Thermophilic Microorganisms,* H. Zuber, ed. Birkhäuser Verlag, Basel, pp. 333–345.

Oshima, T. 1975. Thermine: a new polyamine from an extreme thermophile. *Biochem. Biophys. Res. Commun.* **63**, 1093–1098.

Oshima, T. and K. Imahori. 1974. Description of *Thermus thermophilus* (Yoshida and Oshima) comb. nov., a nonsporulating thermophilic bacterium from a Japanese thermal spa. *Int. J. Syst. Bacteriol.* **24**, 102–112.

Oshima, T., Y. Sakaki, N. Wakayama, K. Watanabe, Z. Ohashi, and S. Nishimura. 1976. Biochemical studies on an extreme thermophile *Thermus thermophilus:* thermal stabilities of cell constituents and a bacteriophage. In *Enzymes and Proteins from Thermophilic Microorganisms,* H. Zuber, ed. Birkhäuser Verlag, Basel, pp. 317–331.

Oshima, M. and T. Yamakawa. 1974. Chemical structure of a novel glycolipid from an extreme thermophile, *Flavobacterium thermophilum. Biochemistry* **13**, 1140–1146.

Quigley, G. S. and A. Rich. 1976. Structural domains of transfer RNA molecules. *Science* **194**, 796–806.

Ramaley, R. F. and J. Hixson. 1970. Isolation of a nonpigmented, thermophilic bacterium similar to *Thermus aquaticus. J. Bacteriol.* **103**, 527–528.

Ray, P. H. and T. D. Brock. 1971. Thermal lysis of bacterial membranes and its prevention by polyamines. *J. Gen. Microbiol.* **66**, 133–135.

Ray, P. H., D. C. White, and T. D. Brock. 1971a. Effect of growth temperature on the lipid composition of *Thermus aquaticus. J. Bacteriol.* **108**, 227–235.

Ray, P. H., D. C. White, and T. D. Brock. 1971b. Effect of temperature on the fatty acid composition of *Thermus aquaticus. J. Bacteriol.* **106**, 25–30.

Rutter, W. J. 1964. Evolution of aldolase. *Fed. Proc.* **23**, 1248–1257.

Sato, S., C. A. Hutchison, and J. I. Harris. 1977. A thermostable sequence-specific endonuclease from *Thermus aquaticus. Proc. Natl. Acad. Sci. USA,* **74**, 542–546.

Singleton, R. and R. A. Amelunxen. 1973. Proteins from thermophilic microorganisms. *Bacteriol. Rev.* **37**, 320–342.

Weeks. O. B. 1974. Genus Flavobacterium. In *Bergey's Manual of Determinative Bacteriology,* 8th ed., R. E. Buchanan and N. E. Gibbons, eds. Williams and Wilkins, Baltimore, pp. 357–364.

Zeikus, J. G. and T. D. Brock. 1971. Protein synthesis at high temperatures: aminoacylation of tRNA. *Biochim. Biophys. Acta* **228**, 736–745.

Zeikus, J. G. and T. D. Brock. 1972. Effects of thermal additions from the Yellowstone geyser basins on the bacteriology of the Firehole River. *Ecology* **53**, 283–290.

Zeikus, J. G., M. W. Taylor, and T. D. Brock. 1970. Thermal stability of ribosomes and RNA from *Thermus aquaticus. Biochim. Biophys. Acta* **204**, 512–520.

Zuber, H. (editor). 1976. *Enzymes and Proteins from Thermophilic Microorganisms.* Birkhäuser Verlag, Basel, 445 pp.

Chapter 5
The Genus *Thermoplasma*

Perhaps the organism we have isolated which has elicited the most interest (and some controversy) has been *Thermoplasma*. The credit for this organism goes first to Gary Darland, who was smart enough to recognize that something others would have considered to be a precipitate was indeed living, and to Robert Belly, who (among other important contributions in my laboratory), showed the significance and distribution of the organism.

I discuss later in this chapter the coal refuse pile as a habitat. Surely it is one of the more fascinating microbial habitats, and one that has yielded to us a variety of interesting material for study. I had always felt that it was pertinent to apply the knowledge we were gaining in Yellowstone to more applied problems, and when we got into work on acid habitats (see Chapter 12), I searched for man-made habitats of similar character. Coal refuse piles provide the most readily accessible and dramatic examples of such acid habitats. But when we first climbed on top of the New Hope South coal refuse pile, I was not prepared for anything thermal as well. I recall my excitement when I saw steam rising from one small part of the pile. A little study showed that there were several steaming sites on this pile, and temperature measurements showed that the temperature ranged from 50°C to 70°C in different locations and at different depths. We collected samples, and Darland set up enrichment cultures. To put this in proper perspective, Darland had been isolating acidophilic thermophiles from acid hot springs for some time, and had a large collection of spore-forming bacilli, now classified as *Bacillus acidocaldarius* (Darland and Brock, 1971). Thus, when we found this steaming coal refuse pile, it was natural to attempt to isolate similar organisms. Instead, *Thermoplasma* was found.

Darland's first notes on this organism call it "The wonder organism." This was not a bad designation, as when we first looked at the enrichment

culture under the microscope, we saw something that looked like a mycoplasma, yet was growing at pH 2–3 and 56°C. The enrichment had been
obtained by adding about 0.5 g of coal refuse material from a site of 56°C
and a pH of 1.96 to a basal salts medium containing glucose and yeast
extract. The medium was adjusted to either pH 2 or 3. After 16 days, the
enrichments were obviously turbid, and microscopic examination revealed
spherical bodies that resembled mycoplasmas. However, our experience
with mycoplasmas was close to zero, and we had to rule out the possibility
that some sort of precipitate had developed from the coal refuse material
during the incubation period. When he transferred from the enrichment,
Darland set up one set using inoculum that had been autoclaved. Only
nonautoclaved material yielded cultures, and of course during successive
transfers the coal refuse material was soon diluted away. At once Darland
was into a major study characterizing this unique organism.

Characteristics of *Thermoplasma*

Two isolates were obtained, 122-1B2 and 122-1B3, by enrichment at pH 2
and 3, respectively. Another isolate, 3-24, was obtained 6 months later from
a different location in the same pile. Later, Belly obtained an even larger
number of isolates (Belly et al., 1973).

Morphologically the isolates were pleomorphic spheres which, by phase
microscopic examination, appeared to reproduce by budding. Although a
sphere appeared to be the basic structural unit, filamentous structures were
often seen, particularly in young cultures. The cells varied in diameter from
about 0.3 to 2 μm.

Electron micrographs revealed the relatively simple structure of a procaryotic organism. Nuclear material was dispersed throughout the cell with
no evidence of a limiting nuclear membrane (Figure 5.1). No indication of
membranous organelles within the cytoplasm was seen. Unlike typical
bacterial cells, however, these isolates lacked a rigid cell wall and were
separated from the surrounding environment by only a double membrane.

The failure to detect a cell wall with the electron microscope was in
keeping with several pieces of indirect evidence. The addition of sodium
lauryl sulfate to a culture of the organism resulted in a very rapid cell lysis,
with a decrease in optical density at 540 nm of about 80% occurring within
30 seconds. We were unable to demonstrate the existence of hexosamine
by the Elson-Morgan assay for amino sugars (Table 5.1). Finally, the
isolates were insensitive to the antibiotic vancomycin at concentrations as
high as 5 mg/ml (Table 5.2). Since this antibiotic is a specific inhibitor of cell
wall synthesis, blocking the addition of the muramic acid-lipid complex to
an acceptor, the inability of this antibiotic to inhibit growth suggested that a
cell wall may not be essential for cell growth. However, the isolates were
inhibited by novobiocin, an antibiotic that inhibits mycoplasmas, at a

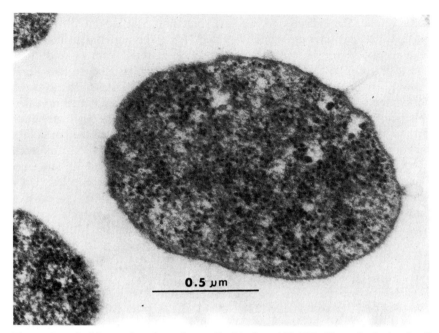

Figure 5.1. A thin section through a cell of isolate 122-1B2. Note the lack of cell wall. (From Darland et al., 1970)

concentration of approximately 0.1 μg/ml. Since the antibiotic tests were performed under rather severe conditions, pH 3 and 56°C, it was necessary to have a control which showed that the antibiotics were indeed active under these conditions. We used another acidophilic thermophile, *Bacillus acidocaldarius* strain 104-1A, a spore-forming rod that grows at pH 3 and 56°C. The inhibition of this organism by both antibiotics showed that the

Table 5.1. Biochemical Composition of Acido-Thermophilic Bacteria[a]

	Protein	RNA	Hexosamine	Carbohydrate
Sulfolobus				
98-3	1	0.18	0.054	0.33
106-3	1	0.16	0.047	0.30
115-2	1	0.16	0.055	0.38
129-1	1	0.20	0.029	0.34
140-5	1	0.26	0.030	0.21
Bacillus acidocaldarius	1	0.10	0.126	0.11
Thermoplasma				
3-24	1	0.24	0	0.09
22-7	1	0.14	0	0.04
122-1B3	1	0.18	0	0.06
R8C	1	0.32	0	0.09

[a]Relative proportions by weight of various components normalized to the protein contents of the cell suspensions. (From Brock et al., 1972.)

Table 5.2. Antibiotic Sensitivity of Acidophilic, Thermophilic Bacteria

Organism[a]	Minimum inhibitory concentration (μg/ml)		
	Novobiocin	Vancomycin	Ristocetin
Bacillus acidocaldarius (peptidoglycan cell wall)	10	50	10
Thermoplasma acidophilum 122-1B2 (no cell wall)	0.1	>5000	—
Thermoplasma acidophilum 122-1B3 (no cell wall)	0.1	>5000	>1000
Sulfolobus acidocaldarius 98-3 (atypical cell wall, no peptidoglycan)	10	>10,000	>1000
Sulfolobus acidocaldarius 115-2	10	>10,000	>1000

[a] Data on *Thermoplasma* from Darland et al. (1970). *Bacillus* and *Sulfolobus,* unpublished data of K. M. Brock. Both vancomycin and ristocetin are specific inhibitors of peptidoglycan synthesis.

antibiotics were active. Because of its acid lability, penicillin, another specific inhibitor of cell wall synthesis, could not be used. The organism was also not inhibited by cycloheximide, a specific inhibitor of eucaryotes.

The thermophilic nature of the isolates is seen in Figure 5.2, in which the growth rate in doublings per hour is plotted against the reciprocal of the absolute temperature. The optimum temperature for growth is about 59°C with a doubling time of about 4 hours. No growth occurred at 65°C or 37°C.

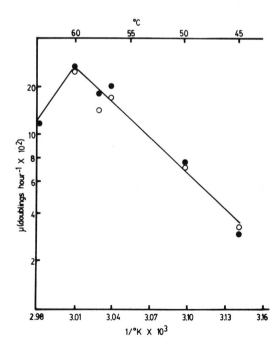

Figure 5.2. Arrhenius plot of the log of the growth rate versus the reciprocal of the absolute temperature. Cultures were grown without aeration in the standard growth medium at pH 2. Growth was followed by measuring the optical density at 540 nm in a Spectronic 20. Open circles, isolate 122-1B2; closed circles, isolate 122-1B3. (From Darland et al., 1970.)

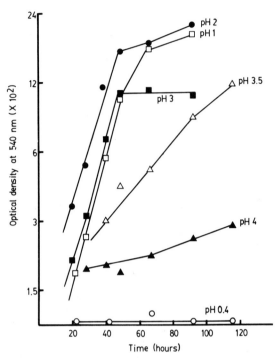

Figure 5.3. Effect of pH on the growth of isolate 122-1B2. The organism was grown at 55°C in the standard growth medium. Growth was followed by measuring the optical density at 540 nm. (From Darland et al., 1970.)

The effect of pH on the growth of one of the isolates, 122-1B2, is seen in Figure 5.3. No growth occurred at pH 0.35, and growth was very slow at pH 4. The organism grew well between pH 0.96 and 3.0, the optimum pH being about 1 to 2. The pH did not vary (within 0.2 unit) throughout this experiment.

DNA Base Composition

One of the characteristics of *Thermoplasma* which we determined was the DNA base composition. We found that it had a low GC content, 25% guanine + cytosine, similar to that of other mycoplasmas. Later we determined the base compositions of three more strains, isolated from geographically diverse areas, and they were also low in GC (Belly et al., 1973). This made me feel a little more confident about the taxonomic relationship of the organism, although in retrospect the GC content alone hardly seems to be a deciding property. At any rate, our GC determinations were apparently erroneous, as subsequently two laboratories have found the base composition to be about 46% G + C (Christiansen et al. 1975; Searcy and Doyle, 1975). The latter laboratory showed that our results may have been erroneous because of the fact that depurination of the DNA of *Thermoplasma* can occur under the culture conditions used. As was shown by Smith et al. (1973), *Thermoplasma* cultures died rapidly after reaching maximum density, so that only 0.1% of the cells remained viable 1 day later, and the fraction of viable cells had decreased to 10^{-8} 2 days after reaching maximum density. If in our work we used nonviable cells as a source for DNA, acid may have entered and caused depurination, which would have lowered its buoyant density, thus leading to an erroneous calculation of its DNA base composition.

Christiansen et al. (1975) found the genome size of *Thermoplasma* to be about 10^9 daltons, whereas Searcy and Doyle (1975) found an even smaller genome size, 8.4×10^8. According to these latter workers, the genome size of *Thermoplasma* DNA is the smallest yet reported for any nonparasitic organism.

Nomenclature

One nomenclatural note should be raised. When I originally coined the name *Thermoplasma acidophila,* I was under the erroneous impression that the "a" ending on *"acidophila"* was correct. As Freundt has pointed out, *Thermoplasma* is a neuter word, so that the correct ending for the species epithet is "acidophilum." I have not bothered to write a paper making this correction, but most of the workers using this organism have used the correct epithet, and this is the epithet that appears in Bergey's 8th edition (Freundt, 1974).

Habitats of *Thermoplasma*

I had from the beginning been concerned about the peculiar habitat of *Thermoplasma,* the self-heated coal refuse pile. This seemed to be a transitory and rather unstable habitat, and so we looked for more likely natural reservoirs of the organism. At first I thought that *Thermoplasma* existed in acid hot springs (Darland et al., 1970), but subsequent work showed that the organism we had isolated from acid hot springs was *Sulfolubus,* an interesting but quite different organism (see Chapter 6). After we had sorted out the differences between *Sulfolubus* and *Thermoplasma,* Belly made a valiant attempt to isolate *Thermoplasma* from Yellowstone hot springs. As seen in Table 5.3, he was completely unable to isolate *Thermoplasma* from appropriate habitats. A very likely habitat in Yellowstone for *Thermoplasma* had seemed to be the *Cyanidium caldarium* mats, which provided a rich source of organic matter for the growth of heterotrophic, thermophilic acidophiles (see the section in Chapter 3 on thermophilic fungi, especially *Dactylaria gallopava*). Indeed, microscopic examination of such mats often revealed mycoplasma-like structures. However, Belly was unable to isolate any *Thermoplasma* cultures, using the methods that worked so well for coal refuse material. He could isolate from the *Cyanidium* mats a number of cultures of the moderately thermo-

Table 5.3. Distribution of *Thermoplasma* in Various Natural Habitats

Habitat	Location[a]	Isolation temperature (°C)	Isolation procedure[b]				Total isolation attempts	Number of isolates
			A2	A1	A1V	A2V		
Acid, hot soils	Y.N.P.	55	149[c]	10	4	149	457	0
		45	145	0	0	0		
Acidic hot springs	Y.N.P.	55	46	58	33	58	223	0
		45	8	8	6	6		
Acidic fumaroles	Y.N.P.	55	18	18	0	0	36	0
Burning coal refuse piles	S. Ind., W. Pa.	55	162	162	0	162	486	113

[a] Samples were obtained from various thermal, acidic regions located in Yellowstone National Park (Y.N.P.) or from coal refuse piles located in Southern Indiana (S. Ind.) or Western Pennsylvania (W. Pa.).

[b] Isolation attempts were made using Allen's basal salts medium containing 0.1% yeast extract at either pH 1.0 (A1) or pH 2.0 (A2) and in the presence (V) or absence of vancomycin (1225 μg/ml).

[c] Figures indicate the number of separate attempts made.

From Belly et al. (1973).

philic acidophile *Bacillus coagulans* (Belly and Brock, 1974), and he also showed that these cultures formed stable protoplasts (or spheroplasts) which in some ways resembled *Thermoplasma* microscopically. But after a total of 716 unsuccessful isolation attempts at Yellowstone, it seemed reasonable to conclude that *Thermoplasma* did *not* live in hot springs.

Thus, at this point, the only habitat we had for *Thermoplasma* was a single coal refuse pile in southern Indiana. This bothered me greatly, so I contacted people at the U.S. Bureau of Mines in Pittsburgh, and arranged for Belly to be taken to a variety of burning coal refuse piles in western Pennsylvania. This trip, made in November of 1971, was wildly successful (Belly et al., 1973). He visited 30 separate coal refuse piles and was able to isolate *Thermoplasma* from 20 of them, obtaining 113 separate isolates. Virtually all samples positive for the presence of *Thermoplasma* were obtained from regions of coal refuse piles that had temperatures of 32–80°C and pH values of 1.17–5.21. A listing of strains isolated which were characterized further is given in Table 5.4. Of some interest is strain 114-1

Table 5.4. Source and Serological Group of Several *Thermoplasma* Isolates

Isolation	Location	Habitat Temperature (°C)	Habitat pH	Serological group[a]
94-1	Winslow, Ind.	35–60	1.17	N.D.[b]
96-2	Baldwin, Pa.	30–55	3.23	V
97-2	Irwin, Pa.	35–62	3.42	II
100-2	Hanker, Pa.	31–58	2.37	N.D.
102-3	New Stanton, Pa.	37–62	3.30	N.D.
105-4	Saltzburg, Pa.	37–59	3.06	IV
107-1	Clymer, Pa.	32–51	2.34	II
110-1	Gypsy, Pa.	30–50	6.78	III
111-1	Carroltown, Pa. (I)	39–65	4.05	IV
112-3	Carroltown, Pa. (II)	32–65	5.14	N.D.
114-1	Carroltown, Pa. (III)	35–58	2.52	V
115-1	Carroltown, Pa. (IV)	32–62	2.33	III
116-1	Watkins, Pa.	37–65	2.10	N.D.
117-2	Ehrenfield, Pa.	45–65	1.65	IV
121-2	Lilly, Pa.	45–65	2.75	V
122-2	Jerome, Pa.	35–65	2.32	N.D.
124-1	Bentleyville, Pa.	37–57	3.44	II
130-2	Mather, Pa.	35–80	3.16	N.D.
133-1	Chartiers, Pa. (I)	35–75	5.21	II
135-1	Chartiers, Pa. (II)	32–55	3.51	I
R8D55	Dugger, Ind.	53–78	1.61	III
122-1B3	Dugger, Ind.	45–56	1.96	I
22-7	Dugger, Ind.	37–52	1.85	IV

[a] See Belly et al. (1973) for details.
[b] N.D.—Not done.

from pile III in Carroltown, Pa. This coal refuse pile had been only burning for 2 years, indicating the short time necessary for the establishment of *Thermoplasma*.

Specific Isolation Procedures for *Thermoplasma*

The procedures used by Belly for isolation of *Thermoplasma* should be briefly described. Since there are rod-shaped organisms (*Bacillus acidocaldarius* and *B coagulans*) that can grow under the same conditions as *Thermoplasma,* it is essential to use some type of selective procedure when carrying out a large number of isolations from nature. The original procedure of Darland was to use a basal salts medium adjusted to pH 2, to which yeast extract at 0.1% was added. This procedure often worked for *Thermoplasma* (see Table 5.3), but to obviate rods, Belly also used the same medium adjusted to pH 1, at which pH rod-shaped bacteria do not grow. Another modification was to use pH 2 medium but to add vancomycin at 1225 μg/ml. This antibiotic, which is stable under warm, acid conditions, inhibits cell wall synthesis, and thus provides conditions selective for wall-less organisms. However, it does present the (probably slight) possibility that it might induce the formation of L-forms of a normal rod-shaped organism. In another procedure, used for water samples, the water was passed through a 0.45-μm membrane filter, and the filtrate obtained then passed through a 0.22-μm filter. The resultant filter was then incubated in culture medium. All enrichment cultures were incubated 4 to 6 weeks at 55°C, and the presence of *Thermoplasma* determined by the occurrence of visible turbidity and confirmed by microscopic examination. Purification of cultures was by dilution in liquid medium, since reproducible colony growth on agar has not been obtained.

Belly characterized the large number of isolates he obtained physiologically, and when Ben Bohlool joined the laboratory, he prepared fluorescent antibodies and grouped the isolates serologically (Belly et al., 1973; Bohlool and Brock, 1974b). All twenty isolates grew over a temperature range of 37–65°C, with an optimum around 55°C. The pH range for growth was between 0.5 and 4.5; no growth was observed at pH 0 or 5. All twenty isolates studied lysed rapidly in the presence of 1% sodium lauryl sulfate, or when at pH 8.5. No lysis was observed when the isolates were heated at pH 2 at 100°C for 30 minutes. Further discussion of the cellular stability of *Thermoplasma* will be found later, but these results indicated that all of the isolates lacked a cell wall.

None of the isolates were able to grow without yeast extract. Although Darland et al. (1970) originally reported that sugars did not stimulate growth, Belly found that growth was stimulated by sugars when the yeast extract concentration was reduced to a growth-limiting concentration of

0.025%. As seen in Table 5.5, sucrose, glucose, galactose, mannose, and fructose stimulated the growth of all strains tested, and a few other substances gave slight stimulation.

To demonstrate the presence of cytochromes, a spectrum of reduced versus oxidized cells was made, using a Cary spectrophotometer. The results showed the presence of a c-type cytochrome, but no absorption bands characteristic of the a- or b-type cytochromes were seen. A spectrum was also performed on a carbon monoxide-treated cell suspension, which showed a strong absorption band at 430 nm characteristic of cytochrome o. Again, no a-type cytochrome was seen in the carbon monoxide spectra. All five strains examined showed similar cytochrome spectra. Langworthy et al. (1972) found naphthoquinones in *Thermoplasma,* suggesting the presence of a complete respiratory chain. Smith et al. (1973) showed that *Thermoplasma* was a facultative anaerobe, when growing in a 1% glucose, 0.1% yeast extract medium. It grew under a N_2 atmosphere, but grew much better in air, and growth was stimulated by forced aeration. Belly had also

Table 5.5. Stimulation of *Thermoplasma* Growth by Addition of Various Compounds to Basal Medium Containing a Growth-Limiting Concentration of Yeast Extract[a]

Addition[b]	Strain 122-1B3	Strain 124-1	Strain 97-2	Strain 110-1	Strain R8D55
Sucrose	+	+	+	+	+
Glucose	+	+	+	+	+
Galactose	+	+	+	+	+
Mannose	+	+	+	+	+
Fructose	+	+	+	+	+
Mannitol	±	±	−	−	−
Lactose	−	−	−	−	−
Glycerol	±	−	−	−	−
Sorbitol	±	−	−	−	−
Inositol	−	−	−	−	−
Ribose	−	−	−	−	−
Aspartate	±	−	±	−	−
Glutamate	−	−	−	−	−
Glycine	−	−	−	−	−
Alanine	−	−	−	−	−
Yeast extract (0.075%)	+	+	+	+	+

[a]Five milliliters of basal salts medium, pH 2.0, supplemented with a growth-limiting concentration of yeast extract of 0.025%, were added to screw-top test tubes. Tubes were incubated at 55°C for 5 days. Growth was determined with an Aminco-Bowman Spectrophotofluorometer and converted into cell counts. + indicates at least one doubling of cell numbers over a control tube with no additions; ± indicates a slight increase in cell numbers; and − indicates no detectable increase in cell number.
[b]Additions were made to a final concentration of 0.1%.
From Belly et al. (1973).

found aeration to be valuable for obtaining higher cell yields, and generally cultured the organism unshaken in large Fernbach flasks to increase the surface area of the medium exposed to the atmosphere. Interestingly, Smith et al. (1973) found that growth was inhibited by CO_2 under anaerobic conditions. Since CO_2 should have no effect on the pH, the nature of this inhibition is not clear.

Serological Studies

Antibodies were prepared against purified cell membranes of *Thermoplasma*. Membranes were prepared by subjecting cell suspensions in acidified distilled water to seven consecutive cycles of freeze and thaw. This resulted in complete disruption of the cells; the membranes were then purified by differential centrifugation. Antisera obtained by injection into rabbits were purified and conjugated with fluoresceine isothiocyanate by standard methods (see Belly et al., 1973). For immunofluorescent staining of cells, cultures were grown in flasks in the presence of microscope slides and the slides were removed after 7–10 days, washed with medium, and fixed with 4% formalin in basal salts medium at pH 2.0. Formalin fixation stabilized the cells so that they could be transferred to neutral phosphate-buffered saline for the fluorescent staining.

Cross-reactions were assessed by visual examination of the intensity of fluorescence, and by adsorption of antibody with formalized cell suspensions of other strains. By means of these assessments, it was possible to group the strains into five antigenic affinity groups, as outlined in Table 5.4. In a subsequent study (Bohlool and Brock, 1974a), immunodiffusion analysis was used to further characterize the *Thermoplasma* strains serologically. Three antisera were prepared and tested for formation of precipitates against cell suspensions lysed at pH 8.0. The lysed preparations were then placed in the peripheral wells of an immunodiffusion plate, with the central well receiving undiluted antiserum. The three antisera tested, which had been prepared against purified cell membranes, exhibited three precipitin bands against their homologous strains. All three isolates showed a certain degree of cross-reaction when tested against each other. One precipitin band was shared by all the isolates tested and formed lines of identity with the homologous system. This indicates that at least one component of the membrane is identical in all the isolates tested. Another precipitin band was also present in the three test isolates but was absent in several of the other isolates. Immunodiffusion absorption was used to further characterize the antigenic differences of several isolates. The data indicated that in addition to the common antigen, there was at least one component specific and unique to the homologous cells.

Immunodiffusion and adsorption provided a much more comprehensive picture of the antigenic relationships of the strains. From the data, the following conclusions could be drawn: (1) isolates 122-1B3, 124-1, and 96-2 are all closely related, but each possesses some unique and individual characteristics; (2) isolate 97-2 is more related to 124-1 than to 122-1B3; (3) isolate R8D55 is antigenically more distinct than 122-1B3 or 124-1; (4) the other isolates tested are related to the three isolates used as sources of antisera only by one component. The overall picture showed that despite the physiological similarities between the various isolates, there was a fair degree of diversity, at least at the antigenic level.

Cellular Stability of *Thermoplasma*

Darland's early results had suggested that the *Thermoplasma* membrane had unusual properties. As part of his study of this organism, Belly determined the effect of various agents on cellular stability. The most striking observation, and one that has been extensively confirmed by others, is that *Thermoplasma* cells are stable at low pH but lyse when brought up to neutral pH (Figure 5.4) (Belly and Brock, 1972). On the other hand, boiling at low pH had no effect on *Thermoplasma* cells (Figure 5.5).

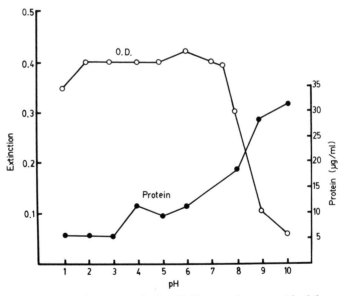

Figure 5.4. Effect of changes in pH on lysis of *Thermoplasma acidophilum*. Twenty micromoles of either HCl-KCl buffer, pH 1 or 2; phthalate-HCl, pH 3 or 4; phthalate-KOH, pH 5; (2-*N*-morpholino) ethane sulphonic acid (MES), pH 6; *N*-tris (hydroxymethyl) methyl-2-aminoethane sulphonic acid (TES), pH 7 or 8; or KCl-borate, pH 9–10, were added to the cell suspension. After 30 minutes, lysis was measured either spectrophotometrically or by protein release. (From Belly and Brock, 1972.)

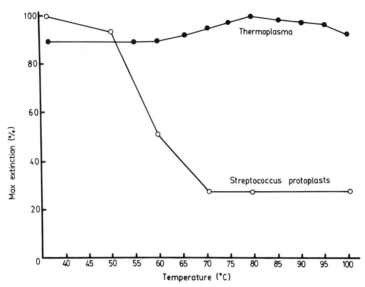

Figure 5.5. The effect of temperature on lysis of *Thermoplasma acidophilum* and *Streptococcus* protoplasts. *Streptococcus* protoplasts were suspended in 0.6 M-sucrose containing 0.075 M-disodium malate (pH 7.3). *Thermoplasma acidophilum* was suspended in distilled water (pH 5.5). Bacteria were heated at temperatures indicated for 30 minutes. After cooling, lysis was measured spectrophotometrically. (From Belly and Brock, 1972.)

What a strange membrane, to be destroyed at "neutral" pH, but to survive boiling! This immediately suggested an unusual membrane structure, which has been confirmed by the chemical studies (see below).

At the time this initial study was done, and before we had any knowledge of the chemistry of the membrane, we attempted to draw an analogy between *Thermoplasma* and *Halobacterium*. *Halobacterium* also has an unusual membrane, which falls apart when the external Na^+ is reduced. We equated H^+, which *Thermoplasma* requires, with Na^+, which *Halobacterium* requires. We then did some experiments attempting to stabilize *Thermoplasma* cells at neutral pH using Na^+ or Li^+ (2.5 *M*), without success. Since the more recent work in other laboratories on membrane chemistry and stability (see below), it appears that the analogy between *Halobacterium* and *Thermoplasma* has some merit, even if monovalent cations and H^+ are not interchangeable.

As opposed to other mycoplasmas and L-forms, *Thermoplasma* cells did not require an osmotic stabilizer. No lysis occurred when cells were suspended in either distilled water or various concentrations of NaCl. On the other hand, surface active agents, which lyse mycoplasma cells readily, also caused rapid lysis of *Thermoplasma* (Belly and Brock, 1972). However, the concentrations of surface active agents required for lysis were about 8 times higher than those required for lysis of other mycoplasmas.

Following up on this work on membrane stability, Smith et al. (1973) showed that although *Thermoplasma* did not lyse at pH 5.5, there was a marked reduction in viable count. If the ionic strength were increased, there was a slight increase in stability at pH 6 to 8, but decreased stability at normal growth pH values. By use of a suspending medium of moderate ionic strength (about 0.05) at a pH of 5, total lysis could be obtained, permitting the recovery and purification of membrane material.

Smith et al. (1973) found that the purified membranes had about 60% protein, 25% lipid, and 10% carbohydrate. The nature of the lipid will be discussed below. The amino acid composition of the total membrane protein did not reveal any major differences from membrane proteins of other mycoplasmas. A difference was noted in the proportion of free amino and carboxyl groups, which were less than half those of another mycoplasma, *Acholeplasma laidlawii.* Smith et al. (1973) showed that when the free carboxyl groups of the *Thermoplasma* membranes were blocked by formation of the glycine methyl ester, which would remove one negative charge, the membranes were relatively stable at all pH values tested. On the other hand, when the carboxyl groups were reacted with ethylene diamine, which would add one positive charge and remove a negative charge, the membranes exhibited the opposite stability to normal membranes, dissociating at low pH and remaining stable at high pH. On the basis of these studies, Smith et al. (1973) concluded that the membranes of *Thermoplasma* normally have a low charge, due in part to the reduced number of chargeable groups, and to the conversion of COO^- to $COOH$ at the low pH used for growth. Increasing the pH thus results in greater ionization of carboxyl groups, resulting in repulsion of negative charges, which leads to disruption of membrane structure and cellular lysis. A similar phenomenon occurs in halobacteria, although here neutralization of negative charges is brought about by Na^+ instead of H^+. Furthermore, if the negative charges of membrane proteins of normal bacteria are increased by succinylation, they disaggregate at neutral pH.

The Lipids of *Thermoplasma*

Because of the interesting membrane properties of *Thermoplasma,* the nature of the membrane lipids is of considerable interest. Two laboratories have been active in studying the chemistry of the lipids of *Thermoplasma,* that of Paul Smith and Thomas Langworthy at the University of South Dakota, and an Italian-English group under the guidance of J. D. Bu'Lock from the University of Manchester, but with the main chemical work done by Mario de Rosa at the CNR Laboratorio per la Chimica Fisica di Molecole di Interesse Biologico, Arco Felice (near Naples), Italy. At Bu'Lock's invitation, I spent 2 weeks at Arco Felice on October 1970, at a time when our work on *Thermoplasma* and *Sulfolobus* was just beginning.

The Italian group had independently isolated *Bacillus acidocaldarius* and I showed them a little microbiology and introduced them to *Thermoplasma*. However, virtually all of their work has been done independently of my guidance.

The lipids of *Thermoplasma* are unique. Perhaps because of this, there has been some difficulty working out the correct structures.

The overall lipid composition of *Thermoplasma* is shown in Table 5.6. Most of the lipid of *Thermoplasma* is present in the membrane (Ruwart and Haug, 1975).

Langworthy et al. (1972) reported that fatty acids are almost completely absent from the *Thermoplasma* lipids, although Ruwart and Haug (1975) found that the neutral lipids had more ester than ether-linked components. (The fatty acids found by Ruwart and Haug were predominantly 16 and 18 carbon saturated and unsaturated fatty acids and could conceivably have come from the yeast extract. No labeling studies were done to show that they were synthesized by the organism.) The hydrophobic side chains are based on the isoprenoid subunit, and are connected to glycerol or carbohydrate via ether linkages. The ether linkage is also found in the lipids of *Halobacterium* and *Sulfolobus* (see Chapter 6 for a discussion of the *Sulfolobus* lipids). The absence of ester linkages could almost be predicted: esters hydrolyze readily in hot acid environments, such as those *Thermoplasma* lives in. Although the internal pH of *Thermoplasma* is not very acid (see later), the pH at the periphery of the membrane is almost certainly acid, so that acid-stable structures are necessary. (Parenthetically, in *Halobacterium* the ether linkages may be necessary to prevent saponification in the alkaline, high-Na^+ environment in which this organism lives.) The group connected to the phosphate of the phospholipid is a monoglycosyl diglyceride, which is similar to the phospholipid of other mycoplasmas.

The precise structure of the nonpolar lipids of the *Thermoplasma* membrane has been worked out (Langworthy, 1977). Despite some controversy (deRosa et al., 1975), the structure is now known to be a 2,3-glycerol tetraether with a saturated C_{40} component which is itself made up from two phytanyl chains linked head-to-head (Figure 5.6). The C_{40} component can itself be acyclic, monocyclic, or bicyclic, and the proportion of these three forms seems to be characteristic for different of the acidophilic thermophiles, as shown in Table 5.7. The nonpolar structure differs from that of

Table 5.6. Lipid Composition of *Thermoplasma acidophilum*[a]

Fraction	Percent dry weight	Percent total lipid
Total lipid	3.1	100
Neutral lipid	0.54	17.5
Glycolipid	0.79	25.1
Phospholipid	1.75	56.6

[a]Data of Langworthy et al. (1972).

Figure 5.6. Structure of the lipid of *Thermoplasma* (Langworthy, 1977). The acyclic form is shown, but both monocyclic and bicyclic hydrocarbon chains are possible (see Table 5.7). In the cyclic forms, one or two of the isoprenoid units undergoes cyclization.

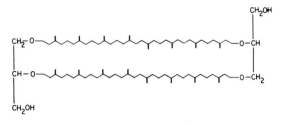

Halobacterium only in that the latter has two C_{20} chains, instead of the C_{40} of the acidophilic thermophiles, which can be viewed as having been formed by a unique head-to-head condensation. Langworthy (1977) has shown that the diglycerol tetraether of *Thermoplasma* has appropriate dimensions (40Å) to completely span the plasma membrane, suggesting a new type of biological membrane assembly which is structurally functional as a lipid monolayer rather than a conventional bilayer. "The lipid domain within the *Thermoplasma* membrane is formed by hydrocarbon chains which are covalently linked across the membrane rather than by intercalation of the hydrocarbon chains from two separate and opposite hydrophobic residues. The diglycerol tetraethers may, therefore, afford required structural stability to these membranes." (Langworthy, 1977).

The similarities between the lipids of *Thermoplasma* and *Sulfolobus* provide strong evidence for the importance of these structures in the stabilization of the membranes of these organisms under hot, acid conditions. Because these two organisms are clearly unrelated (despite the rather tortured reasoning of deRosa et al., 1974a), we have here an excellent example of convergent evolution. Interestingly, *Bacillus acidocaldarius,* which grows under similar environmental conditions to *Thermoplasma* (although it is slightly less acidophilic), but has a typical gram-positive cell wall, possesses typical fatty acids ester-linked to glycerol. This suggests that the cell wall in some way is able to afford protection from the onslaught of the H^+ ion. [Although *Sulfolobus* does have a cell wall (see later), it is an atypical cell wall, and provides little rigidity to the cell.]

Thermoplasma also forms a lipopolysaccharide-like substance which consists of a polymer of mannose, glucose, and glycerol and 40-carbon hydrocarbon in a molar ratio of 24:1:1:2 (Mayberry-Carson et al., 1974). This polymer has a monomeric weight of about 5300 and a molecular weight of greater than 1,200,000. Although extractable from lipid-free cells by procedures used to extract lipopolysaccharide (hot aqueous phenol), this polymer differs radically in chemical structure from typical lipopolysaccharide. Despite this, its fine structure is morphologically similar to lipopolysaccharide isolated from gram-negative bacteria (Mayberry-Carson et al., 1975). The precise cellular location of this lipopolysaccharide has apparently not been determined.

Table 5.7. Composition of Hydrocarbons in Thermophilic Acidophilic Bacteria

Organism	Component, % total hydrocarbons		
	$C_{40}H_{82}$ (acyclic)	$C_{40}H_{80}$ (monocyclic)	$C_{40}H_{78}$ (bicyclic)
Thermoplasma acidophilum	65.0	32.5	2.5
Sulfolobus acidocaldarius 98-3	29.9	32.3	37.8
S. acidocaldarius (ex-98-3)	trace	14.4	85.6
Strain MT-3[a]	3.6	18.2	71.6
Strain MT-4	2.2	8.6	70.8

[a]Strains MT-3 and MT-4 are related to *Sulfolobus* (see Chapter 6). (From deRosa et al., 1975.)

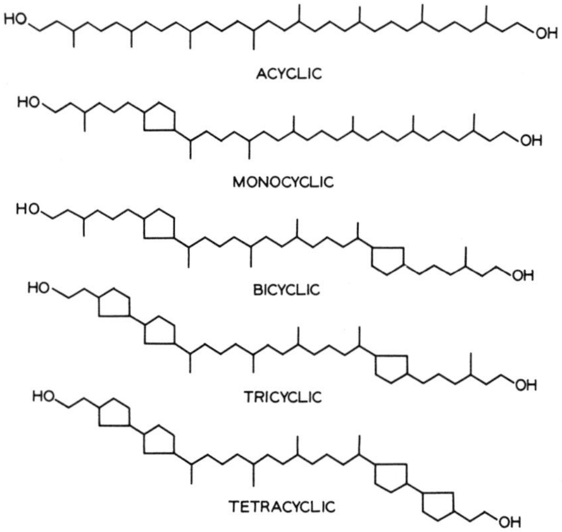

ACYCLIC

MONOCYCLIC

BICYCLIC

TRICYCLIC

TETRACYCLIC

Membrane Vesicles and Spin Label Studies

As first shown by P. F. Smith et al. (1973), membrane vesicles of *Thermoplasma* can be prepared by lysing cells at neutral pH, purifying the membrane components by centrifugation, and allowing formation of vesicles by reacidification. G. G. Smith et al. (1974) used this procedure to incorporate spin labels into *Thermoplasma* vesicles, in order to study lipid phase transitions. They found that the lipid regions of *Thermoplasma* were highly rigid, even more rigid than those of *Halobacterium*, which had the most rigid membrane previously known. Depending on the spin label used, two lipid phase transitions were seen, at 45°C and 60°C. Although perhaps fortuitous, these temperatures define approximately the temperature limits for growth of the organism. As already noted (see above), Langworthy (1977) has concluded that the *Thermoplasma* membrane is composed of a new type of assembly in which the hydrocarbon chain completely spans the membrane to form a rigid monolayer. This structure probably accounts for the extreme rigidity of the *Thermoplasma* membrane.

Intracellular pH

The intracellular pH of acidophilic organisms is always of interest. As noted in Chapter 1, there are many acid-labile molecules in cells, and proteins are also precipitated by acid. It is unlikely that life is compatible with intracellular pH values below 5. Two studies have been done to determine the intracellular pH of *Thermoplasma*. Hsung and Haug (1975) used a technique which is based on measuring the distribution of ^{14}C-labeled 5,5-dimethyl-2,4-oxazolidinedione (DMO) between the cells and the extracellular medium. Because DMO has a pK value of 6.1 at 56°C, almost all DMO molecules are in the nonionized form at pH 2 or 4. If DMO permeates the cell and the intracellular pH is higher than the pK values, some of the molecules will dissociate into the ionized form. Furthermore, if DMO enters passively, the intracellular concentration of the nonionized form should be the same as that of the extracellular form, so that if the intracellular pH is higher than the pK, there will be a net accumulation of DMO within the cell. By measuring the intracellular volume (using ^{14}C-dextran as an impermeant), and the amount of DMO label that is present in the cells at equilibrium, the intracellular pH can be calculated. A series of measurements on living cells, boiled cells, and cells poisoned with dinitrophenol or sodium azide showed that the intracellular pH was generally in the range of 6.5. This result is consistent with the observation that if a pellet of *Thermoplasma* cells is broken by sonification, then the pH of the resulting extract varied from 6.3 to 6.8. The fact that the intracellular pH of boiled cells (which were still morphologically intact) was about the same as that of living cells suggested to Hsung and Haug (1975) that the almost neutral intracellular pH is due to an impermeability of the plasma membrane to H^+, rather than to an active pumping mechanism. This result is also consistent with the lack of effect of metabolic inhibitors on the intracellular pH. According to them, "a Donnan potential across the membrane generated by an internal charged macromolecule impermeant to the cell membrane, can account for maintenance of the hugh pH gradient without participation of active mechanisms."

On the other hand, Searcy (1976) has used two different techniques and has found the internal pH of *Thermoplasma* to be near 5.5; he has also found that in boiled cells the internal pH is the same as the external pH. In his first procedure, Searcy (1976) titrated a dense suspension of *Thermoplasma* cells in 0.25 M sucrose to neutrality, at which point lysis occurred. He then back-titrated with acid, and obtained a different curve. In the uptitration, the cells showed little buffering capacity until a pH of 6.0–6.5 was reached, at which point lysis began and the release of cytoplasmic constituents occurred, providing buffering for the titration. Once this buffering capacity had been exhausted, further additions of alkali resulted in progressive increases in pH again. Back-titrating this suspension with acid gave a smoothly buffered curve, with the buffering capacity now being greater in the region of low pH. At pH 5.6 the lines crossed, and he interprets this crossing pH as the intracellular pH of the cells. His reasoning follows: "At some point in the titration curve, the external and internal pH will be both the same (assuming this can occur within the range of cellular stability). At this point, if there were cellular lysis, no change in pH would be observed. Lysis was obtained by adding equal amounts of first base and then acid. The only point . . . satisfying the condition of returning to the original pH after equal additions of base and acid is where the lines cross, and so this must be the intracellular pH." In nine replicates of the titration experiment, the estimated intracellular pH was 5.4 ± 0.2 (standard deviation).

Searcy's second approach to the measurement of intracellular pH resembles the sonication procedure of Hsung and Haug (1975). Washed cells were adjusted to different pH values between 4.0 and 6.0 and then sheared in a French pressure cell. When the pH of the broken cell suspensions was measured, it was found that the pH after breakage was different from before, except when the pH of the original suspension was at a value that was 5.5. He interprets this value as the intracellular pH in the following way: "If the original extracellular pH was lower than the pH inside the cells, rupturing the cells caused an increase in the apparent pH. Conversely, a high extracellular pH was lowered when the cells were ruptured. Thus the intracellular pH is that value which was unaffected by rupturing the cells." He found that if the cell suspension had been boiled before the extracellular pH was adjusted, then the pH of the suspension after rupturing was always the same as the original extracellular pH, suggesting that boiled cells cannot exclude H^+ ions. In a way, this result is more satisfying than that of Hsung and Haug (1975), since a biologist is uncomfortable with the observation that boiling has no effect on the membrane function. (Note that boiling *does* kill *Thermoplasma!*) Searcy suggests that a possible explanation for the results of Hsung and Haug is that DMO is adsorbed by the cells, or becomes dissolved in the membrane lipids, thus giving an overestimation of the true intracellular pH. This possibility could be tested by use of the equilibrium dialysis technique.

Osmotic Relations of *Thermoplasma*

Searcy (1976) also studied the osmotic relationships of *Thermoplasma*. As he points out, since *Thermoplasma* lacks a cell wall, it must be in osmotic equilibrium with its environment (no possibility for the development of turgor pressure). Belly and Brock (1972) had also commented on the ability of *Thermoplasma* to remain stable in triple distilled water, whereas other mycoplasmas lyse. Since *Thermoplasma* grows in very dilute culture media, and must contain some osmotically active materials in the cells (enzymes, amino acids, etc.), in order for cellular stability to be maintained the concentrations of inorganic ions must be kept low in order to achieve osmotic equilibrium. Since potassium is the most common counterion in cells, Searcy measured the intracellular potassium concentrations of cells grown at different osmolarities. When growing in the most dilute medium (about 50 mOsm), the intracellular K^+ concentration was only 17 mM, which is the lowest intracellular concentration of this ion yet reported. (*Escherichia coli* and most other bacteria have about 100–600 mM K^+.) As the osmolarity of the medium was increased with sucrose or sorbitol, the intracellular concentration of K^+ increased until it reached a value of 150 mM. Since Na^+ was not present in the culture medium, the only other major cation which might be present would be Mg^{2+}, which was present at about 30 mM (although much of the Mg^{2+} would probably be bound to cell constituents).

Possibly related to this low intracellular ion concentration is the presence in *Thermoplasma* of a histone-like protein associated with its DNA (Searcy, 1975). Conceivably this protein replaces some of the osmotically active cations which would normally be involved in neutralization of the negative charges on cellular nucleic acid.

The Nature of the Yeast Extract Requirement

Although *Thermoplasma* grows in media much simpler than other mycoplasmas, it has not been possible to replace the yeast extract in the medium. Some detailed studies on the nature of the yeast extract requirement were carried out by P. F. Smith et al. (1975). A culture medium was prepared containing salts and glucose, but no yeast extract, and then various additions were made. Certain peptones possessed some activity, but yeast extract provided by far the greatest growth stimulation. Therefore, yeast extract was subjected to a variety of purification procedures, in an attempt to isolate a specific growth factor(s). The results suggested that the active component(s) is a polypeptide. A fraction was isolated which had a molecular weight of about 1000 and had 8 to 10 amino acids. This polypeptide contained primarily basic and dicarboxylic amino acids, although only one amino group per molecule was free and the polypeptide bound cations avidly. The requirement for this polypeptide appeared absolute, although since a relatively large concentration was required, it should perhaps be viewed not as a simple nutrient. Smith et al. (1975) postulate that the requirement for this polypeptide may be related in some way to the high H^+ concentration required for growth. They speculate that the polypeptide may act as an ion scavenger for some trace metal requirement, protection

Figure 5.7. A typical self-heating coal refuse pile (right side of picture). This was a pile at the Blackfoot #5 Mine, near Winslow, Indiana, photographed in 1969. Note the smoke and steam. The large pile on the left is not coal refuse, but a spoil bank from the strip mining operation. Spoil banks do not self-heat and are much less acidic.

of the organisms from high H^+, involvement in ion transport, or as a supply of essential amino acids in a form available to the cell. Conceivably, in the acidic environment, uptake of free amino acids might be difficult.

Nutrition in the Natural Habitat

A typical self-heating coal refuse pile is shown in Figure 5.7. *Thermoplasma* is *not* found in frankly burning parts of such piles, but in areas that are merely smoldering, with temperatures around 50–55°C. The yeast extract requirement of *Thermoplasma* raises immediately the question of nutrient sources in coal refuse piles. Immunofluorescence data of Bohlool and Brock (1974b) show that *Thermoplasma* does grow in the coal refuse pile habitat. Furthermore, growth also occurred with an aqueous extract of coal refuse material, prepared by incubating at 60°C a 1:1 mixture of coal refuse and distilled water, followed by centrifugation (Bohlool and Brock, 1974b). This indicates that coal refuse provides the necessary nutrients for *Thermoplasma* growth.

What does coal refuse contain that might support the growth of *Thermoplasma?* Although the bulk of the organic material (coal) has been removed, the coal refuse always contains coal fragments and other organic materials. In the absence of self-heating, it is unlikely that this complex, high molecular weight material would serve directly as a source of nutrients for *Thermoplasma,* but the self-heating process will induce a series of pyrolytic reactions which should lead to the formation of lower molecular weight materials. As noted earlier, *Thermoplasma* is not the only organism able to grow in self-heated coal refuse piles. The thermophilic fungus *Dactylaria gallopava* has also been isolated. Although a study of the chemistry of self-heated coal refuse would probably be even more compli- cated than a study of the chemistry of yeast extract, it could conceivably lead to the discovery of the real *Thermoplasma* growth factors, which might, in fact, be new to science.

Evolution of *Thermoplasma*

It is quite clear that *Thermoplasma* is a unique organism and is not a stable L-form of some rod-shaped bacterium. The fine structure (Figure 5.1) and the colony morphology (Figure 5.8) suggest some relationship to the myco-

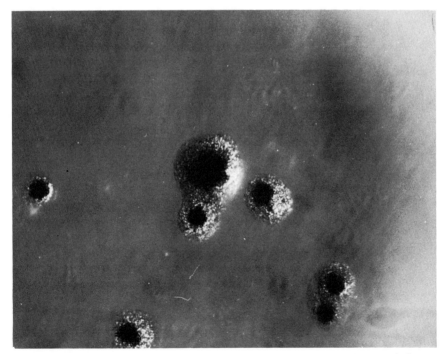

Figure 5.8. Colonies of *Thermoplasma,* showing the "fried-egg" morphology. These colonies are about 1 mm in diameter. Colonies have never been obtained reproducibly, and in most cases growth on acid agar ceases after colonies reach about 100 cells.

plasma group, but the membrane and lipid biochemistry show that the organism is unique. The only known organism that could conceivably be related is *Bacillus acidocaldarius,* which grows over the same temperature and pH range. However, the lipids of *B. acidocaldarius,* although interesting, are normal fatty acyl esters, and it is hard to imagine *Thermoplasma* losing a cell wall and acquiring its isoprenoid ether lipids all at the same time.

Given the fact that *Thermoplasma* is a unique organism, where has it come from? As I have pointed out earlier, so far its habitat seems to be quite restricted. Although there are hundreds of burning coal refuse piles, on a geographical basis they are few and far between. Is it possible that this is the primary habitat of the organism? In this respect, I might note that although the coal refuse pile is a man-made object, analogous natural habitats do occur. Coal seams can get exposed to the atmosphere through natural causes, such as earthquakes and erosion, and self-heating of such coal seams can occur (See Chapter 2 and Burns, 1969). Thus it is possible for habitats suitable for *Thermoplasma* to have been generated over long periods of time. Dispersal from one habitat to another would still be difficult to explain, but no more difficult than explaining the dispersal of any hot spring organism.

One other type of habitat which we considered might be suitable for *Thermoplasma* is the stomach of the warm-blooded animal. Stomach pH values range from 1 to 2, and the temperature is 37°C, a temperature at which our strains of *Thermoplasma* do grow, although slowly. Belly et al. (1973) attempted to establish *Thermoplasma* in the stomachs of rabbits by force-feeding cultures over a 5-week period. Microscopic examination, both by phase-contrast and by immunofluorescence, failed to reveal any evidence of *Thermoplasma,* either in normal or inoculated rabbits. The predominant organism in the stomach of rabbits is the yeast *Saccharomycopsis guttulatus.* In another study, we attempted to culture *Thermoplasma,* or any other acidophilic bacterium, from the human stomach, using as inocula biopsy material obtained from the University Hospitals, University of Wisconsin. This was also unsuccessful, as has been a large number of attempts to isolate *Thermoplasma* from normal soils and waters.

Further studies on the ecology of *Thermoplasma* might give some insight into the habitat and evolution of the organism. Unfortunately, coal refuse piles are not easy to study ecologically (Bohlool and Brock, 1974b), and in this day of active environmental protection, no new piles are being created and old ones are being covered. It is ironic that environmental activities of this sort may result in the eventual destruction of all *Thermoplasma* habitats. Is *Thermoplasma* an endangered species?

References

Belly, R. T., B. B. Bohlool, and T. D. Brock. 1973. The genus *Thermoplasma. Ann. N.Y. Acad. Sci.* **225,** 94–107.

Belly, R. T. and T. D. Brock. 1972. Cellular stability of a thermophilic, acidophilic mycoplasma. *J. Gen. Microbiol.* **73**, 465–469.

Belly, R. T. and T. D. Brock. 1974. Widespread occurrence of acidophilic strains of *Bacillus coagulans* in hot springs. *J. Appl. Bacteriol.* **37**, 175–177.

Bohlool, B. B. and T. D. Brock. 1974a Immunodiffusion analysis of membranes of *Thermoplasma acidophilum*. *Infect. Immunol.* **10**, 280–281.

Bohlool, B. B. and T. D. Brock. 1974b. Immunofluorescence approach to the study of the ecology of *Thermoplasma acidophilum* in coal refuse material. *Appl. Microbiol.* **28**, 11–16.

Brock, T. D., K. M. Brock, R. T. Belly, and R. L. Weiss. 1972. *Sulfolobus:* a new genus of sulfur-oxidizing bacteria living at low pH and high temperature. *Arch. Mikrobiol.* **84**, 54–68.

Burns, F. H. 1969. The smoking pillars of Horton River. *The Beaver,* Spring 1969, pp. 40–43.

Christiansen, C., E. A. Freundt, and F. T. Black. 1975. Genome size and deoxyribonucleic acid base composition of *Thermoplasma acidophilum*. *Int. J. System. Bacteriol.* **25**, 99–101.

Darland, G., and T. D. Brock. 1971. *Bacillus acidocaldarius* sp. nov., an acidophilic thermophilic spore-forming bacterium. *J. Gen. Microbiol.* **67**, 9–15.

Darland, G., T. D. Brock, W. Samsonoff, and S. F. Conti. 1970. A thermophilic, acidophilic mycoplasma isolated from a coal refuse pile. *Science* **170**, 1416–1418.

deRosa, M., A. Gambacorta, and J. D. Bu'Lock. 1975. The Caldariella group of extreme thermoacidophile bacteria: direct comparison of lipids in *Sulfolobus, Thermoplasma,* and the MT strains. *Phytochemistry* **15**, 143–145.

deRosa, M., A. Gambacorta, G. Millonig, and J. D. Bu'Lock. 1974. Convergent characters of extremely thermophilic acidophilic bacteria. *Experientia* **30**, 866.

Freundt, E. A. 1974. The mycoplasmas. In *Bergey's Manual of Determinative Bacteriology,* 8th ed., R. E. Buchanan and N. E. Gibbons, eds. Williams and Wilkins, Baltimore, pp. 952–953.

Hsung, J. C. and A. Haug. 1975. Intracellular pH of *Thermoplasma acidophila*. *Biochim. Biophys. Acta* **389**, 477–482.

Langworthy, T. A. 1977. Long-chain diglycerol tetraethers from *Thermoplasma acidophilum*. *Biochim. Biophys. Acta* **487**, 37–50.

Langworthy, T. A., P. F. Smith, and W. R. Mayberry. 1972. Lipids of *Thermoplasma acidophilum*. *J. Bacteriol.* **112**, 1193–1200.

Mayberry-Carson, K. J., T. A. Langworthy, W. R. Mayberry, and P. F. Smith. 1974. A new class of lipopolysaccharide from *Thermoplasma acidophilum*. *Biochim. Biophys. Acta* **360**, 217–229.

Mayberry-Carson, K. J., I. L. Roth, and P. F. Smith. 1975 Ultrastructure of lipopolysaccharide isolated from *Thermoplasma acidophilum*. *J. Bacteriol.* **121**, 700–703.

Ruwart, M. J. and A. Haug. 1975. Membrane properties of *Thermoplasma acidophila*. *Biochemistry* **14**, 860–866.

Searcy, D. G. 1976. *Thermoplasma acidophilum:* intracellular pH and potassium concentration. *Biochim. Biophys. Acta* **451**, 278–286.

Searcy, D. G. 1975. Histone-like protein in the prokaryote *Thermoplasma acidophilum*. *Biochim. Biophys. Acta* **395**, 535–547.

Searcy, D. G. and E. K. Doyle. 1975. Characterization of *Thermoplasma acidophilum* deoxyribonucleic acid. *Int. J. System. Bacteriol.* **25**, 286–289.

Smith, G. G., M. J. Ruwart, and A. Haug. 1974. Lipid phase transitions in membrane vesicles from *Thermoplasma acidophila. FEBS Lett.* **45**, 96–98.

Smith, P. F., T. A. Langworthy, W. R. Mayberry, and A. E. Hougland. 1973. Characterization of the membranes of *Thermoplasma acidophilum. J. Bacteriol.* **116**, 1019–1028.

Smith, P. F., T. A. Langworthy, and M. R. Smith. 1975. Polypeptide nature of growth requirement in yeast extract for *Thermoplasma acidophilum. J. Bacteriol.* **124**, 884–892.

Note added in proof. Further indication of the uniqueness of *Thermoplasma,* and its relationship to *Sulfolobus* and *Halobacterium,* is shown by the fact that all three organisms show unique and remarkably similar structures of their 16S ribosomal RNA. On the basis of this fundamental molecular property, these three bacterial genera also show close relationship to the methanogenic bacteria, and all of these organisms have been grouped together in a new kingdom called the Archebacteria (Carl Woese, University of Illinois, personal communication). If *Thermoplasma* is an ancient organism, as this grouping suggests, then it is even more vital to find a natural habitat for it other than the self-heating coal refuse pile.

Chapter 6
The Genus *Sulfolobus*

Our work on *Sulfolobus* originally arose out of a determination to find a natural habitat for *Thermoplasma*. As discussed in the previous chapter, *Thermoplasma* was originally discovered as an acidophilic thermophile living in self-heated coal refuse piles. As this seemed an unlikely permanent habitat of the organism, I naturally thought about the hot, acid environments of Yellowstone. My determination to find *Thermoplasma* in Yellowstone was strengthened by an observation that Gary Darland had made during his attempt to define the upper temperature for life in acidic environments (see Brock and Darland, 1970). He had immersed microscope slides in the effluents of a number of acid, thermal streams, and by a quantification of the organisms that developed on the slides he could define the upper temperature limits. At that time, we were aware only of rod-shaped and filamentous bacteria, but in several of the effluents, at temperatures at which rods and filaments did not appear, there were odd-shaped spherical structures. Since most of these acid streams are sulfur-rich, I interpreted these structures as some sort of amorphous sulfur deposit. In retrospect, I should have realized that they were living organisms, since they not only were phase dark, but appeared on the slides in a distribution that could have been interpreted as microcolonies. As we had already concluded earlier (Brock, 1967; Bott and Brock, 1969), development of microcolonies was a good indication of growth at *in situ* temperatures.

In the summer of 1970, Doemel and I went to Iceland for the last 2 weeks of June on a NASA-sponsored trip. Bill Samsonoff, then a graduate student of Sam Conti, was spending the summer with us, looking for material for electron microscopic study. While I was gone, he had placed microscope slides in the effluent of Roaring Mountain, and when I returned I examined the slides microscopically. The development of spheres on these slides was

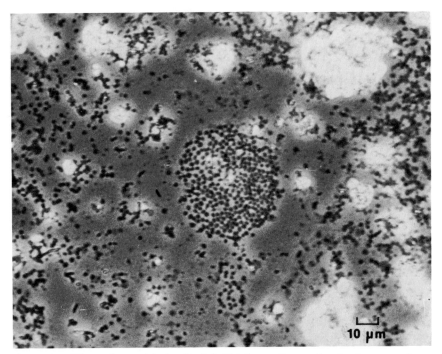

Figure 6.1. Phase photomicrograph of *Sulfolobus* cells growing *in situ* on a micro-scope slide immersed in a small pool of the effluent of the spring on the south side of Roaring Mountain. Temperature of the pool 84°C, pH 2.0. The slide had been immersed for 13 days. Photo with water-immersion lens.

very good (Figure 6.1) and I became interested. By then, we had character-ized *Thermoplasma,* and knew how to isolate it. Since we knew about *Thermoplasma,* I was convinced that this was what we were seeing. I then began a detailed survey of hot, acid springs. Now that I knew what I was looking for, I was seeing spheres everywhere. In addition to the attached forms in Roaring Mountain, some of the best examples were free-floating forms in the large pool at Sulfur Caldron. It was a simple matter to set up our *Thermoplasma* enrichment procedure, using basal salts at pH 2, 0.1% yeast extract, with and without glucose and elemental sulfur. Because *Thermoplasma* grew at 55°C, I set up one set at that temperature, but because the habitat temperatures where I was seeing the spheres were 70–85°C, I set up a second set at 70°C, the most convenient high temperature we had in our W. Yellowstone laboratory. No growth was obtained at 55°C, but in the 70°C enrichments, extensive developments of spheres were seen. These spheres were somewhat more refractile than those of *Thermo-plasma,* but otherwise seemed similar.

At the same time, Gary Darland went back over his notes on observa-tions of the immersion slides, and informed me of the locations where the odd-shaped spheres had been seen. All of the habitats were at high temperatures, and all except two were at low pH (the two high pH sites probably had something else, not further investigated). With this background, it was now a simple matter to do an extensive field investigation and to set up a large number of enrichment cultures. Soon I had many cultures and was getting some idea of the temperature range for growth. It was clear that all of the enrichments were growing at considerably higher temperatures than *Thermoplasma*. In mid-August I went to Mexico City for the International Congress of Microbiology, and afterward went down to El Salvador, where there were some interesting thermal areas and a new geothermal energy project. I found a number of hot, acid pools there and brought back material for more cultures. Then in October, I went to Italy under a trip arranged by Prof. Bu'Lock and sponsored by the Italian research organization CNR. In the classic acid thermal area, Solfatara, in the Bay of Naples, I found more hot acid springs, and obtained more cultures. Meanwhile, Bob Belly was isolating more *Thermoplasma* strains and my present wife, Kathie, began characterizing all our isolates. When we returned to Bloomington she, Bob Belly, Rich Weiss, and Carl Fliermans all were involved, one way or another, in sorting out the distinctions between *Sulfolobus* and *Thermoplasma*.

At this time we were still sure we were working with hot spring enrichments of *Thermoplasma*. Aside from the temperature of growth, the identical conditions were being used for coal refuse pile and hot spring isolates. Remember that all of our Yellowstone isolates had been derived from yeast extract enrichments aiming for *Thermoplasma*. Kathie took the cultures we had available, both hot spring and coal refuse isolates, and began to characterize them for growth parameters, temperature, pH, and nutrition. The isolates quickly grouped themselves: the coal refuse isolates grew optimally at 55°C; the hot spring isolates at 70°C [subsequently we found a variety of temperature strains of *Sulfolobus* (see later)]. The coal refuse isolates would only grow when yeast extract was present in the medium, whereas the hot spring isolates would grow with casein hydrolysate, peptone, and amino acid mixes. Soon we were also able to grow the hot spring isolates in completely synthetic media, with amino acids or pentose sugars. However, as already shown in Table 5.2, both the hot spring and coal refuse isolates were resistant to ristocetin and vancomycin, two antibiotics that inhibit cell wall synthesis.

By early 1971, we had sufficiently characterized the two groups of isolates to know that they were different, but as yet we did not know how different. About this time, Rich Weiss began electron microscopic studies, using negative staining, thin sections, and freeze-etching. He quickly showed that the hot spring isolates had some sort of cell wall, although definitely not a typical bacterial cell wall. Thus, we knew that we did not have just another species of *Thermoplasma*, but something quite different. When Kathie performed hexosamine assays on acid hydrolysates of cell suspensions from the two groups, she was easily able to show that the hot spring isolates all had hexosamine, whereas the coal refuse isolates did not.

During this period, Bob Belly was doing the DNA base compositions of a variety of hot spring and coal refuse isolates. Although the values for the coal refuse isolates turned out to be erroneously low (see Chapter 5), the differences between the two groups of isolates are still striking. All of the *Thermoplasma* isolates had low GC values, whereas the hot spring isolates all had GC values of 60% or above.[1] At this time I began toying with names, and came up with *Acidoglobus,* to indicate that we had an acidophilic organism that was global in shape. Then, because of the lobed-shaped structures that Rich was seeing in the electron microscope, I fixed on the name *Acidolobus,* which I thought was a little more euphonious. This was the name that we used when we gave our paper at the ASM meeting in Minneapolis in 1971, although fortunately we did not use the name in the abstract (*Bacteriological Proceedings,* 1971, paper G34, p. 29).

Strangely, we did not test our hot spring isolates for growth on sulfur until May 1971. We were probably so fixed on the idea that the isolates were similar to *Thermoplasma,* which we knew didn't grow on sulfur, that we never thought of autotrophic growth. This despite the fact that the main habitats from which we had isolated the organism were sulfur springs. In defense of our lack of action, it might be noted that since all of the isolates had come from yeast extract enrichments, it seemed unlikely that they would grow on sulfur. However, Carl Fliermans was working in the laboratory on hot, acid soils, and from a Roaring Mountain soil he had isolated a spherically shaped organism which used elemental sulfur as sole energy source. His isolate grew very slowly, and he nursed it throughout the fall of 1970 and spring of 1971 while we were carrying out the work outlined above. Bob Belly had also isolated on elemental sulfur some acidophilic thermophiles from Yellowstone habitats. Finally, in early May 1971, Kathie tested our hot spring isolates enriched in yeast extract, and Bob Belley's and Carl Flierman's isolates enriched on elemental sulfur. As shown by continued growth on elemental sulfur, and by the appearance of sulfate in the medium (assayed turbidimetrically using $BaCl_2$), all of the hot spring isolates could grow on elemental sulfur. *Thermoplasma,* on the other hand, did not grow on elemental sulfur, although in the presence of yeast extract it reduced sulfur to H_2S.

By this time we were back at Yellowstone for the summer of 1971, after a hectic spring that not only involved a lot of research, teaching, and packing up to move the whole laboratory (five graduate students) to Yellowstone, but a complete move to Wisconsin of my own household plus all of the laboratory equipment that Indiana University would let loose. (I had accepted a position as E. B. Fred Professor of Natural Sciences in the Department of Bacteriology.) We went to Yellowstone for the summer

---

[1]Although all of our isolates had high G + C values, both de Rosa et al. (1974) and Furuya et al. (1977) have isolated *Sulfolobus*-like organisms with G + C values around 45%. That these differences are real and not due to experimental error is shown by the fact that de Rosa et al. (1974) did confirm our DNA base determination on one strain which we sent them.

employed by Indiana University, and returned to Madison employed by the University of Wisconsin. I owe a lot of gratitude to Prof. Dean Fraser, then chairman of the department at Indiana, for being so cooperative during this rather disruptive move.

Now that we were back in Yellowstone, and knew that our yeast-extract enriched isolates would grow on sulfur, we were naturally interested to isolate new strains on elemental sulfur alone. Although growth was much slower, we eventually obtained isolates, and all of them would also grow heterotrophically. It was also possible, using fluorescence microscopy and acridine orange staining, to observe spherically shaped cells attached to sulfur crystals, both in culture and in natural material (Figure 6.6). It was quite clear that the best name for our organism was *Sulfolobus*. It is a nice name, and meaningful.

The derivation of the name *Sulfolobus acidocaldarius* is straightforward. "Sulfo" is Latin for sulfur, "lobus" Latin for lobe, "acido" is Latin for acid, and "caldarius" is Latin for hot. Thus, *Sulfolobus acidocaldarius* is the lobed-shaped organism that uses sulfur and grows in hot, acid environments.

When first thinking of names for a sulfur oxidizer, I thought of *Thiolobus*. Thinking it would be useful to have the name checked by an expert, I wrote Prof. R. E. Buchanan, now deceased, who was probably the world expert on bacterial nomenclature. He pointed out that although correct, *Thiolobus* was inelegant because it combined the Greek word for sulfur, "thio," with the Latin word for lobed. He suggested *Sulfolobus,* which I immediately liked.

One final note before getting into a more precise characterization of *Sulfolobus*. I had been aware for some time of a thesis by James Brierley at Montana State University which had a brief description of a spherically shaped organism he had isolated from an acid hot spring (Brierley, 1966). However, I had completely forgotten about this description, since it was embodied in a few lines buried in the middle of an extensive study of iron- and sulfur-oxidizing thiobacilli. His organism, strain 8-3-65-2 (1), was isolated on elemental sulfur but also oxidized ferrous iron. It could grow at 60°C, but not at 65°C, although it resisted heating to 80°C. He interpreted this heat resistance, plus the presence of a structure in the cell which stained with a bacterial spore stain, as an indication that the organism formed a spore. Although Brierely had never published on his organism, about the time that we were completing our work on *Sulfolobus,* his wife, Corale, was just completing her Master's thesis (New Mexico Tech.) on his isolate. Although they have preferred not to use the designation *Sulfolobus* for their isolate (Brierley and Brierley, 1973), the organisms are clearly closely related, especially since we have shown that iron oxidation is quite common in natural populations of *Sulfolobus* (Brock et al., 1976). Since Jerry Mosser later isolated several strains of *Sulfolobus* able to grow optimally around 60°C (Mosser et al., 1974b), it is now fairly certain that Brierley's organism is a low-temperature strain of *Sulfolobus*.

Morphology of *Sulfolobus*

Under the phase microscope, *Sulfolobus* cells appear phase dark and refractile, and although generally spherical in shape, they often look irregular (Figure 6.2). *Sulfolobus* can be distinguished from *Thermoplasma* by careful microscopic examination: *Thermoplasma* cells are completely round whereas *Sulfolobus* cells are more irregular; and when cells are tumbling through the field, distinct lobes can be seen. The diameter of *Sulfolobus* cells are about 0.8–1.0 μm and little size variation is seen. *Thermoplasma* cells seem to reproduce by budding or by narrow hyphae whereas *Sulfolobus* cells show no obvious signs of budding or hyphae. *Thermoplasma* cells are quite variable in diameter, ranging from the limit of resolution at 0.1–0.2 μm to large cells 5–10 μm in diameter. In *Thermoplasma* cultures large empty cells 5–10 μm in diameter are often seen, within which denser structures can be seen; when empty cells are seen in *Sulfolobus* cultures they are no larger than normal cells and lack the denser interior structures seen in *Thermoplasma* cultures. Probably the best characteristic for distinguishing the two organisms is sphericity, *Thermoplasma* cells always being evenly spherical and *Sulfolobus* cells being lobed spheres.

The most striking difference between *Sulfolobus* and *Thermoplasma* under the electron microscope is the presence of a wall in the former and the absence of a wall in the latter. However, the wall structure of *Sulfolo-*

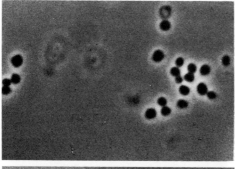
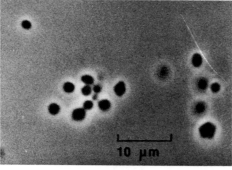

Figure 6.2. Phase photomicrograph of *Sulfolobus* cells from a pure culture, strain 98-3.

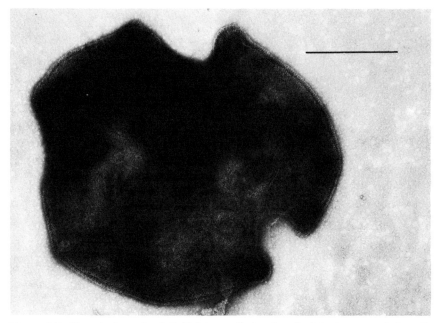

Figure 6.3. Negatively stained *Sulfolobus* cell showing the lobed nature of the cell and the subunit structure of the cell wall. Bar represents 0.5 μm. (From Brock et al., 1972.)

bus is atypical; the characteristic peptidoglycan layer seen in other bacteria is absent.

Under the electron microscope, *Sulfolobus* and *Thermoplasma* are easy to distinguish by negative staining with phosphotungstic acid, as shown in Figures 6.3 and 6.4. The lobed shape and characteristic wall structure of the *Sulfolobus* cell are easily seen, whereas the *Thermoplasma* cells are essentially spherical and show no sign of a wall. Perhaps because of its wall, *Sulfolobus* cells are easier to prepare for negative staining than *Thermoplasma;* the latter frequently lyse so that one needs to search extensively on grids to find any whole cells.

Nature of the Cell Wall

An enlarged portion of the cell wall (Figure 6.5) shows the presence of a diffuse electron transparent layer of subunits, and the absence of a typical peptidoglycan layer. deRosa et al. (1974) have advanced the idea that the lobes of *Sulfolobus* are artifacts induced by centrifugation. However, as Rich Weiss has pointed out (personal communication, August 17, 1974):

> I would regard the claim that the lobate structure of *Sulfolobus acidocaldarius* is a centrifugation artifact, as highly unlikely. First of all, the work I published in *J. Bacteriol.* [Weiss, 1974] shows that centrifuged cells are not lobate, but

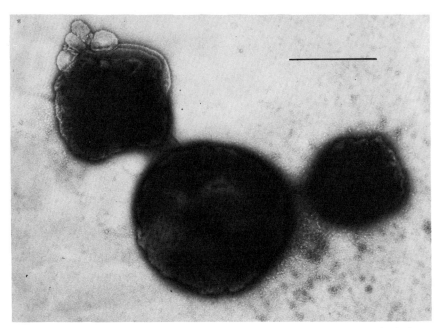

Figure 6.4. Negatively stained *Thermoplasma* cells showing the spherical shape and absence of cell wall. Bar represents 1 μm. (From Brock et al., 1972.)

irregular spheres [see his Figure 1]. Secondly, all the work we did at Yellowstone Park shows that both fixed and unfixed cells of *Sulfolobus acidocaldarius* when prepared without centrifugation by the method of negative staining are highly lobed. This does not mean that all uncentrifuged cells are lobed; quite the contrary, many cells are spherical. However, this characteristic does not depend on centrifugation. We found that the lobed nature of this organism is the result of its developmental cycle (growth curve). That is, organisms that are sampled in the log phase are generally irregular spheres whether centrifuged and prepared by thin sectioning or uncentrifuged and seen by the negative staining method. Also, the highly lobed structure can be seen quite clearly using the light microscope and comparison of cells in various stages of the growth cycle can be made at this level. Such comparisons are in complete agreement with data derived from thin sectioned and negatively stained material. By way of summary I would say that cells are generally spherical under normal conditions; centrifugation does not change this characteristic. In continuous culture in the hot springs in the park, cells are highly lobed, and this characteristic is present in uncentrifuged and centrifuged cells. Finally, I have found no changes in the cell wall structure of fractionated cells after centrifugation. If one fragments log phase cells (which are spherical) and negatively stains these (without centrifugation) the fragments are spherical. If the same treatment is carried out on stationary phase cells (lobed shaped cells), the fragments are lobed. The characteristics of the fragments do not change with centrifugation; the same is true for intact cells. The cell wall of *Sulfolobus acidocaldarius* is an extremely rigid structure that is able to withstand physical stress and it appears to be unaffected by centrifugation.

Weiss (1974) worked out methods for the isolation and purification of the cell wall of *Sulfolobus*. Cells were disrupted with a French press and the membrane removed by extracting the fragments with Triton X-100. The Triton-insoluble material retained the characteristic subunit structure of the cell wall. The diameter of the subunits as measured in negatively stained preparations was 15.5 nm with a center-to-center spacing of about 20 nm. Similar measurements were found for material prepared by freeze-fracturing. Chemical analysis of the cell wall revealed a protein-hexosamine-carbohydrate ratio of 1:0.026:0.140 (wt/wt/wt). Thus, the wall contains a large proportion of protein. On the basis of studies on the effects of various enzymes and extraction agents on cell walls, Weiss concluded that the cell wall of *Sulfolobus* appears to be a protein-lipid complex. Since the wall was sensitive to dithiothreitol, a disulfide bond-breaking reagent, he hypothesized that the lipoprotein component is responsible for cell wall integrity, perhaps by hydrophobic protein-protein interaction.

The cell wall protein was found by Weiss (1974) to be enriched with charged amino acids (aspartate, glutamate, lysine) as well as with branched-chain hydrophobic amino acids (valine, leucine, and isoleucine). Weiss made a special attempt to detect the cell wall components typical of bacteria: diamino pimelic and muramic acids. Using special procedures on the amino acid analyzer, he did not find significant amounts of these substances. Weiss (1974) also showed that the *Sulfolobus* cell wall was not affected by lysozyme-EDTA. Brock et al. (1972) had suggested on the basis of the high resistance of *Sulfolobus* to antibiotics that inhibit bacterial cell wall synthesis that the organism may lack a typical peptidoglycan cell wall. Brock et al. (1972) showed that whole cells of *Sulfolobus* did contain hexosamine, and Weiss (1974) has shown that hexosamine is associated both with the cell wall and the cell membrane. Although perhaps fortuitous, the bacteria living at 90°C in the alkaline Boulder Spring also lack a peptidoglycan cell wall, as determined electron microscopically, and seem to have a distinct simple wall with a distinct subunit structure. Perhaps even more interesting, in view of the similarities in lipids (see Table 5.7), the cell wall of *Halobacterium* is morphologically similar to that of *Sulfolo-*

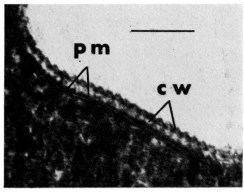

Figure 6.5. Enlargement of the cell envelope of *Sulfolobus,* showing the plasma membrane (pm) and the cell wall (cw). Bar represents 0.1 μm. (From Brock et al., 1972.)

bus (Cho et al., 1967; Steensland and Larsen, 1969). *Halobacterium* also lacks a peptidoglycan cell wall.[2]

Pili and Attachment to Sulfur

As noted earlier (Figure 6.6), *Sulfolobus* attaches to sulfur crystals (see also Shivvers and Brock, 1973). Attachment is much more common in flowing than in nonflowing springs. Weiss (1973) detected pili or pilus-like structures on *Sulfolobus* cells attached to sulfur, and suggested that these structures may be involved in the attachment of the bacteria to the sulfur particles. He also suggested that the same structures may be involved in the attachment of the bacteria to glass slides or to other solid surfaces present in the environment. Evidence for the importance of such structures in the habitat is suggested by the fact that when *Sulfolobus* cells which are attached to sulfur crystals are examined carefully in the light microscope, they appear to be separated from the crystal face by a short distance, which perhaps represents the distance occupied by the pili. As also pointed out by Weiss, large clusters of cells are often seen attached to the crystals, and most of the cells in these clusters are not attached to the sulfur directly, but to other cells (Fig. 6.6). Conceivably, the pili could serve to anchor cells tightly to each other.

Further work on this interesting phenomenon would be of interest, since in culture *Sulfolobus* cells are rarely seen attached to each other. The development of this attachment mechanism in flowing springs may be a response to the generally lower nutrient levels in spring waters as opposed to culture media.

Taxonomy

When it came to deciding where *Sulfolobus* should be placed in Bergey's *Manual*, 8th edition, I had some correspondence with the co-editor N. E. Gibbons. He said: "I have been going over the MSS for Bergey 8 again and wondering what to do about *Sulfolobus*. As it is a lobed organism it might go in Part IV Prosthecate and Budding Bacteria. There has been a lot of discussion as to just what goes in this part... If the lobes meet the definition of prosthecae, *Sulfolobus* may go under the group "prosthecae have no reproductive function" rather than under the GUA's (genera of uncertain affiliation) at the end. Another possibility is to put it in Part XII Gram negative chemolithotrophs along with the other organisms that metabolize

---

[2]I have also been informed (Carl Woese, University of Illinois, personal communication, 1977) that *Sulfolobus, Halobacterium,* and the methanogenic bacteria are all similar in the structure of their 16S ribosomal RNA. The methanogenic bacteria also lack a peptidoglycan cell wall.

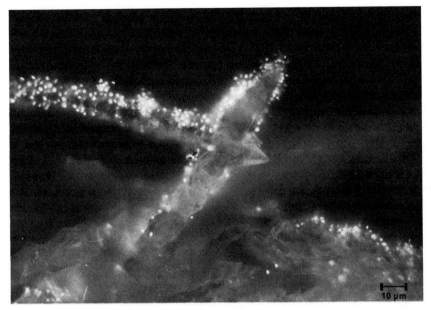

Figure 6.6. Photomicrograph by acridine orange fluorescence microscopy of *Sulfolobus* cells attached to crystals of elemental sulfur. The material was collected from a flowing sulfur spring in the Amphitheater Springs area.

sulfur. I believe the present definition of chemolithotroph would include *Sulfolobus* as very few organisms meet the criteria originally laid down for autotrophs. As we have no Odds and Sods Part, which of these possibilities do you prefer?'' My answer was simple: ''I would definitely prefer *Sulfolobus* to be placed with the Gram negative chemolithotrophs which oxidize sulfur. This placement will avoid the controversial question of what constitutes a prostheca.'' And this is where the genus ended up (Brock, 1974).

Specific Isolation Procedures

I have already described in the introduction some of the procedures we used in the initial isolation of *Sulfolobus*. It is much easier to isolate cultures mixotrophically or heterotrophically than autotrophically. A procedure adopted in later work was to take a sample of spring water and add 0.1 volume of a 1% yeast extract solution which had been acidified to pH 2. Incubation at 70°C should lead to the development of faint turbidity in a few days, and microscopic examination will reveal the presence of *Sulfolobus*. The turbidity reached in cultures is never as great as it would be in normal bacteria. Since cell numbers can rise over 10^9/ml, this is probably because the fairly nonrefractile spherical cells do not scatter light as much as normal bacteria. (The same is true for *Thermoplasma*.)

If autotrophic cultures are desired, addition is made of 0.05–0.1 volume of spring water to a basal salts medium containing sterile elemental sulfur. Enrichment on sulfur is always better if a little yeast extract is added. The yeast extract may partly stimulate growth by wetting the sulfur, so that bacterial attachment is better [Weiss (1973) showed better attachment of *Sulfolobus* to sulfur in the presence than in the absence of yeast extract], but yeast extract may also provide growth factors. On yeast extract alone, the pH of the culture will rise, but with sulfur and yeast extract together, the pH remains low, and this may be another reason why growth is better on sulfur + yeast extract than on yeast extract alone. Purification of *Sulfolobus* cultures from enrichments must be done by end-point dilution, since reproductible growth on agar is difficult. (Colonies on agar can be obtained, far easier than was the case with *Thermoplasma,* but because of the nuisance involved in making up acid agar media, limiting dilution is probably easier.) To be absolutely certain of culture purity, I prefer a three-tube or five-tube most-probable-number procedure, checking all tubes microscopically for presence of *Sulfolobus*. Incubation must be for several weeks, to be certain that growth has occurred at the highest dilution possible. From the dilution at which some tubes are positive and other negative, a positive tube is selected and used as the source of inoculum for a stock culture.

All incubations must be in covered water baths to prevent evaporation. With culture tubes, stainless steel capped tubes are used (cotton must be avoided as it gets very wet and also undergoes acid hydrolysis). For stocks, we generally use screw-capped medicine bottles filled to only 10% of their volume, and enrich the atmosphere with CO_2 occasionally.

Although *Sulfolobus* can be freeze-dried or stored under liquid N_2, I do not know how long viability remains. We continued to maintain a few stocks by successive transfer at weekly intervals on yeast extract medium until 1977, when we finally stopped.

Habitats

As I have indicated in the introduction to this chapter, once we knew how to look for it, we found *Sulfolobus* to be widespread in hot, acid environments. The organism is often present in such high numbers that it can be readily detected microscopically, either in sulfur pools (Figure 6.7) or in flowing sulfur springs. In the former, the organism is generally found free-floating in the water, and a little care must be taken in microscopy to be certain that one is looking at *Sulfolobus* and not at debris or precipitate. In flowing springs, sulfur crystals are frequently found deposited on the bottom of the outflow channel (Figure 6.8), and microscopic examination of these crystals will usually reveal the presence of *Sulfolobus* (Figure 6.6) if the temperature and pH are appropriate.

In 1969–1971, I did an extensive survey of Yellowstone springs for the presence of *Sulfolobus*. I also surveyed springs in El Salvador and Italy. Later Kathie and I visited Iceland and Dominica and found *Sulfolobus* there also. As seen in Table 6.1, almost all of the habitats had pH values under 4 and temperatures over 60°C. Cultures were obtained from most of these habitats, as shown in Table 6.2. In subsequent work, Bohlool (1975) has found *Sulfolobus* in New Zealand, and Furuya et al. (1977) in Japan.

The chemistry of some of the Yellowstone springs is given in detail in Table 6.3 and it can be seen that all of those analyzed had detectable amounts of sulfur, sulfide, elemental sulfur, and/or ferrous iron. From the turnover times of the springs (see p. 160), we can calculate that there are sufficient concentrations of these energy sources to support the population growth rates that have been measured.

Table 6.1. Habitats in Which *Sulfolobus*-type Cells Have Been Observed[a]

Location[b]	pH	Temperature (°C)
Yellowstone National Park		
Norris Geyser Basin		
1. Congress Pool	2.5	78.5
2. Growler Spring	1.55–2.2	82–89
3. Locomotive Spring	1.8–2.7	83–88.6
4. Spring 98-4 (mixes with Locomotive Spring)	2.25–2.5	80.5–82
5. Sulfur Spring 40-2 at Norris Junction	2.4	80
6. Spring 84-3	2.0	87–88
7. Spring 98-5	2.2–2.45	85–89
8. Spring 98-6	1.8–1.95	83–89
9. Spring 120-2	2.9	89
10. Spring 120-5	3.5	88.5
Roaring Mountain (Southern Effluent), outflow channel	2.0–2.2	76–84
Amphitheater Springs		
1. Sulfur spring 38-1 at western end, outflow channel	2.1	77.5
2. Sulfur spring 38-2 at western end, outflow channel	2.2	76.5
Sylvan Springs (Gibbon Geyser Basin)		
1. Pool 7a	2.5	78.2
2. Pool 15a	1.8	88.8
3. Spring 87-7	1.8	83
4. Spring 89-1	2.65	83
5. Spring 89-3	1.7	79
6. Spring 89-6	1.55	88–90

(For continuation see next page.)

Table 6.1. Habitats in Which *Sulfolobus*-type Cells Have Been Observed[a] (*continued*)

Location[b]	pH	Temperature (°C)
Canyon Area		
1. Spring 111-3	2.5	62
2. Spring 111-5	2.75	73
Mud Volcano Area		
1. Sulfur Caldron	1.35–1.65	65–68.8
2. Mud Geyser	1.7–2.05	60–65.5
Shoshone Geyser Basin		
1. Spring 115-1	2.3	83
2. Spring 115-2	2.1	80–85
3. Spring 115-3	2.1	80
4. Spring 116-1	2.65	75–80
El Salvadore-Sauce area (near Ahuachipan)		
1. Spring 106-2	2.0	87–88
2. Spring 106-3	2.0	64.5
3. Spring 106-4	2.2	85
Italy-Agnano area		
1. Spring 140-4	1.85	79.5
2. Spring 140-5	1.95	84
3. Spring 140-6	1.85	81
4. Spring 142-2	1.7	80
Iceland		
Kerlingarfjoll area		
1. Spring 74-1	2.6	95
2. Spring 75-1	2.4	97.5
Geysir area		
1. Spring 82-5	4.0	67
2. Spring 82-6	3.5	65
Krisuvik area		
1. Spring 108-1	2.3	90.5
2. Spring 109-4	2.3	64
3. Spring 109-5	2.5	84
4. Spring 109-6	2.55	91.5
5. Spring 110-2	2.0	64
Dominica, West Indies		
Valley of Desolation		
1. 21-4, warm soil	1.5	30
2. 21-8, hot pool	3.85	70
3. 22-1, mud pot	4.15	99
4. 22-3, hot pool	3.1	94
Wotten Waven		
1. 32-2, hot soil	2.1	60
2. 32-4, hot soil	3.0	75

[a]Mostly unpublished data of T. D. Brock.
[b]See also Bohlool (1975) for data on New Zealand habitats.

Table 6.2. Sources of *Sulfolobus* Cultures

Designation	Location	pH	Source Temperature	pH	Isolation Temperature	Medium
	Aquatic Habitats					
98-3	Locomotive Spring	2.4	83	3.0	70	yeast extract
106-3	El Salvador	2.0	64.5	2.6	70	yeast extract
111-3	Canyon Area	2.5	62	3.0	55	yeast extract
115-2	Shoshone Basin	2.1	80–85	2.0	70	yeast extract
129-1	Roaring Mountain	2.1	84.5	2.0	70	yeast extract
132-1	Sulfur Caldron	1.5	68.8	1.5	70	yeast extract
136-1	Mud Geyser	1.7	65.5	2.0	65	yeast extract
140-5	Italy	1.95	84	2.0	70	yeast extract
140-6	Italy	1.85	81	2.0	70	yeast extract
74-1[a]	Iceland	2.6	95	3.0	70	S^0, S^0 + yeast extract
75-1	Iceland	2.4	97.5	3.0	70	S^0, S^0 + yeast extract
82-5	Iceland	4.0	67	3.0	70	S^0, S^0 + yeast extract
82-6	Iceland	3.5	65	3.0	70	S^0, S^0 + yeast extract
108-1	Iceland	2.3	90.5	3.0	70	S^0, S^0 + yeast extract
109-4	Iceland	2.3	64	3.0	70	S^0, S^0 + yeast extract
109-5	Iceland	2.5	84	3.0	70	S^0, S^0 + yeast extract
109-6	Iceland	2.55	91.5	3.0	70	S^0, S^0 + yeast extract
110-2	Iceland	2.0	64	3.0	70	S^0, S^0 + yeast extract
21-8	Dominica	3.85	70	3.0	70	S^0, S^0 + yeast extract
22-1	Dominica	4.15	99	3.0	70	S^0, S^0 + yeast extract
22-3	Dominica	3.1	94	3.0	70	S^0, S^0 + yeast extract
	Terrestrial Habitats					
71-5[b]	Amphitheater Springs soil	2.7	83	3.0	55	sulfur
88-4	Roaring Mountain soil	1.4	63	3.0	55	sulfur
125-7	Roaring Mountain soil	1.65	75–85	3.0	70	sulfur
125-14	Roaring Mountain soil	1.45	68–77	3.0	55	sulfur
125-18	Steamboat Springs, Nev.	1.80	46	3.0	55	sulfur
21-4	Dominica	1.5	30	3.0	55	sulfur
32-2	Dominica	2.1	60	3.0	70	S^0, S^0 + yeast extract
32-4	Dominica	3.0	75	3.0	70	S^0, S^0 + yeast extract

[a] At a later time, an attempt was made to grow the enrichment cultures of 74-1, 75-1, and 108-1 at 90°C, but no growth was obtained, although excellent growth was obtained at 70°C.
[b] Cultures 71-5 and 88-4 were isolated by C. B. Fliermans, the rest by T. D. Brock. For more details on locations of habitats, see Table 6.1.

Temperature Relations

The studies on the effect of temperature and pH of pure cultures confirm the observations on natural habitats. As seen in Table 6.4, most strains of *Sulfolobus* tested have temperature ranges from 55°C to 80°C with one strain (140-5 from Naples) growing at 85°C. Some isolates of the Italian group (labeled by them the MT strains) have temperature optima a little higher than our isolates (although our Neapolitan isolate had the highest temperature optimum of our original group of cultures). The temperature optima of the strains listed in Table 6.4 may be similar in part because they were isolated under similar conditions, with incubations at 70°C. In a subsequent study we showed that temperature strains existed (see below),

a

Figure 6.7a,b. Two hot acid pools, rich in elemental sulfur, which had good *Sulfolobus* populations. (a) Sulfur Caldron. Beer cans occasionally dropped from the parking lot above, but the spring quickly disposed of such objects. Temperature over many years has not varied much from 65°C, and the pH has always been quite low, around 1.5. (b) Evening Primrose, in the Sylvan Springs area. The most acidic spring in Yellowstone, pH 0.9. In 1971 this spring had a higher pH, 1.3, and a temperature of 90°C, and was devoid of *Sulfolobus*. Sometime during the 1971–72 winter, the temperature of the spring dropped to below 60°C. A massive *Sulfolobus* population colonized the spring, most of the elemental sulfur disappeared, and the pH dropped to 0.9. These observations provide some of the best evidence that *Sulfolobus* does not grow above 90°C in Yellowstone.

and the temperature optimum depended both on the temperature of the habitat and the temperature of isolation. Although *Sulfolobus* was seen in an Iceland spring at 97.5°C (Table 6.1), the culture obtained by 70°C enrichment would not grow at 90°C.

In the study of temperature strains, Mosser et al. (1974b) studied *Sulfolobus* both in its natural habitat and in culture. The temperature optimum for the oxidation of ^{35}S-elemental sulfur was measured in natural populations. At the same time, enrichment cultures were prepared with inoculum from each spring using several incubation temperatures, and the temperature optima of pure culture isolates were then studied in the laboratory. A

b

Figure 6.7b. Legend see opposite page.

Figure 6.8a. Photograph of a flowing sulfur spring in the Amphitheater Springs area. The spring is in a small crevice toward the bottom of the picture, and flows toward Lemonade Creek, which is in the background.

Figure 6.8b. Close-up of the sulfur deposit, which has been laid down as a result of the oxidation of hydrogen sulfide to elemental sulfur. The sulfur deposit decreases in amount downstream as the H_2S is exhausted. *Sulfolobus* is found attached to the sulfur crystals in the area (see Figure 6.6) with appropriate temperature.

Table 6.3. Characteristics of Solfataric Springs and Half-Times for Dilution of Added Chloride[a]

Location	Temperature (°C)	pH	Volume (liters)	Cells/ml	Natural Cl⁻	Concentrations in ppm					Half-time for chloride dilution
						S^0	$S^=$	$SO_4^=$	Fe^{2+}	Total iron	
Norris Geyser Basin											
Locomotive	90–92	2.3	2422	5.4×10^6	6.2	0.3	2.5	231	17.4	18.2	16 hr
White Bubbler	90–92	2.1	52	2.0×10^6	8.7	<0.04	1.4	324	1.2	1.7	10 hr
Vermillion	91–92	2.3	273	6.4×10^6	7.4	15.0	1.1	400	11.6	11.9	17 hr
26-5	89–92	2.2	257	7.1×10^5	6.1	0.02	0.8	396	2.1	3.2	13 hr
21-1	90–92	2.3	97	3.0×10^6	23.8	2.2	2.6	151	3.0	3.3	14 hr
Sylvan Springs											
59-1	76–81	2.1	453	4.0×10^6	1.1	72.0	5.7	344	2.8	4.4	24 hr
59-2	70–79	1.5	22	1.3×10^8	2.8	697	4.0	1516	7.4	14.2	24 hr
54-5	82–86	1.8	595	2.1×10^6	13.9	1.0	1.1	472	3.2	3.6	24 hr
56-3	63–67	2.1	95	2.4×10^6	31.0	3.2	5.0	291	3.6	4.5	34 hr
58-3	82–87	1.8	80	6.4×10^6	14.7	0.2	1.3	506	15.4	16.8	13 hr
Mud Volcano											
Moose Pool	72–80	1.6	1.1×10^6	2.5×10^7	31.9	1110	15.5	2260	120.3	119.8	28 days
Sulfur Caldron	65–68	1.5	7.6×10^5	7.4×10^7	1.7	993	22.4	2260	—	—	35 days

[a]From Brock and Mosser (1975).

Table 6.4. Temperature and pH Range for Growth of *Sulfolobus*[a]

Strain	Temperature (°C)		pH	
	Minimum	Maximum	Minimum	Maximum
98-3	55	80	1	5.9
106-3	55	80	1	5.9
111-3	55	75	1	4.5
115-2	55	80	1	5.9
129-1	60	80	1.5	5.8
132-1	55	80	1	5.3
136-1	55	80	1	4.6
140-5	55	85	1	5.8
140-6	60	80	1	5.8

[a]Tested in basal salts with 0.1% yeast extract added. For temperature series, all media were adjusted to pH 2. For pH studies, incubation was at 70°C. Growth was recorded after 4–6 days incubation. (From Brock et al., 1972.)

series of springs was selected all of which had good populations of *Sulfolobus,* but with different temperatures. The temperature optima of the natural populations, as determined with the ^{35}S technique, varied considerably. As seen in Figure 6.9, three distinct temperature optima were exhibited: (1) the optimum was about 63°C at Mud Geyser (site temperature 59°C; (2) the optimum was 70°C at Evening Primrose (site temperature 57°C); (3) the optimum was 80°C at Sulfur Caldron (site temperature 67°C), Vermillion (site temperature 92°C), Moose Pool (site temperature 76°C), and Spring 24-

| Optimal temperature of population
■ Range of temperatures in spring

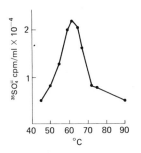 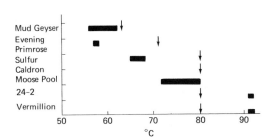

Figure 6.9. Optimum temperatures for [35]S⁰ oxidation by natural populations of *Sulfolobus* in several springs. The bar represents the range of temperature for each spring and the arrow indicates the temperature optimum for the population. The left graph shows data from Mud Geyser, typical of those from which the temperature optima were estimated. (From Mosser et al., 1974b.)

2 (site temperature 92°C). These results indicate that populations distinct in their temperature requirements can develop at different sites, but in spite of this, only at Mud Geyser and Moose Pool did the bacteria appear to be well adapted to the temperature of their environment. At the other sites, the maximum rate of S^0 oxidation did not occur at the environmental temperature. In Evening Primrose and Sulfur Caldron, sites with relatively low environmental temperatures, the optimum temperature was higher than the environmental temperature, whereas in Vermillion and 24-2, sites with the highest environmental temperatures, the optimum was lower than the environmental temperature.

In the same study, cultures were attempted using incubation temperatures of 60°C, 75°C, and 90°C. No cultures could be obtained using 90°C, but cultures were obtained at 60°C and 75°C. After several transfers at the enrichment temperature, clones were obtained by a most-probable-number dilution method. The temperature optima of the clones corresponded closely, although not exactly, with the incubation temperature. From several springs, two strains with distinct temperature optima were obtained, indicating that mixtures of strains of *Sulfolobus* occur in these springs. The temperature optima of the isolates were determined using $^{35}S^0$ oxidation, $^{14}CO_2$ fixation, and from growth curves, and the optima determined by these three methods in general agreed. This suggests that the optima determined in the natural habitat using the $^{35}S^0$ method are correct.

Thus, in only two of the six springs studied were the *Sulfolobus* populations well adapted to the temperature of their habitats. This might be explained by assuming that there are only a limited number of temperature strains of *Sulfolobus,* and that modification of temperature optimum by mutation does not occur readily in this organism. More difficult to explain is why springs are not colonized by strains optimally adapted to the habitat temperature. Our studies of these springs over about 4–5 years indicate that the temperatures of these springs are nearly constant. We have never observed variations in temperature of more than 10°C. Growth rate measurements (see later) indicate that the organisms are multiplying in the springs, with doubling times varying from about 1 day, in the small springs, to 30–40 days, in the larger ones. Even the slow growth rates in the larger springs are fast enough to permit strains optimally adapted to their habitat to develop during the period under study (2–3 years).

The explanation for the lack of agreement between optimal and environmental temperatures is probably not the same for all the springs. For the springs with the highest temperatures, Vermillion and 24-2, both near boiling, it seems reasonable to conclude that strains with temperature optima greater than 80°C do not occur in Yellowstone Park. Very low rates of sulfur oxidation were obtained at 90°C with natural populations, and no growth occurred in enrichment cultures incubated at that temperature. In spite of this, the bacteria do maintain themselves at these sites, as shown by the *in situ* growth rate measurements, but it seems unlikely that they could do so at much higher temperatures. Conceivably, the bacteria at these sites are multiplying in the sediments where the temperatures may be slightly lower. Since temperature optima around 90°C have been found for bacteria colonizing springs of neutral pH (see later), the high acidity of its habitat may prevent *Sulfolobus* from better adapting to high temperatures. It should be noted that *Sulfolobus* populations do exist in acid springs at 95–99°C in Iceland (Brock, unpublished), where the boiling point is higher due to lower elevation, so it is possible that high temperature strains exist there.

It is of some interest that temperature strains of the thermal blue-green alga *Synechococcus* occur (Peary and Castenholz, 1964), but that temperature strains of the acidophilic thermal alga *Cyanidium caldarium* do not occur. Like *Sulfolobus, Cyanidium* occupies acid environments, whereas *Synechococcus* grows in neutral to slightly alkaline environments. It may be that the added stress of high acidity is a factor restricting the adaptability of acid-dwelling organisms. Another factor may be that since such acidophilic thermophiles are the sole occupants of their environmental niches, they do not meet competition from other organisms, and hense there is less selection pressure for optimal adaptation.

Serology

Ben Bohlool prepared antisera against purified cell walls of *Sulfolobus,* and used these antisera in immunofluorescence and immunodiffusion studies (Bohlool and Brock, 1974). The procedures used were similar to those already described for *Thermoplasma* (Chapter 5). Antisera were prepared against four strains, and were then tested for cross-reaction against other strains (Table 6.5). The antisera for strains 98-3 and V-75 showed a high degree of specificity, for their fluorescent antibodies did not react with any of the other strains. Cells of the two other strains, EP-60 and SC-60, did react moderately well with each other's antisera, but not with those from strains 98-3 and V-75. Cross-reactions tested by immunodiffusion corroborated those obtained with fluorescent antibody staining. Because of the specificity of the 98-3 antiserum, it was tested against 20 separate strains, of which 4 reacted strongly, 3 weakly, and the rest were negative.

Because EP-60 and 98-3 antisera did not cross-react, they were used to observe the geographical distribution of the two serotypes in a variety of Yellowstone hot springs. The reactive cells were quantified by a membrane filter technique which permits efficient and nonselective removal of *Sulfolobus* cells from the spring waters. As seen in Table 6.6, EP-60 reactive cells were present in all the hot springs studied, but the number and percentage of EP-60 reactive cells varied widely. Cells reactive with 98-3 were present in many of the springs, but not in all, and the numbers were relatively lower than those of EP-60 reactive cells. In most springs, both serotypes were present. In a related study, microscope slides were immersed in several

Table 6.5. Immunofluorescence and Immunodiffusion Specificities of *Sulfolobus* Antisera

	Antiserum						
	98-3		EP-60		SC-60		V-75 FA Reaction
Organisms	FA Reaction	Number of precipitin bands	FA Reaction	Number of precipitin bands	FA Reaction	Number of precipitin bands	
98-3	4+ [a]	1	–	0	–	0	–
EP-60	–	0	4+	2	3+	2	–
SC-60	–	0	2+	1	4+	3	–
V-75	–	n.t.	–	n.t.	–	n.t.	4+
A-60	2+	1					
MP-60	2+	0					
MP-75	–	n.t.					
115-2	2+	1					
129-3	4+	n.t.					

[a]Symbols: n.t., not tested; (–), no reaction; 1 to 4+, increasing intensity of fluorescence; FA, fluorescent antibody. (From Bohlool and Brock, 1974.)

springs, both flowing and nonflowing, and cells reactive with the two antibodies that developed on the microscope slides were determined. In this study it was found that EP-60 cells were much more widespread and abundant in nonflowing springs, whereas 98-3 cells predominated in flowing springs. Microcolony development could be observed on the slides (Figure 6.10) and from the rate of increase in number of cells per microcolony with time it was possible to estimate growth rates. In Locomotive Spring, the growth rate was estimated in this way to be approximately one generation per 36 hours. Further discussion of growth rates of *Sulfolobus* in nature will be found later in this chapter, but this measured rate is of the same order of magnitude as that measured for the total population in the spring water itself.

In a later study, Bohlool (1975) isolated a number of *Sulfolobus* strains from New Zealand hot springs and tested them for cross-reactivity with EP-60 and 98-3 antisera by immunofluorescence. Most of the strains showed no cross-reaction, but two New Zealand isolates showed strong cross-reaction with 98-3 and two isolates showed moderate cross-reaction with EP-60. The serological relationship between some of the New Zealand and Yellowstone strains raises some interesting questions of dispersal and evolution of *Sulfolobus,* which will be discussed later in this chapter.

Perhaps the most significant result to come from the Yellowstone studies was that individual hot springs generally harbored more than one serotype of *Sulfolobus.* This was true even of some hot springs that were quite small in size. For instance, the volumes of springs 59-1 and 59-2 are only 453 and 22 liters, respectively. As seen in Table 6.6, both of these springs had populations of both serotypes, and there may have been other serotypes as well, since not all of the *Sulfolobus* cells were stained with both antisera. The ability of single small springs to harbor several serotypes may only reflect the fact that competitive interactions between *Sulfolobus* strains are not strong, and with so many *Sulfolobus* habitats in these geyser basins, there is large opportunity for dispersal and colonization of different strains. This conclusion agrees with the observations of the presence of more than one temperature strain in a single spring, as discussed above.

Cellular Stability

No extensive studies on the cellular stability of *Sulfolobus* have been done paralleling those done with *Thermoplasma* (Chapter 5). However, general impressions are that *Sulfolobus* is more stable to various environmental influences than is *Thermoplasma,* probably because of the presence of the cell wall. Weiss (1974) showed that the cell wall remained intact at neutral pH, with the preservation of the subunit structure, although the wall could be solubilized by guanidine hydrochloride, sodium dodecyl sulfate (SDS), and partially by pH 10.5. Treatment of walls with dithiothreitol rendered

Table 6.6. Geographical Distribution of *Sulfolobus* Serological Types in the Water of Hot Springs of Yellowstone National Park[a]

	Temperature (°C)	pH	Total count[b]	FA[c] count EP-60		FA count 98-3	
				No. reactive cells/ml	Percent of total	No. reactive cells/ml	Percent of total
Nonflowing springs							
I. Mud Volcano							
Moose Pool	72.4	1.6	1.2×10^8	4.9×10^7	41	2.1×10^6	1.8
Mud Geyser	60.2	1.7	2.3×10^7	2.4×10^7	104	0	0
Sulfur Caldron	65.7	1.6	1.8×10^8	3.1×10^7	21	2.6×10^6	1.4
II. Norris							
Locomotive	88.4	2.3	2.6×10^7	6.3×10^6	24	0	0
White Bubbler	91.2	2.1	1.3×10^7	4.9×10^5	4	0	0
21-1	91.8	2.3	2.4×10^7	4.3×10^6	18	0	0
24-2	91.4	2.9	2.0×10^7	1.2×10^6	6.0	1.8×10^7	90
26-2	89.2	2.2	1.0×10^7	4.3×10^5	0.4	8.5×10^4	0.9
III. Sylvan							
Evening Primrose	56	1.1	4.6×10^7	2.4×10^7	52	1.3×10^6	3.0
54-5	82.7	1.9	2.5×10^6	2.1×10^6	84	3.7×10^3	0.16
58-3	83.3	1.7	3.6×10^7	1.5×10^7	42	8.5×10^4	0.25
59-1	80.5	2.1	1.4×10^7	1.1×10^6	8	1.2×10^4	0.1

59-2	75.2	1.6	2.0×10^8	1.2×10^8	60	6.5×10^5	0.32
59-3	80.5	2.4	1.9×10^7	9.8×10^5	5	3.6×10^5	2.0

Flowing springs

I. Amphitheater

Site I

Water	75.5	2.2	2.5×10^3	2.3×10^3	92.0	3.7×10^1	1.6
Sediment			3.7×10^4	3.2×10^4	86.0	1.2×10^4	32

Site II

Water	74.0	2.3	2.2×10^3	1.5×10^3	68.0	1.1×10^2	5
Sediment			4.0×10^4	3.1×10^4	77.0	2.7×10^4	67

II. Roaring Mountain, Southern Effluent

Site I

Water	69	2.2	2.3×10^3	1.9×10^3	83	6.1×10^1	2.6
Sediment			1.5×10^7	4.0×10^5	2.7	1.6×10^6	10.7

Site II

Water	65	2.2	1.2×10^3	9.0×10^2	75	2.5×10^1	2.0
Sediment			5.6×10^6	3.0×10^5	5.0	3.3×10^5	6.0

[a] From Bohlool and Brock (1974).

[b] By Petroff-Hauser count for populations above 10^7/ml and membrane filter acridine orange count for numbers below 10^7/ml.

[c] FA, fluorescent antibody.

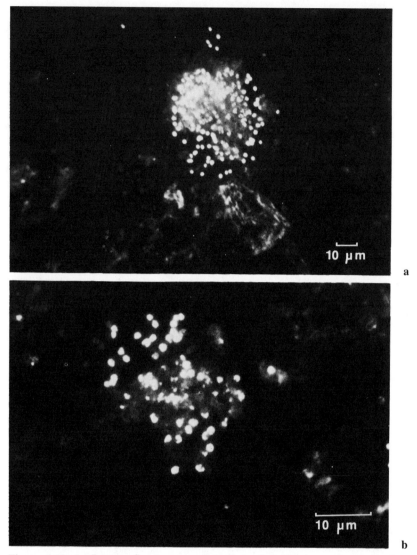

Figure 6.10a,b. Immunofluorescence detection of *Sulfolobus* colonies on immersion slides from a hot spring. Six-day immersion slides from Locomotive Spring stained with EP-60 fluorescent antibody. (a) A relatively large colony with intact cells. (b) A higher magnification of a smaller colony showing evidence of cell degradation in the center. (From Bohlool and Brock, 1974.)

the cell walls more sensitive to SDS, so that complete dissolution occurred. Nothing is known about the stability of the cell membrane of *Sulfolobus*, except that it can be solubilized by Triton-X (Weiss, 1974).

Lipids of *Sulfolobus*

The lipids of *Sulfolobus* have been studied by Langworthy et al. (1974) and deRosa et al. (1974). The latter group has concentrated primarily on the structure of the hydrocarbon chains, whereas the former group has done a more detailed characterization of the lipids. Langworthy et al. (1974) found that *Sulfolobus* cells contained about 2.5% total lipid, of which 10.5% was neutral lipid, 67.6% glucolipid, and 21.7% polar lipid. The lipids contained a $C_{40}H_{80}$ isoprenoid similar to that of *Thermoplasma*. The structural relationships between these two isoprenoids were discussed in Chapter 5 (see Table 5.7). As in *Thermoplasma*, *Sulfolobus* contained almost no fatty acids. The glycolipids were composed of about equal amounts of the glycerol diether analogue of glycosyl galactosyl diglyceride and a glucosyl polyol glycerol diether. The latter compound contained an unidentified polyol attached by an ether bond to the glycerol diether. About 40% of the lipid phosphorus was in the diether analogue of phosphatidyl inositol, and the rest was in approximately equal amounts of two inositol monophosphate-containing phosphoglycolipids, inositolphosphoryl glycosyl galactosyl glycerol diether and inositolphosphoryl glucosyl polyol glycerol diether. The glycolipids appear similar to those found in *Halobacterium cutirubrum*, but the phospholipids appear to be unique. The presence of only inositol-containing phospholipids is quite unusual as these components are not commonly present in bacteria.

The long-chain isoprenoid ethers found both in *Thermoplasma* and *Sulfolobus* are almost certainly related in some way to the survival of these organisms in hot, acid environments. This has been discussed in Chapter 5.

Langworthy has also found (personal communication) that a new species of phospholipid appears in autotrophically grown *Sulfolobus*, as compared with heterotrophically grown cells. Further work on the relationship of these lipids to cellular stability and metabolism would be of considerable interest. Furuya et al. (1977) reported that their *Sulfolobus*-like organism from a Japanese hot spring also had ether-linked lipids.

Sulfur Oxidation

The ability of *Sulfolobus* to oxidize sulfur was first indicated by the ability of the organism to grow autotrophically on elemental sulfur as the sole energy source. Under these conditions, sulfate was formed, and the pH

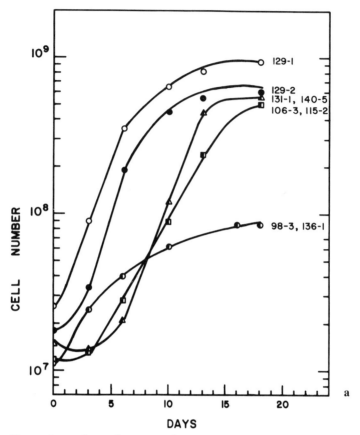

Figure 6.11a. Legend see opposite page.

dropped. In order to study the process of sulfur oxidation in detail, Shivvers and Brock (1973) developed a method using ^{35}S-labeled elemental sulfur. Because elemental sulfur is insoluble, the oxidation could be measured by filtering samples of cultures and measuring the radioactivity that entered the filtrate. All of the radioactivity in the filtrate was $^{35}SO_4^{2-}$, as shown by the fact that it was insoluble in CS_2 and was quantitatively precipitated by $BaCl_2$. Controls have shown that no oxidation of $^{35}S^0$ occurs in uninoculated bottles, or when the organisms are poisoned with formaldehyde. The relationship between growth and oxidation of $^{35}S^0$ for a number of strains of *Sulfolobus* is shown in Figure 6.11. Quantification of growth in the later phases of the growth cycle is difficult because most of the cells are attached to the sulfur crystals. This is shown clearly in the data of Table 6.7, in which the relative numbers of cells attached and unattached to sulfur crystals were counted by acridine orange fluorescence microscopy. In this experiment, the cells were grown unshaken, except for occasional mixing of the culture during sampling. Experience has shown that growth is not

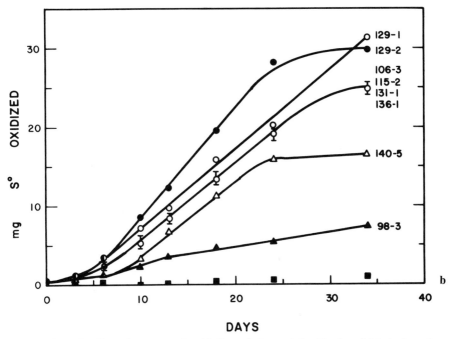

Figure 6.11a,b. Growth rates and oxidation of elemental sulfur by eight strains of *Sulfolobus*. (a) Growth rates with elemental sulfur as sole energy source. The numbers represent only free-floating cells. (b) Oxidation of $^{35}S^0$. Temperature of incubation 70°C. Strain 98-3 has an optimum between 75°C and 80°C, which probably explains its low rate of oxidation at 70°C. (From Shivvers and Brock, 1973.)

Table 6.7. Attachment of *Sulfolobus* to Elemental Sulfur[a]

			Relative cell numbers		
Time (days)	Cells/ml unattached	Optical density	Unattached (U)	Attached (A)	Ratio U:A
0	2.6×10^7	0.009	407	0	—
3	9.0×10^7	0.018	525	6	87:1
6	3.5×10^8	0.056	1011	83	12:1
10	5.6×10^8	0.072	129	342	1:3
13	8.1×10^8	0.091	273	1658	1:6
17	2.1×10^9	0.112	88	1560	1:18

[a]Strain 129-1, incubation at 70°C. (From Shivvers and Brock, 1973.)

Table 6.8. Sulfur Balance, *Sulfolobus acidocaldarius* Strain 129-1[a]

Distribution of sulfur	Percentage of total sulfur			
	10-day culture	18-day culture	34-day culture	Control
S^0	79	58	36	94
SO_4^{2-}	12	30	51	2
Incorporated into organisms	1	3	4	0
Total	92	91	91	96

[a]From Shivvers and Brock (1973). Control, uninoculated.

any better if cultures are shaken or aerated, although growth is stimulated if the gas phase is enriched with 5% CO_2 in air.

A sulfur balance was determined for two strains of *Sulfolobus* growing autotrophically on elemental sulfur. As seen in Table 6.8, virtually all of the oxidized sulfur can be accounted for as sulfate, and only very little sulfur is incorporated into the organisms. This would be predicted: chemolithotrophs must oxidize large amounts of substrate in order to obtain the energy necessary for growth. The fact that only sulfate is formed agrees with the fact that at the temperature and pH involved, other sulfur oxidation products, such as thiosulfate or sulfite, are unstable.

Because *Sulfolobus* will grow both autotrophically and heterotrophically, it was of interest to determine the effect of organic materials on sulfur oxidation. As seen in Table 6.9, yeast extract markedly stimulated growth in sulfur medium, but inhibited sulfur oxidation. At a concentration of 0.1%, which would normally be considered quite low, yeast extract inhibited sulfur oxidation by about 30-fold. Yet, because of the increased growth, the total quantity of sulfur oxidized was only reduced by one-third. Thus, it seems that *Sulfolobus* is getting energy from the oxidation of both

Table 6.9. Effect of Yeast Extract Addition to the Culture Medium on Growth and Rate of Sulfur Oxidation[a]

Yeast extract concentration (%)	Cell no./ml	Sulfur oxidized (μg/ml)	Sulfur oxidized (μg per ml per 10^7 organisms)
0	3.48×10^8	95.13	2.74
0.01	1.42×10^9	104.05	0.73
0.05	4.9×10^9	47.84	0.098
0.1	5.5×10^9	39.19	0.071

[a]Initial inoculum, 2.6×10^7. Incubation time, 6 days. (From Shivvers and Brock, 1973.)

sulfur and organic matter, although yeast extract probably represses the enzymes of the sulfur oxidation pathway.

CO_2 Fixation

Since *Sulfolobus* cultures growing autotrophically obtain all of their carbon by CO_2 fixation, it was of interest to measure rates of CO_2 fixation during growth. Much less work has been done on CO_2 fixation than on sulfur oxidation, because the latter process is of more ecological and biogeochemical interest. As compared with the ^{35}S experiments, the rates of CO_2 fixation using ^{14}C could be measured in quite short incubation periods, of the order of an hour or so. Mosser et al. (1974b) showed that the temperature optimum for CO_2 fixation by *Sulfolobus* cultures was the same as was the temperature optimum for sulfur oxidation and for growth. Shivvers and Brock (1973) showed that yeast extract inhibited CO_2 fixation.

Because of the ability of *Sulfolobus* to attach to elemental sulfur crystals, it was of interest to measure $^{14}CO_2$ fixation by attached and unattached cells. As shown in Table 6.10, CO_2 fixation was considerably higher for the attached cells than for cells freely suspended in the medium. Because in the experiment shown, most of the cells would be free rather than attached (see Table 6.7), it can be concluded that in mixed suspensions, most of the bacterial activity resides with cells attached to sulfur.

Sulfide Oxidation

Hydrogen sulfide is a fugitive compound in the *Sulfolobus* environment. As Zinder and Brock (1977) have shown, it is the dominant sulfur gas emanating from hot, acid springs, but although HS^- is quite soluble in water, the

Table 6.10. CO_2 Fixation of Sulfur-Grown Cells[a]

Strain	Location of bacteria	cpm	
		Control	Experimental
98-3	Free	71	172
98-3	Attached to sulfur	37	1064
140-5	Free	46	475
140-5	Attached to sulfur	40	928

[a]Cultures grown 4 days at 70°C on basal salts at pH 3 plus 1% elemental sulfur. The sulfur was allowed to settle and aliquots of the freely suspended cells removed. Then the cultures were shaken and aliquots of sulfur plus cells removed. The latter samples had most bacteria attached to sulfur crystals. Incubation was 3.5 hr at 70°C. In the controls, formaldehyde was added before adding isotope at zero time. Cell counts for unattached bacteria were 98-3, 8.9×10^7/ml; 140-5, 1.1×10^8/ml; cpm, counts per minute.

pK of the H_2S/HS^- pair is around neutrality, so that at the acid pH values of the springs, only H_2S is present. The solubility of H_2S in water at these temperatures is low, so that the gas quickly escapes into the atmosphere. In some of the springs, fairly large amounts of iron are present (Brock et al., 1976), and if ferric ions are present they will react with and rapidly oxidize sulfide (Brock and Gustafson, 1976). If there is low iron and high sulfide, then the sulfide will reduce all of the iron to the ferrous form, and some ferrous sulfide may be present.

Although HS^- reacts fairly rapidly with O_2 (Brock and O'Dea, 1977), H_2S hardly reacts at all, and in the absence of ferric ions H_2S and O_2 can coexist. Thus, in low-iron springs that are high in sulfide, sufficient H_2S may be present to serve as a significant energy source. Zinder and Brock (unpublished) have shown that *Sulfolobus* can catalyze the oxidation of H_2S with O_2 under these specialized, low-iron conditions. To avoid volatilization of H_2S, the experiment must be done in sealed, completely filled tubes. Because under these conditions continuous aeration is difficult, the water was enriched in O_2 by oxygenating with pure O_2 at room temperature. The filled oxygenated tubes were then equilibrated at temperature, and the reaction started by removing the caps, and quickly injecting with a long needle to the bottom of the tube small amounts of a stock solution of sulfide to give a final concentration of around 2 $\mu g/ml$. The capped tubes were incubated and replicates removed at intervals. A sample was quickly removed with a syringe and the sulfide fixed in zinc acetate, following standard procedures (Brock and Mosser, 1975). The sulfide was then assayed colorimetrically.

As seen in Figure 6.12, *Sulfolobus* populations rapidly oxidized sulfide, and there was a sharp temperature optimum at 80°C. The experiment illustrated was done with cells from Moose Pool, where it had earlier been

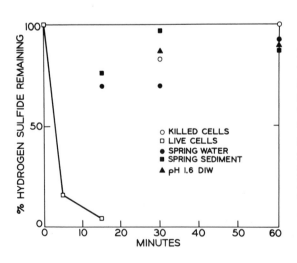

Figure 6.12. Rate of oxidation of hydrogen sulfide by natural population of *Sulfolobus* from Moose Pool. The cells were washed three times and suspended in oxygenated pH 1.6 sulfuric acid in completely filled, sealed tubes (to avoid loss of H_2S by volitization). There is a sharp optimum at 80°C. (Unpublished data of S. D. Zinder and T. D. Brock.)

Figure 6.13. Time course of oxidation of hydrogen sulfide by *Sulfolobus* from Moose Pool and appearance of elemental sulfur. Procedure as in Figure 6.12. Temperature of incubation 65°C. (Unpublished data of S. H. Zinder and T. D. Brock.)

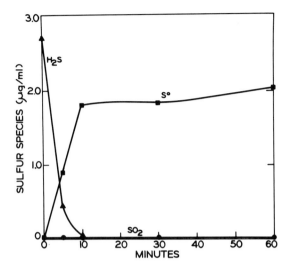

shown (Mosser et al., 1973) that the temperature optimum for elemental sulfur oxidation was also 80°C.

Assays were made of sulfur oxidation products to determine what the sulfide was oxidized to. Sulfite was assayed colorimetrically and gas chromatographically as described by Zinder and Brock (1977) and elemental sulfur was assayed by the trichloroethylene method of Pachmayr (1960). No sulfite was detected, and the oxidation appeared to stop at the level of elemental sulfur. As shown in Figure 6.13, there was a direct correlation between the disappearance of sulfide and the appearance of elemental sulfur. Although because of the high sulfate content of these spring waters it is not possible to assay for an increase in sulfate, it seems reasonable to conclude that no sulfate is formed from sulfide, as there was a quantitative conversion of sulfide to sulfur.

Since *Sulfolobus* oxidizes elemental sulfur to sulfate, it might be thought paradoxical that it oxidizes sulfide only to elemental sulfur. However, the rate of oxidation of sulfide is very rapid, the process being completed, at optimal temperature, in 15 minutes, whereas the rate of oxidation of elemental sulfur is much slower, being reckoned in terms of hours or days (see previous section). The much slower rate of elemental sulfur oxidation is probably due to the insolubility of this material. It might be thought, however, that if elemental sulfur were formed as an obligate intermediate of the oxidation of sulfide to sulfate, then it would be present in the cell either in soluble form or in small colloidal droplets, and that in this form further oxidation to sulfate would occur rapidly. This seemed not to be the case, and the only other explanation I can think of is that the enzyme(s) for the oxidation of elemental sulfur is on the outside of the cell membrane, whereas when elemental sulfur is formed from sulfide, the enzyme (NAD or flavin-linked) for the oxidation of sulfide is on the inside, so that if elemental sulfur is formed within the cell from sulfide oxidation it does not reach the sulfur oxidase system and thus accumulates.

Ferrous Iron Oxidation

In his original work, Brierley (1966) showed that his isolate of *Sulfolobus* could oxidize ferrous iron, and this was subsequently confirmed by Corale Brierley (Brierley and Brierley, 1973). She demonstrated ferrous iron oxidation both by the Fe^{2+}-dependent uptake of O_2 and by the appearance of Fe^{3+}. Douglas Shivvers in my laboratory showed that *Sulfolobus* strain 136-1 (sulfur-grown) would oxidize ferrous iron at rates considerably higher than those expected from autooxidation, although little growth was obtained (Figure 6.14). Obtaining growth on ferrous iron presents some technical difficulties, because at the temperature of growth ferrous iron autooxidizes even at pH 2, and an extensive precipitate of jarosite (ferric sulfate hydroxide) forms. It is fairly easy to add small amounts of ferrous iron to cultures and demonstrate oxidation over short-term incubations, but to obtain significant growth, large amounts of ferrous iron must be added

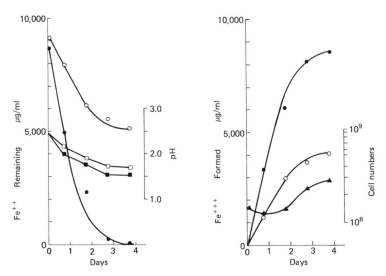

Figure 6.14. Oxidation of ferrous iron by *Sulfolobus* strain 136-1, grown on elemental sulfur, transferred to mineral salts medium (pH 2.5), and incubated with gentle bubbling with CO_2 and air at 70°C. Assays for ferrous and ferric iron were made periodically using the phenanthroline method. Cell counts were made with the Petroff-Hausser counting chamber. The control (open symbols) was uninoculated. Circles: ferrous and ferric iron; squares: pH; triangles: cell count (numbers per ml). (Unpublished data of Douglas Shivvers and T. D. Brock.)

and incubation must be continued for a number of days. So far as I know, no one has provided data on growth of *Sulfolobus* on ferrous iron such as we have provided for growth on elemental sulfur.

We have carried out an extensive study on ferrous iron oxidation in natural populations of *Sulfolobus* taken from Yellowstone springs (Brock et al., 1976). Many *Sulfolobus* springs are fairly high in Fe^{2+}, Fe^{3+}, or both (Table 6.3), the iron presumably derived by acid attack on iron-rich minerals in the volcanic rocks of the basins. Wide variations in iron contents were found (Table 6.3), and within a given spring there were also fairly wide temporal variations. I discuss the hydrology of these acid springs in more detail in Chapter 12. Two sources of soluble iron can be envisaged, the minerals of the immediate spring basin, and the water entering the spring by underground seepage. That the latter source is significant was shown by a simple experiment in which two small springs were drained completely by siphoning, and samples of the first water reentering the springs were collected for iron assay. Ferric iron concentrations of the first water reentering drained springs were two- to fourfold higher than the steady-state concentration, and ferrous concentrations were lower. Because of the low pH of the underground water, both ferrous and ferric iron are soluble, so that the relative proportions of ferrous and ferric iron in the entering water are probably determined partly by the possibilities for bacterial oxidation in the upslope seepage area, and partly by the possibilities for contact of the water with reducing agents, primarily hydrogen sulfide. Because the vents in these springs are generally rich in H_2S (Zinder and Brock, 1977), the results of these drainage experiments suggest that soluble iron enters the springs via underground seepage and is reduced to the ferrous state by the H_2S present [as shown by Brock and Gustafson (1976), the reduction of ferric iron by H_2S is an extremely rapid process at acid pH]. Since the hydrologic studies have shown that the springs are in steady state, with water seeping out as fast as it is flowing in (see Chapter 12), it seems reasonable to conclude that the steady-state concentrations of ferrous and ferric ions are determined by the relative rates of the reduction reaction brought about by H_2S and of the oxidation reaction brought about by the bacteria.

The role of *Sulfolobus* in ferrous iron oxidation in these springs was studied by incubating samples of spring water in the laboratory and assaying for the disappearance of ferrous iron with time. If the natural concentration of ferrous iron were low, ferrous sulfate was added to raise the concentration to around 50–60 $\mu g/ml$, whereas if the ferrous concentration were already of that magnitude, enrichment was not done. To show that the ferrous iron oxidation was due to bacterial action, controls were used in which the cells were poisoned by the addition of 10% NaCl (formaldehyde, our most commonly used poison, could not be used because it interferred with the ferrous iron assay, as did mercuric bichloride). As shown in Figure 6.15, 10% NaCl completely inhibited ferrous iron oxidation by Locomotive

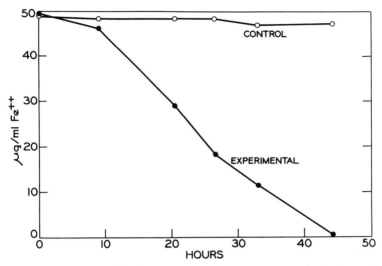

Figure 6.15. Oxidation of ferrous iron by *Sulfolobus* population from Locomotive Spring. The concentration of ferrous iron was raised from 15 to 49 ppm at 0 time. The control had 10% NaCl added to inhibit bacterial activity. Temperature of incubation 80°C. (From Brock et al., 1976.)

Spring bacteria, whereas in the absence of NaCl the ferrous iron concentration was rapidly depleted.

To determine how widespread bacterial oxidation of ferrous iron was in acidic hot springs, a survey of a number of high-iron springs was carried out. The results are summarized in Figure 6.16 and it can be seen that the process occurred in most of the springs studied. The inability to detect ferrous iron oxidation in several of the springs will be explained below. Studies on the effect of temperature show that in different springs there were distinct temperature optima for ferrous iron oxidation. The temperature optimum was 80–85°C in Locomotive Spring, 70°C in spring 58-3, and 60°C in Moose Pool. In all three springs, no significant oxidation occurred at 90°C during the time of the experiment. The absence of oxidation at 90°C and the distinct temperature optimum are the best evidence that the process of ferrous iron oxidation is a biological one. As has been discussed earlier, distinct temperature strains of *Sulfolobus* exist, and the present data show that temperature strains can probably also be detected by measuring ferrous iron oxidation. However, in the case of Moose Pool, the temperature optimum for elemental sulfur oxidation by the natural population was 80°C (Mosser et al., 1973), whereas for ferrous iron oxidation an optimum in the same spring of 60°C was measured. We must conclude that the dominant strain oxidizing sulfur is different from the strain oxidizing ferrous iron. The temperature of Moose Pool has ranged from about 70°C to 80°C during the years we have had this spring under observation, but it has never fallen as low as 60°C. Thus, I have no explanation for the 60°C optimum for ferrous iron oxidation, especially since in other pools strains with higher temperature optima exist, as indicated by the temperature optima of the natural populations.

One of the striking observations which we made when initiating studies on ferrous iron oxidation was that the process did not seem to occur in several springs in which there was a large amount of suspended sediment. This is clearly shown by the results for Moose Pool and Sulfur Caldron in Fig. 6.16. When the sediment was allowed to settle for about 1 hr, the bacteria, present primarily in the supernatant water, readily oxidized ferrous iron. It was then shown that the sediment did not, in fact, inhibit ferrous iron oxidation, but in some way reduced the ferric iron formed back to the ferrous state. Thus, ferrous iron was being oxidized at a significant rate in the presence of sediment, but was being reduced back to the ferrous form, so that there was no net loss of ferrous iron.

The interpretation of these results in the paper by Brock et al. (1976) was somewhat complicated because it was not known at the time that *Sulfolobus* would itself *reduce* ferric iron using elemental sulfur as an electron donor (see next section). Since the sediment had large amounts of elemen-

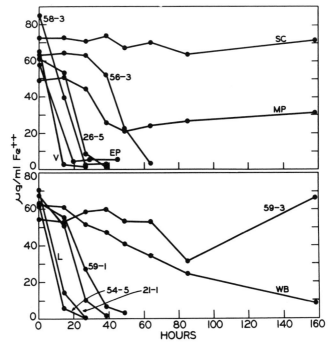

Figure 6.16. Bacterial oxidation of ferrous iron in waters from various Yellowstone hot springs. Temperature of incubation 80°C, except 56-3 and Evening Primrose, which were 70°C. Controls which had 10% NaCl added to inhibit bacterial activity showed no oxidation in any case. Ferrous sulfate was added to all waters except Moose Pool to give the 0 time values reported. (From Brock et al., 1976.)

tal sulfur, it seems likely that the bacteria were carrying on two processes simultaneously (not necessarily the same bacterial cells, but the same species), ferrous iron oxidation using O_2 as electron acceptor and elemental sulfur oxidation using ferric iron as electron acceptor. Although this does not explain all of the findings in the Brock et al. (1976) paper, it explains most of them. The use of ferric iron as an electron acceptor in an aerobic system such as this may seem surprising, but it should be pointed out that at the incubation temperatures involved, there is really very little O_2 in solution.

Ferric Iron Reduction

It was shown by Brock and Gustafson (1976) that ferric iron was not reduced chemically by elemental sulfur at low pH. Brock and Gustafson (1976) then showed that *Thiobacillus* and *Sulfolobus* could catalyze this reduction, using ferric iron as an electron acceptor. In the case of *Sulfolobus,* ferric iron reduction was shown using both elemental sulfur and glutamic acid as electron donors. Because of the more rapid growth on glutamic acid, it was easier to demonstrate the reduction process. As seen in Figure 6.17, growth and ferric iron reduction occurred during the 3-day incubation period under aerobic (really microaerophilic) conditions. Experiments were also set up to see whether *Sulfolobus* could grow using ferric

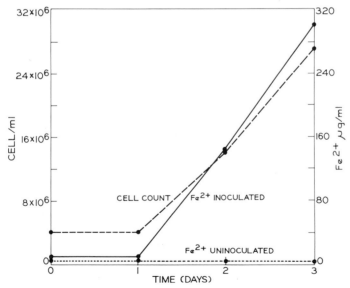

Figure 6.17. Ferric iron reduction and cell growth of *Sulfolobus* strain 98-3 in glutamate-containing medium. Aerobic incubation, 70°C. (From Brock and Gustafson, 1976.)

iron as an electron acceptor anaerobically with glutamate as energy source, but these experiments were thwarted when it was discovered that under anaerobic conditions, glutamate (and several other organic compounds used by *Sulfolobus* as energy sources) reduced ferric iron nonbiologically. Because of this it is conceivable that the reduction of ferric iron seen heterotrophically under aerobic conditions could be nonspecific, resulting from the utilization of dissolved O_2 by the bacteria, with the subsequent chemical reduction of ferric iron by residual glutamate. However, this could not explain the results with elemental sulfur, since elemental sulfur and ferric iron do not react under anaerobic conditions in the absence of bacteria. Although no details were presented, Brierley (1974a) reported that *Sulfolobus* could reduce hexavalent molybdenum $(MoO_4)^{2-}$ to the pentavalent form (molybdenum blue; Mo_2O_5) in the presence of elemental sulfur and inferred that the sulfur was oxidized. This is analogous to the ferric iron reduction reported by Brock and Gustafson (1976).

Heterotrophic Nutrition

All strains of *Sulfolobus* that we have tested are able to grow heterotrophically as well as autotrophically, although the strain studied by Brierley and Brierley (1973) grows only autotrophically. As seen in Table 6.11, our strains (from Yellowstone, El Salvador and Italy) grew on a variety of

Table 6.11. Heterotrophic Nutrition of *Sulfolobus* Strains[a]

Nutrient	Strain				
	98-3	106-3	115-2	129-1	140-5
Yeast extract	+	+	+	+	+
Tryptone	+	+	+	+	+
Peptone	+	+	+	+	+
Casamino acids	+	+	+	+	+
Casein hydrolysate	+	+	+	+	+
Glutamate	+	+	+	±	+
Glutamine	+	±	+	±	+
Alanine	+	+	+	±	+
Asparatate	+	+	+	−	+
Fructose	−	−	−	−	−
Glucose	−	−	−	+	−
Galactose	−	−	−	+	−
Sucrose	+	+	−	+	+
Lactose	−	−	−	+	−
Ribose	+	+	+	+	+

[a]All tested at 0.1% concentrations in basal salts at pH 2, incubation temperature 70°C. + good growth; ± fair growth; − no growth. No growth was obtained with any strain with the amino acids phenylalanine, histidine, proline, or leucine. (From Brock et al., 1972.)

complex organic substrates, as well as on amino acids and certain sugars. The MT isolates of deRosa et al. (1974) are said to grow on sugars but not on amino acids.

Although all of our isolates would grow on amino acids, the variety of amino acids used was fairly restricted. All strains used alanine and glutamine and most strains used glutamate and aspartate. None of the strains tested grew on phenylalanine, histidine, proline, or leucine, Of the sugars tested, fructose was not used at all and glucose, galactose, and lactose were used only by one strain. Sucrose, on the other hand, was used by most strains, and ribose by all. In a later study, other pentose sugars were tested and it was found that growth also occurred on xylose and arabinose. The ability to use sucrose but not glucose and fructose may seem surprising, but this is not unprecedented, as certain pseudomonads show the same phenomenon (Doudoroff et al., 1949). (Our sucrose solution was sterilized separately from the acid medium; it apparently does not acid hydrolyze during the 2–3 days of incubation.)

The ability to use pentoses in preference to hexoses may seem surprising until it is recalled that the springs in which these organisms live are frequently surrounded by wooded land, and logs and bits of tree material often fall in the springs. The pentosans present in wood hydrolyze readily under the hot acid conditions in the springs, liberating free pentose sugars. If such sugars are common in its environment, it may not be too surprising that *Sulfolobus* has evolved the ability to utilize them.

It may be that *Sulfolobus* grows heterotrophically in certain of the springs that are low in ferrous iron and reduced sulfur compounds, such as White Bubbler (see Table 6.3). No simple way to test this point comes to mind.

Leaching and Oxidation of Sulfide Minerals

The solubilization of minerals from low-grade sulfide ores by percolation of acid through a large pile of ore is called leaching. In most cases, this is a microbial process, carried out by the organisms *Thiobacillus ferrooxidans* and *Thiobacillus thiooxidans*. Although this is a widely used process in the mining industry, it is very poorly understood ecologically. The little knowledge available has been summarized by Beck (1967). Because the oxidation of sulfide minerals is an exothermic reaction, heat builds up in the piles, and Beck (1967) reported temperatures as high as 80°C. I attempted at one time to obtain samples of material from the depths of copper leaching piles, to look for the presence of *Sulfolobus,* but was unable to obtain any significant cooperation from copper companies. There is no problem obtaining samples of the leach liquid draining out of the bottom of the piles, but what is needed are samples from within the piles themselves, where the temperature is high, and such samples could only be obtained by drilling. It seems to me that it would be of considerable importance to determine whether *Sulfolobus* was present in these piles, since this might lead to some knowledge about how to increase the efficiency of the process.

So far, what has been done is to determine that *Sulfolobus* will promote the leaching of low-grade ores in flasks or in percolation columns in the laboratory (Brierley and Murr, 1973; Brierley, 1974a,b). Brierley showed that *Sulfolobus* was able to strongly promote the solubilization of molybdenum from molybdenum sulfide (molybdenite), and scanning electron microscopy showed that the organism had attached to and was growing on the mineral crystals. According to her, *Sulfolobus* is unique in being tolerant to high concentrations of molybdenum; it will grow at a concentration of 750 ppm molybdenum whereas *Thiobacillus ferrooxidans* has a maximum tolerance of 5–90 ppm.

Sulfolobus could also be used to leach low-grade copper ores (Brierley, 1974b), although the results were not as dramatic as with molybdenite. Respiration and growth were inhibited by concentrations of Cu^{2+} of over 1000 ppm, and this copper tolerance is similar to that of the thiobacilli. Since the feasibility of leaching sulfide ores with *Sulfolobus* is now established, further work on the temperatures in the leach dumps would be desirable, in order to locate habitats suitable for *Sulfolobus*. Since *Sulfolobus* does apparently disperse throughout the world (see section on Biogeography), it could conceivably colonize thermal sites in leach dumps spontaneously, but it would appear likely that the rate of development could be greatly increased by inoculation.

As noted in the previous section, *Sulfolobus* effectively reduces ferric iron to the ferrous form. This reduction process may itself be significant in leaching. Sato (1960) has shown that the oxidation of sulfide minerals is a two-step process. There is a selective movement of metal atoms from the crystal into the surrounding solution, thus enriching the remaining solid phase with sulfur atoms. When solid sulfur is left over, it becomes oxidized to sulfate. In the case of sulfides of copper, lead, silver, and zinc, Sato concluded that the solid sulfur remaining would be at the oxidation state of elemental sulfur, S^0. Chemically, such solid sulfur reacts slowly with ferric iron, but the rate is greatly speeded up by bacteria such as *Sulfolobus* (Brock and Gustafson, 1976). Thus, the utilization by *Sulfolobus* of ferric iron as an electron acceptor could be expected to speed up the oxidation of solid sulfur on the mineral, and thus speed up the rate of oxidation of the mineral as a whole. Since in most natural situations, sulfides of copper, etc., are accompanied by sulfides of iron (pyrite, marcasite), the iron of which becomes oxidized to the ferric state, it can be assumed that in any leaching dump there will be ferric iron present to act as an electron acceptor. Through this series of steps, *Sulfolobus* may thus markedly increase the solubilization of the sulfide mineral. It seems likely that O_2 is limiting in most leach dumps, since the piles are quite large and poorly aerated, and since the acid leach solution entering the top of the pile is high in ferric iron but the effluent contains almost exclusively ferrous iron. It would be of considerable importance to measure O_2 concentrations within leach piles, and to measure the rate of bacterial oxidation of sulfide minerals anaerobically in the presence of ferric iron. If the proposed mechanism is correct, it suggests that more rapid or effective leaching with ferric iron would be obtained if care were taken to develop and maintain a large active population of bacteria within the leach dump and to regulate the concentration of ferric iron.

Growth Rates of *Sulfolobus* in Nature

The steady-state nature of the hot springs in which *Sulfolobus* lives has made it possible to develop procedures for measuring *in situ* growth rate of the organism (Mosser et al., 1974a). The details of procedures for determining the turnover rates of the springs will be described in Chapter 12. Essentially, what was done was to enrich the springs with sodium chloride and measure the rate at which the chloride ion was diluted. In many of the springs the dilution rates were surprisingly rapid (Mosser et al., 1974a). Since none of the springs studied had any surface inflow or outflow, it must be concluded that water enters and leaves by underground seepage. An example of the kind of results obtained in a small spring is shown in Figure 6.18. It can be seen that there was a rapid disappearance of chloride with time, whereas sulfate, temperature, pH, and concentration of *Sulfolobus* remained constant. Since control experiments had shown that chloride (a very water soluble ion which does not adsorb to particles) moved with the water, it could be hypothesized that the water turnover rate was quite rapid. Sulfate concentration remained constant because sulfate was entering the spring as rapidly as it was leaving. But why did *Sulfolobus* numbers remain constant? Two possibilities existed: (1) cells were not entering or leaving the spring with the water seepage, but were lysing, and the lysed cells were replaced by growth; and (2) cells were moving out of the spring at the same rate as the water but the loss was being balanced by growth. If the latter were the case, then the growth rate of the organism could be calculated from the chloride dilution rate.

To determine whether the cells were being diluted at the same rate as the water, we used one of Ben Bohlool's serologically distinct strains. Strain 98-3 was chosen because there were several springs in which this serotype was absent from the springs (see Serology section above). The procedure was to add large numbers of formaldehyde-fixed cells of 98-3 to the water, and then measure concentration with time, using a membrane filter technique to concentrate the cells for fluorescent antibody staining. In preliminary experiments, the stability of formaldehyde-killed cells was tested by incubating killed cells at pH 2 at 80°C for 3 days. Fluorescence counts indicated no loss of stainable cells during this 3-day period. Since the half-

Figure 6.18. Change in concentration of chloride with time after addition to White Bubbler Spring. The estimated half-time for chloride dilution is shown in the parentheses. Data are also presented showing that sulfate, temperature, pH, and concentration of *Sulfolobus* cells do not change with time. (From Mosser et al., 1974a.)

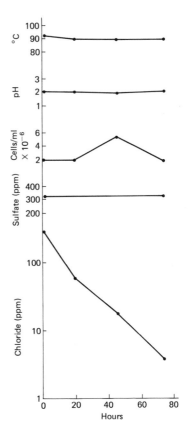

dilution rates in the springs are considerably faster than 3 days, instability of the killed cells should not be a limitation on the use of this technique.

Figure 6.19 shows the results of an experiment in which serologically distinct cells were added to springs and also shows the chloride dilution rates for the same springs. As seen, the loss rates for the formaldehyde-fixed cells agreed reasonably well with those for chloride dilution. Thus, we can conclude that *Sulfolobus* cells seep out of the spring at the same rate as the water. Since the walls of these springs are composed of a quartz-rich gravel-like material, which is very porous, it is not surprising that the bacteria move out rapidly.

There was still some concern that *Sulfolobus* cells might move *into* the spring via seepage from uphill, and complicate interpretation of the results. To test this, we completely drained three springs by siphoning the water downhill, and then assayed the first water that entered the springs. The drainage was complete in each spring within less than 0.5 hour, and fresh water began to fill the pools almost immediately. When the springs were drained, the bottoms were washed out thoroughly with *Sulfolobus*-free water to eliminate the bacterial cells from the sediment as much as possible.

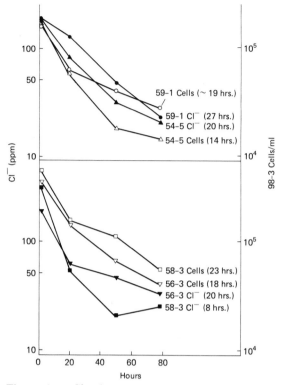

Figure 6.19. Simultaneous chloride dilution and loss of serologically distinct for-maldehyde-killed cells added to four different springs. Half-times are based on least-squares analyses and are given in the parentheses. Correlation between the loga-rithm and time was −0.92 to −0.98 in all cases except for cell dilution at site 54-5, which was −0.76. The chloride concentration at site 58-3 had reached the natural background level by the final sampling time, and the dilution rate was calculated with the data from the first three sampling times. (From Mosser et al., 1974a.)

At the time the first water reappeared in the bottom of the pools, a few minutes after draining, samples were taken, and further samples were taken at intervals over the next few days. Two of the springs had completely recovered their volumes within a day, and the larger springs recovered in several days.

The first water which entered the springs after drainage was quite low in bacterial numbers, virtually at the limits of detection (an acridine orange fluorescent microscopic method was used). This shows that the steady-state concentration of cells must be maintained by growth within the water in the spring pool rather than by inflow of bacteria with the source water. Once the springs were drained and refilled, inoculation would inevitably occur. As seen in Figure 6.20, there was a rapid increase in cell numbers, with exponential growth occurring. The doubling times during this period of exponential growth in *Sulfolobus*-free water were much more rapid than growth rates in the steady state, as would be predicted. The reason for the slower growth rate in the steady state is not known but could be due to limiting levels of an essential nutrient or to the presence of accumulated toxic metabolic products. However, the exponential growth rates in the springs that had been drained and allowed to refill were considerably greater than the growth rates obtained on elemental sulfur in the laboratory, suggesting that optimal conditions have not been used for the laboratory cultures, or that sulfur is not the energy source for growth in nature.

A summary of the steady-state growth rates is given in Table 6.3. It can be seen that there are essentially two classes of growth rates, those in the springs of small volume, which are fairly rapid, and those in the large-volume springs, which are much slower. In the large-volume springs, Moose Pool and Sulfur Caldron, the growth rates calculated as necessary to maintain the steady-state concentrations are almost 2 orders of magnitude less than those at the other sites. However, it may be that the bacteria in these large springs are actually dividing faster than required by the steady-state dilution rates. Rates of sulfur oxidation by samples from these springs

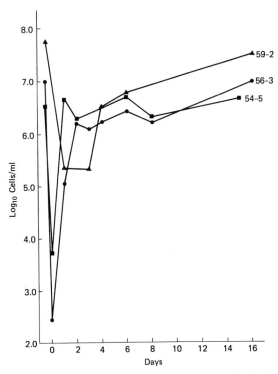

Figure 6.20. Rate of increase of cell numbers in three springs which were drained, cleaned, and allowed to refill. Spring 56-3 did not contain *Sulfolobus* but contained a high-temperature *Thiobacillus*. Data for this spring are given for comparison. (From Mosser et al., 1974a.)

are the highest of all the *Sulfolobus*-containing springs examined, and cultures derived from these springs exhibit the same growth rates as isolates from other springs. Conceivably, a cyclic process of cell division, death, and dissolution could permit growth rates faster than those required to maintain the steady state. Whether such a cyclic process could permit growth rates as high as those in the smaller springs is not known.

Ecology of *Sulfolobus* in Hot Acid Soils

I discuss the solfatara environment in some detail in Chapter 12. The dominant feature of solfataras are not hot springs, which are really rather restricted in size and number, but hot soils. Solfatara areas generally occupy hillsides, plateaus, small ravines, and shallow hollows. In much of a solfatara area springs are absent, and the area is characterized by acidic soils of various temperatures, the soils being heated by steam rising to the surface. Solfatara soils abound with crystals, veins, and lumps of elemental sulfur. In these sulfur-rich soils, there is usually sufficient moisture for the development of bacteria, although cell densities are never as high as they are in hot springs. Even with careful search under the microscope, it is almost never possible to see bacteria, either attached to or separated from the sulfur crystals. Because of this, different techniques were necessary to study the distribution and activity of these organisms (Fliermans and Brock, 1972). The most important technique involved measurement of $^{14}CO_2$ fixation (Smith et al., 1972). In these acid, organic-poor soils, virtually all $^{14}CO_2$ is due to the activity of sulfur-oxidizing autotrophs, so that the distribution of $^{14}CO_2$-fixing ability correlates well with the distribution of bacteria. Preliminary work showed that the temperature optimum for $^{14}CO_2$ fixation was near the temperature of the soil. To actually quantify *Thiobacillus* and *Sulfolobus,* Fliermans set up most-probable-number (MPN) dilutions, using an autotrophic medium with elemental sulfur at pH 3. At the end of the incubation period (14 days), the tubes were examined microscopically for the presence of organisms and the MPN calculated. At the higher dilutions, only one type is usually present, either *Thiobacillus* or *Sulfolobus,* and because these two organisms differ greatly morphologically, it is a simple matter to quantify the two organisms in a soil sample.

For his field work, Fliermans selected sites on solfataras in which thermal gradients were present. Hot soils can easily be located in the winter by the absence of snow (Figure 6.21), but can generally be located even in summer by the presence of steam. Observations of steaming ground is easiest early in the morning, since then (in Yellowstone) the air temperatures are always low and steam condenses readily (see Figure 2.6). The terminus of a thermal gradient is also easily revealed by the presence of grasses, small trees, and other higher vegetation. One of the best examples of a thermal gradient is that shown in Figure 6.22, at Roaring Mountain.

Figure 6.21. Area of hot ground in Yellowstone Park, as revealed by absence of snow. The surrounding normal ground had about 2 meters of snow on the level at the time this photograph was taken.

The measurements of temperature, pH, and other parameters at this thermal gradient are given in Table 6.12. At every station along the transect, the temperature is lower on the surface, and decreases considerably with depth. At station 88, which is the hottest, the temperature at 20 cm is close to the boiling point. Fliermans intentionally avoided establishing stations at sites where there were active fumaroles, but in such fumaroles the temperature even at the surface is boiling. (Temperatures above boiling are rare or absent in Yellowstone furnaroles, although quite common in Italian, Hawaiian, and Icelandic fumaroles.)

It can also be seen in Table 6.12 that at some stations the concentrations of elemental sulfur are enormous. Indeed, it was for just such high-sulfur soils that Fliermans developed his very simple spectrophotometric assay for elemental sulfur (Fliermans and Brock, 1973). At each site and at each depth, Fliermans collected soil for measurements of $^{14}CO_2$ fixation and for most-probable-number determination of sulfur-oxidizing bacteria. The relationship between bacterial numbers and temperature along two transects, one at Roaring Mountain and the other at Amphitheater Springs, is shown in Figures 6.23 and 6.24. It can be seen that there is a clear separation between *Thiobacillus* and *Sulfolobus,* with *Thiobacillus* present at low temperatures and *Sulfolobus* at high temperatures. Only in the temperature range of about 55°C do the two organisms overlap.

The relationship between $^{14}CO_2$ fixation and numbers of sulfur-oxidizing

Table 6.12. Characteristics of the Solfatara Habitat at Roaring Mountain

Station	Distance from station 87 (m)	Depth[a] (cm)	Temperature 4 Aug. 1971	Moisture (%)	Soil pH	SO_4^{2-} (ppm)	S^0 (ppm)
87	0	0	28.0	14.3	2.7	260	
		5	32.0	14.4	0.7	4,100	
		10	44.9	15.4	1.6	3,000	
		15	51.1	15.1	1.3	2,590	1,000
		20	57.0				
88	5	0	35.0	12.2	1.2	730	5,000
		5	42.0	13.2	1.0	950	156,000
		10	67.0	16.7	1.3	470	110,000
		15	81.0	12.4	1.1	245	24,000
		20	89.9				
89	10	0	36.0	13.9	2.6	50	
		5	47.0	15.2	1.5	480	2,200
		10	59.0	22.3	1.5	400	152,000
		15	79.0	25.9	1.2	690	34,000
		20	83.9				
90	15	0	33.0	17.3	3.1	50	4,000
		5	44.0	23.0	3.1	35	
		10	56.0	22.0	3.1	23	1,000
		15	71.0	16.6	2.9	50	2,000
		20	75.0				
91	16	0*	26.0	17.6	4.2	<10	
		5	36.0	27.0	3.7		
		10	44.0	21.7			
		15	51.0	23.4			

92	21	20	60.0				2,000
		0	30.0	20.4	3.8	20	2,000
		5	34.0	15.2	4.5	30	
		10	38.0	20.3	4.0	40	
		15	42.0	23.8	4.8	20	
		20	49.0				
93	26	0	34.0	14.3	3.5	<10	5,000
		5	42.0	14.1	2.6	20	
		10	50.0	20.0	3.7	40	
		15	60.0	20.9	3.7	30	
		20	65.0				
94	27	0	34.0	20.9	4.7	<10	1,000
		5	35.2	18.9	4.0		
		10	46.0	14.5	4.9		
		15	50.0	25.6	5.0		
		20	58.0				
95	32	0	30.0	13.2	4.0	10	1,000
		5	40.0	19.2	4.5	20	
		10	45.0	16.4	3.7	28	
		15	46.0	18.8	3.3	20	
		20	52.6				
96	37	0	30.0	17.1	4.7	20	2,000
		5	34.0	15.8	4.2	20	
		10	37.0	14.7	4.6	25	
		15	40.0	25.7	4.1	20	
		20	48.0				
97	42	0	29.0	18.6	3.8	20	1,000
		5	35.0	19.1	3.6	20	1,000
		10	38.5	18.8	3.0	18	<200
		15	42.0	19.7	4.9	20	5,000

(See next page for continuation)

Table 6.12. Characteristics of the Solfatara Habitat at Roaring Mountain (*continued*)

Station	Distance from station 87 (m)	Depth[a] (cm)	Temperature 4 Aug. 1971	Moisture (%)	Soil pH	SO_4^{2-} (ppm)	S^0 (ppm)
98	47	20	45.0				
		0	30.0	19.8	4.1	10	2,000
		5	34.0	15.2	4.4	10	2,000
		10	35.5	18.5	4.1	10	<200
		15	35.0	18.2	4.3	15	1,000
		20	41.0				
99	52	0	30.0	14.3	4.3	10	15,000
		5	29.0	19.4	3.5	30	<200
		10	30.5	21.2	3.9	20	6,000
		15	31.0	18.8	4.7	25	<200
		20	32.0				
100	57	0	30.0	36.5	4.8	20	35,000
		5	29.0	18.9	4.8	38	15,000
		10	30.0	14.1	4.4	15	6,000
		15	30.0	16.3	4.4	20	
		20	30.0				
101	62	0	28.0	55.9	4.3	15	
		5	28.3	20.5	4.6	10	
		10	29.0	17.7	5.1	<10	
		15	30.0	16.2	4.1	10	
		20	28.1				

[a]Moisture, pH, SO_4^{2-}, and S^0 determined on samples collected from 5-cm segments of soil cores. Thus, 0 represents soil collected from the interval 0 to 5 cm, 15 represents soil collected from the interval 15 to 20 cm, etc. Station 87 was established near the base of the mountain and the transect extended in a westerly direction (from Fliermans and Brock, 1972). (For further details, see C. B. Fliermans, Ph.D. thesis, Indiana University, 1972.)

Figure 6.22. A transect at Roaring Mountain, showing the progression from hot ground at the base of the mountain (foreground), to normal temperatures where trees are present in the background. Intermediate temperatures are indicated by the presence of mosses and grasses, and where even these plants are absent the temperatures are above 40°C. About 2 meters of snow were on the ground in nonthermal areas.

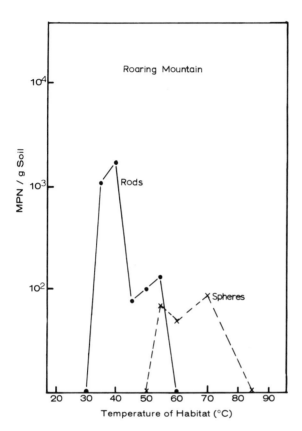

Figure 6.23. Numbers of sulfur-oxidizing bacteria in soils collected at different temperatures along the Roaring Mountain transect, incubations being carried out at the temperature of the habitat. MPN, most-probable-number. (From Fliermans and Brock, 1972.)

bacteria is shown in Figure 6.25. This is a composite graph collecting the results from a wide variety of samples, and it can be seen that if the soil has more than 300 to 350 sulfur-oxidizing bacteria per gram, bacterial count correlates well with $^{14}CO_2$ uptake. Also, there seems to be a lower limit of about 200 to 250 cpm/g soil, below which there is no correlation between numbers of sulfur oxidizers and rate of CO_2 fixation. At this low level, it is possible that CO_2 fixation is due to heterotrophs or to nonspecific absorption. Thus, in the analysis of data, a value of 200–250 cpm/g soil was used, above which incorporation was considered to reflect the activity of sulfur-oxidizing bacteria. Figure 6.26 presents a composite graph in which the mean levels of incorporation of $^{14}CO_2$ are plotted against temperature. It can be seen that there are two peaks of activity, one at a temperature around 20–30°C and another around 70°C. The low-temperature peak is probably due to the activity of *Thiobacillus* and the high-temperature peak to *Sulfolobus*. Although a direct connection between CO_2 fixation and sulfur oxidation may not occur, it may be hypothesized that sulfur oxidation occurs predominantly at low temperatures in these solfatara soils.

Fliermans's work emphasizes the considerable spatial and temporal variability of the hot-acid soil environment. However, although the environment is spatially heterogeneous, the conditions at a single location are sufficiently stable so that bacteria with temperature optima similar to the habitat temperature can develop. However, the detailed data (see Ph.D. thesis, C. B. Fliermans, Indiana University, Bloomington) showed variation of temperature with time over a 2-year period of as much as 30°C at some depths, whereas at other locations variations were less than 5°C. At Roaring Mountain, temperatures were also measured once in the winter and were not cooler than in the summer, and in some cases were even hotter. However, at station 88, one of the hotter stations at Roaring Mountain, temperature variation as great as 20°C was found over a single weekly period, and diurnal variation in temperature as great as 10°C was found. Since the specific heat of soil is low, it responds fairly quickly to temperature changes.

The problem of moisture in these hot acid soils is of some interest. Because they are hot, they might be thought to be dry, yet the source of

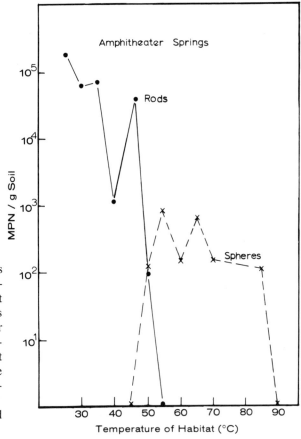

Figure 6.24. Numbers of sulfur-oxidizing bacteria in soils collected at different temperatures along the Amphitheater Springs transect, incubations being carried out at the temperature of the habitat. MPN, most-probable-number.
(From Fliermans and Brock, 1972.)

heat is steam. Fliermans reported that moisture contents are around 15–20% for most of his sites. However, moisture content does not reveal the water availability at a site, since water availability also depends on how tightly the water is bound to soil particles. The most precise way of expressing water availability is by measurement of water potential. To examine the water potentials of some of these sites, Charlene Knaack and I made *in situ* determinations of water potential, using the filter paper method described by Fawcett and Collis-George (1967). In this procedure, a filter paper of known water absorptive capacity is buried in the soil at the desired site, allowed to equilibrate, and the moisture content measured. From a calibration curve relating water potential to water content, the water potential of the site can be calculated. In the case of these hot acid soils, it is not possible, of course, to immerse the paper in direct contact with the soil, because the cellulose will hydrolyze in a week. What we did was to construct plastic chambers with a number of holes through which moisture could move, and suspend the filter papers within these chambers. The filter paper-chamber combinations were then immersed at the desired sites, allowed to equilibrate for a week, removed and placed in tared, moisture-proof containers, returned to the laboratory, and moisture contents determined. In all cases studied, the water potentials were less than 1 bar, which means they were very close to saturation. Thus, it is probably unlikely that in these hot, acid soils the organisms are under any moisture

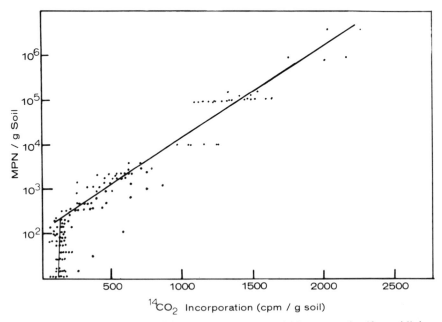

Figure 6.25. Relationship of $^{14}CO_2$ uptake rate and viable count of sulfur-oxidizing bacteria. MPN, most-probable-number. (From Fliermans and Brock, 1972.)

Figure 6.26. Rate of uptake of $^{14}CO_2$ in soils of different temperatures. The number of determinations at each point is indicated in parentheses on the graph and each point is the average of the values obtained. (From Fliermans and Brock, 1972.)

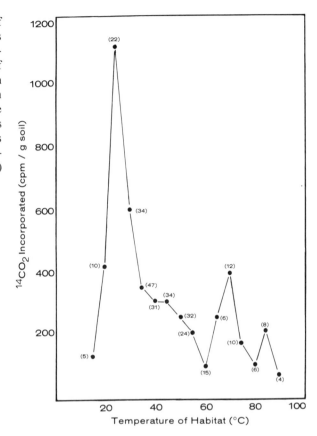

stress. The high water potentials of these soils is not surprising when it is considered that clay minerals, a major source of water-binding capacity, are virtually absent, and the source of heat, steam, is also a source of moisture. In many cases, the filter papers were completely saturated with water, suggesting that water had condensed on them. This probably occurred because of small variations in temperature at the sites, as a result of changes in air temperature. If a saturated atmosphere at a certain temperature is cooled, moisture will condense as the dew point is reached. Thus, neither *Thiobacillus* nor *Sulfolobus* should have any major difficulty finding sufficient moisture to grow in these hot acid soils.

Biogeography and Dispersal

Despite its restricted habitat, *Sulfolobus* is world-wide in its distribution. I have isolated typical strains from Yellowstone Park, Iceland, Italy, and Dominica (West Indies), Bohlool (1975) has isolated it from New Zealand

and Furuya et al. (1977) from Japan. Several of the New Zealand isolates were of the same serotypes as the Yellowstone strains.

Hot, acid environments are even more restricted than hot, neutral environments. Thus, dispersal is even more difficult for *Sulfolobus* than for *Thermus,* since the latter can find a much larger number of natural hot springs, as well as many man-made thermal habitats. If *Sulfolobus* disperses by hot-spring hopping, it must be able to resist drying for long periods of time, and to land with impressive accuracy. We have no real information on the processes involved.

It was of interest to determine whether *Sulfolobus* grew in the self-heating coal refuse piles where *Thermoplasma* was found. Many sites on these coal refuse piles are of proper temperature for *Sulfolobus.* Bob Belly made a number of attempts to isolate *Sulfolobus,* both heterotrophically and autotrophically, but only once was he successful, and we are not certain whether the one isolate obtained might have been a contaminant. If *Sulfolobus* does live in hot coal refuse piles, it is not very common. Because of the short life of these piles, it may not be surprising that *Sulfolobus* is absent, until it is remembered how common *Thermoplasma* is.

Evolution of *Sulfolobus*

Despite the obvious similarities of *Sulfolobus* and *Thermoplasma,* these organisms are much more different than they are alike. *Sulfolobus* is an autotroph, *Thermoplasma* a heterotroph. *Sulfolobus* will grow on a variety of simple carbon compounds, *Thermoplasma* requires a complex medium containing a factor from yeast extract. *Sulfolobus* has a DNA base composition of high GC, *Thermoplasma* has a DNA of low GC (the MT strains of deRosa et al., which are more like *Sulfolobus* than *Thermoplasma,* have DNA base compositions that overlap those of *Thermoplasma,* however). *Sulfolobus* has a cell wall, *Thermoplasma* does not. The only similarity between these two organisms lies in their requirements for hot, acid environments, but even here, there are differences, since *Thermoplasma* is not able to grow at nearly as high a temperature as *Sulfolobus* (although certain temperature strains of *Sulfolobus* have optima similar to those of *Thermoplasma*).

The fact that *Sulfolobus* and *Thermoplasma* have similar lipids is of interest, but almost certainly this can be explained by convergent evolution. This hypothesis is strengthened by the fact that *Halobacterium,* another quite different organism, also has lipids similar to those of the two acidophilic thermophiles.

There is good reason to believe that acid environments did not arise on earth until the atmosphere became oxidizing. Sulfuric acid can only form under oxidizing conditions, since the only significant source of oxidizing power is O_2. It is true that sulfuric acid can be formed anaerobically by the oxidation of sulfide with ferric iron, but ferric iron itself can only be formed if O_2 is present, because of the high oxidation-reduction potential of the Fe^{3+}/Fe^{2+} pair. Thus, it seems unlikely that *Sulfolobus* was an early form of life.

The existence of *Sulfolobus* reveals the remarkable adaptability of living protoplasm. The ability to grow in hot environments is impressive enough, but the ability to grow in hot *and* acid environments is truly amazing. It seems unlikely that such an evolutionary jump occurred very often, perhaps only once, and the fact that similar or identical organisms can be found as far apart as Yellowstone and New Zealand suggests that the available hot, acid habitats on earth have been colonized by a single organism.

The cell wall of *Sulfolobus,* although different from that of other bacteria, probably provides some element of stability to the cell. As Weiss has noted, the wall of *Sulfolobus* seems to be fairly rigid. In view of Searcy's observations on the osmotic relations of *Thermoplasma,* discussed in the previous chapter, it may be that the *Sulfolobus* wall plays some osmotic role, making it possible for the organism to grow in the exceedingly dilute waters in which it lives. This cell wall may also be involved in some way in the ability of *Sulfolobus* to attach to and oxidize elemental sulfur, although Weiss (1973) thinks that pilus-like structures may play that role.

Finally, *Sulfolobus* is the most thermophilic autotroph known, and it would be of considerable interest to study the CO_2 fixation enzymes. Unfortunately, it is fairly difficult to obtain large cell yields of *Sulfolobus* when growing the organism autotrophically, so that enzymatic studies might be difficult. However, a concerted effort could lead to interesting results, especially in light of McFadden's consideration (McFadden, 1973) regarding the evolution of the key enzyme in the Calvin cycle, ribulosediphosphate carboxylase.

Note

The original paper describing the genus *Sulfolobus* was rejected twice by the *Journal of Bacteriology* and was then published in *Archives for Microbiology.* In considering the great interest that this organism has elicited, I went back and reviewed the correspondence with the *Journal of Bacteriology* to determine why the paper was rejected. It may be of some interest to quote some of the reviewer's comments (since they are anonymous, they should not mind). One reviewer thought the organism was interesting but questioned the electron microscopy, and especially the angularity of the organism. He raised the question of fixation artifact and was also concerned about the lack of nucleic fibrillar areas. He also thought he saw pili on one micrograph (which Weiss later provided evidence for), and he was concerned that we did not establish a holotype. Since a large number of genera have been described without *any* electron microscopic evidence, it is not clear why some question about interpretation should be cause for *rejection* of the paper.

The second reviewer for the first submission did not believe that we had provided any evidence that the organism uses elemental sulfur as an energy source. I quote: "If the authors were naming the new genus, *Lobus,* then the manuscript would be acceptable. Since they name it '*Sulfolobus;* a new genus of sulfur oxidizing bacteria . . . ,' I do not believe that sufficient evidence has been presented that the organism does in fact oxidize sulfur. A new genus should be defined on solid rock. I suggest that the manuscript be retained by the authors until they provide numbers for sulfur oxidized and sulfate formed. They make no mention of sulfide oxidation (and this seems a most logical substrate) or of thiosulfate oxidation." He also questioned whether certain tables should be included.

Considering the second review to be the most damning, I went back in the literature to see what had been done when describing other sulfur-oxidizing bacteria. The case of *Thiobacillus thiooxidans* was most pertinent, since this bacterium grew primarily on elemental sulfur. In the original Waksman and Joffe paper (1922, *J. Bacteriol.* **7**, 235–256), all that was done was to state (without data) that the organism grew on elemental sulfur. Although we had transferred the organism continuously in elemental sulfur, and thought that enough, we also had extensive data showing the formation of sulfate during the growth of *Sulfolobus* on elemental sulfur. Thus, I prepared another table giving the data (we had already stated them in text form) and resubmitted the paper. My covering letter said: "Concerning the main point at issue, evidence for sulfur oxidation, the first sentence at the top of page 17 states that sulfate was produced. To provide the solid evidence requested by the reviewer, we have added the data in Table 7A. However, the request for data on sulfide and thiosulfate oxidation shows a lack of understanding on the part of your reviewer concerning the stability of sulfur compounds at high temperature and low pH. Neither sulfide nor thiosulfate would be stable and hence cannot be studied."

The resubmitted paper went to the two original reviewers plus *three* new reviewers. The old reviewer No. 2, who was concerned about sulfur oxidation, had the following new comments: "The addition of Table 7A does provide evidence that sulfur has been oxidized to sulfate; out of about 10,000 μg of sulfur added per ml 65 μg of sulfate was produced in the best cultures in 6 days. Sulfur appears to be a poor source of energy, and it is apparent why the authors were not eager to display these data." (I wonder whether this reviewer has grown organisms on elemental sulfur. They grow *slowly*. Elemental sulfur is not a poor energy source, it is just not easily available.) "The rebuttal statement by the authors that sulfide is unstable at high temperature and low pH and hence cannot be studied is completely unacceptable. H_2S is the main source of sulfur in the solfataras (as stated on page 3); it is the most available form of sulfur in terms of solubility, and it is continually supplied geochemically from much hotter areas. Experimentally it *can* be supplied continuously to shaken cultures at high temperature and low pH in amounts comparable to that found in natural solfataras." My response to this was as follows: "If H_2S autooxidizes quickly to sulfur, then whether it is added continuously or discontinuously makes no difference, one will still be dealing with a mixture of H_2S and S^0, and which is being oxidized cannot be determined. However, this is really a minor point. Even with *Thiobacillus* it has not been shown that H_2S functions as an energy source. Why should we be required to show this for a new and barely discovered organism?'' The new reviewer No. 1 was concerned that *Sulfolobus* grew better on sucrose than on glucose and fructose (not being aware of *Pseudomonas saccharo-philia*, etc.), and was not convinced that truly autotrophic growth was occurring. He was concerned about the small amounts of elemental sulfur oxidized to sulfate in our Table 7A, plus the fact that there were large amounts of sulfur in the medium and only small numbers of bacteria. The new reviewer No. 2 also did not believe that we had proved sulfur oxidation. Among other things, he felt we should demonstrate the presence of CO_2 fixation enzymes (this for the first description of a new genus?). New reviewer No. 3 wanted to see generation times of autotrophic and heterotrophically grown cells, and was concerned about a statement we had made about the possible significance of the lobes in cell division. He also wanted to see phase and Nomarski photomicrographs of the organism.

At this stage, I had no energy to try to bull this paper through the *Journal of Bacteriology* and sent it to *Archives for Microbiology*, which has an excellent reputation for publishing papers on topics in general bacteriology. It was accepted with minor corrections, but it must be admitted that I modified the paper before sending it to *Archives*. I deleted several tables, added a phase photomicrograph, and made the point strongly that all strains grow on elemental sulfur as sole energy source. Later, when Douglas Shivvers joined my laboratory, he spent a year studying elemental sulfur oxidation with ^{35}S, and put the autotrophic nutrition on a stronger footing. It is almost certain that the doubts of the *Journal of Bacteriology* reviewers stimulated me to ask him to work on this subject, and my pride was sufficient that I wanted his paper on this subject be published in the *Journal of Bacteriology* (Shivvers and Brock, 1973). Considerably later, after we had learned to stabilize H_2S under acid conditions by eliminating all of the ferric iron, we did study briefly the oxidation of H_2S by *Sulfolobus*, and showed that this was indeed a process that it carried out. However, these were short-term experiments, and not the continuous-bubbling growth experiments one of the reviewers wanted. (I defy him to do what he wanted us to do, and make any sense out of it!)

Why am I going over this ancient history? First, it did affect our subsequent work on *Sulfolobus*, and made it considerably stronger. Second, it is a good example of why there should be more than one journal in any field. Third, it makes one begin to question whether one should ever publish on a new genus, because there will always be more things that could be done. How nice it would have been to have worked in 1880, when every organism cultured was a new genus, and little was needed to characterize it but a morphological description. Every period of development in bacteriology brings new requirements, without eliminating the old ones. Now we cannot publish on a new genus, or even a new species, without a DNA base composition. Electron micrographs are *de rigeur*, although most of the genera in *Bergey's Manual* were described in the days before the electron microscope existed. And one of our reviewers even wanted us to prove the existence of the CO_2 fixation enzymes! In the days of old, a new genus stood up and shouted. Our acidophilic, thermophilic organisms come as close to doing this today as any organisms will. It should be clear from the discussion of the past two chapters that our work hardly ever proceeded in a direct line from habitat to organism. There were lots of stumbles and misdirections. I have no apologies for this: after all, we had no precedent on which to base our work. I am sure there are more new organisms out there. Where? Mars? No, here at home. Look for an extreme environment and study it.

References

Beck, J. V. 1967. The role of bacteria in copper mining operations. *Biotechn. Bioeng.* **9**, 487–497.

Bohlool, B. B. 1975. Occurrence of *Sulfolobus acidocaldarius,* an extremely thermophilic acidophilic bacterium, in New Zealand hot springs. Isolation and immunofluorescence characterization. *Arch. Microbiol.* **106**, 171–174.

Bohlool, B. B. and T. D. Brock. 1974. Population ecology of *Sulfolobus acidocaldarius.* II. Immunoecological studies. *Arch. Microbiol.* **97**, 181–194.

Bott, T. L. and T. D. Brock. 1969. Bacterial growth rates above 90°C in Yellowstone hot springs. *Science* **164**, 1411–1412.

Brierley, C. L. 1974a. Molybdenite-leaching: use of a high-temperature microbe. *J. Less-Common Metals* **36**, 237–247.

Brierley, C. L. 1974b. Extraction of copper from sulfide ores using a thermophilic microorganism. Final Technical Report to U.S. Bureau of Mines, Socorro, New Mexico.

Brierley, C. L. and J. A. Brierley. 1973. A chemoautotrophic and thermophilic microorganism isolated from an acid hot spring. *Can. J. Microbiol.* **19**, 183–188.

Brierley, C. L. and L. E. Murr. 1973. Leaching: use of a thermophilic and chemoautotrophic microbe. *Science* **179**, 488–489.

Brierley, J. 1966. Contribution of chemoautotrophic bacteria to the acid thermal waters of the Geyser Springs group in Yellowstone National Park. Ph.D. thesis, Montana State University, Bozeman. (The material on the *Sulfolobus*-like organism is on pp. 58–60.)

Brock, T. D. 1967. Life at high temperatures. *Science* **158**, 1012–1019.

Brock. T. D. 1974. *Sulfolobus.* In *Bergey's Manual of Determinative Bacteriology,* 8th ed., R. E. Buchanan and N. E. Gibbons, eds. Williams and Wilkins, Baltimore, pp. 461–462.

Brock, T. D., K. M. Brock, R. T. Belly, and R. L. Weiss. 1972. *Sulfolobus:* a new genus of sulfur-oxidizing bacteria living at low pH and high temperature. *Arch. Mikrobiol.* **84**, 54–68.

Brock, T. D., S. Cook, S. Petersen, and J. L. Mosser. 1976. Biogeochemistry and bacteriology of ferrous iron oxidation in geothermal habitats. *Geochim. Cosmochim. Acta* **40**, 493–500.

Brock, T. D. and G. K. Darland. 1970. Limits of microbial existence: temperature and pH. *Science* **169**, 1316–1318.

Brock, T. D. and J. Gustafson. 1976. Ferric iron reduction by sulfur- and iron-oxidizing bacteria. *Appl. Environ. Microbiol.* **32**, 567–571.

Brock, T. D. and J. L. Mosser. 1975. Rate of sulfuric-acid production in Yellowstone National Park. *Geol. Soc. Am. Bull.* **86**, 194–198.

Brock, T. D. and K. O'Dea. 1977. Amorphous ferrous sulfide as a reducing agent for the culturing of anaerobes. *Appl. Environ. Microbiol.* **33**, 254–256.

Cho, K. Y., C. H. Doy, and E. H. Mercer. 1967. Ultrastructure of the obligate halophilic bacterium *Halobacterium halobium.* *J. Bacteriol.* **94**, 196–201.

deRosa, M., A. Gambacorta, G. Millonig, and J. D. Bu'Lock. 1974. Convergent characters of extremely thermophilic acidophilic bacteria. *Experientia* **30**, 866.

Doudoroff, M., J. M. Wiame, and H. Wolochow. 1949. Phosphorolysis of sucrose by *Pseudomonas putrefaciens.* *J. Bacteriol.* **57**, 423–427.

Fawcett, R. G. and N. Collis-George. 1967. A filter-paper method for determining the moisture characteristic of soil. *Aust. J. Exp. Agr. Anim. Husb.* **7**, 162–167.

Fliermans, C. B. and T. D. Brock. 1972. Ecology of sulfur-oxidizing bacteria in hot acid soils. *J. Bacteriol.* **111**, 343–350.

Fliermans, C. B. and T. D. Brock. 1973. Assay of elemental sulfur in soil. *Soil Sci.* **115**, 120–122.

Furuya, T., T. Nagumo, T. Itoh, and H. Kaneko. 1977. A thermophilic acidophilic bacterium from hot springs. *Agric. Biol. Chem.* **41**, 1607–1612.

Langworthy, T. A., W. R. Mayberry, and P. F. Smith. 1974. Long chain glycerol diether and polyol dialkyl glycerol triether lipids of *Sulfolobus acidocaldarius*. *J. Bacteriol.* **119**, 106–116.

McFadden, B. A. 1973. Autotrophic CO_2 assimilation and the evolution of ribulose diphosphate carboxylase. *Bacteriol. Rev.* **37**, 289–319.

Mosser, J. L., B. B. Bohlool, and T. D. Brock. 1974a. Growth rates of *Sulfolobus acidocaldarius* in nature. *J. Bacteriol.* **118**, 1075–1081.

Mosser, J. L., A. G. Mosser, and T. D. Brock. 1973. Bacterial origin of sulfuric acid in geothermal habitats. *Science* **179**, 1323–1324.

Mosser, J. L., A. G. Mosser, and T. D. Brock. 1974b. Population ecology of *Sulfolobus acidocaldarius*. I. Temperature strains. *Arch. Microbiol.* **97**, 169–179.

Pachmayr, F. 1960. Vorkommen und Bestimmung von Schwefelverbindungen im Mineralwasser. Doctoral disseration, Ludwig-Maximillians Universität, Munich, Germany.

Peary, J. A. and R. W. Castenholz. 1964. Temperature strains of a thermophilic blue-green alga. *Nature* **202**, 720–721.

Sato, M. 1960. Oxidation of sulfide ore bodies. II. Oxidation mechanisms of sulfide minerals at 25°C. *Econ. Geol.* **55**, 1202–1231.

Shivvers, D. W. and T. D. Brock. 1973. Oxidation of elemental sulfur by *Sulfolobus acidocaldarius*. *J. Bacteriol.* **114**, 706–710.

Smith, D. W., C. G. Fliermans, and T. D. Brock. 1972. Technique for measuring $^{14}CO_2$ uptake by soil microorganisms *in situ*. *Appl. Microbiol.* **23**, 595–600.

Steensland, H. and H. Larsen. 1969. A study of the cell envelope of the Halobacteria. *J. Gen. Microbiol.* **55**, 325–336.

Weiss, R. L. 1973. Attachment of bacteria to sulphur in extreme environments. *J. Gen. Microbiol.* **77**, 501–507.

Weiss, R. L. 1974. Subunit cell wall of *Sulfolobus acidocaldarius*. *J. Bacteriol.* **118**, 275–284.

Zinder, S. H. and T. D. Brock. 1977. Sulfur dioxide in geothermal waters and gases. *Geochim. Cosmochim. Acta* **41**, 73–79.

Chapter 7
The Genus *Chloroflexus*[1]

Until the discovery of *Chloroflexus,* no photosynthetic bacteria were known that were filamentous or that possessed gliding motility. Since these attributes are found in the blue-green algae, considerable interest was aroused in the discovery of an organism that possessed bacteriochlorophylls and carried out a typical bacterial (rather than plant) type photosynthesis. The credit for the discovery of *Chloroflexus* goes to Beverly Pierson, a student of Richard Castenholz (Pierson, 1973), who deduced the likelihood that this organism was a photosynthetic bacterium and then proceeded to isolate it in pure culture and study some of its characteristics. At about the same time, John Bauld in my laboratory was working on this organism without knowing that it was photosynthetic, and when Pierson's results became known, he was able to quickly confirm this finding. Later in my laboratory, Mike Madigan carried out detailed nutritional and physiological studies, and was able to place on a firmer footing the taxonomic status of this organism (Trüper, 1976).

Among other things, the discovery of *Chloroflexus* placed a new light on the interpretation of filamentous microfossils found in Precambrian rocks, which had previously been assumed to be the remains of blue-green algae. Since it is thought that the rise in atmospheric O_2 in the late Precambrian was due to blue-green algae, it is important to be able to specify when blue-green algae first evolved. *Chloroflexus* clearly tells us that the presence of filamentous microfossils, or laminated rocks (see Chapter 11), is not in itself sufficient evidence for the existence of blue-green algae.

---

[1]In some early work on this organism, the spelling *Chloroflexis* was used (Castenholz, 1973; Bauld and Brock, 1973, 1974). Castenholz later changed the spelling to that now used.

As will be discussed in more detail later in this chapter, *Chloroflexus* is related to the green sulfur bacteria in terms of pigments and photosynthetic apparatus, but is much more versatile than other green sulfur bacteria, being able to grow aerobically in the dark as well as phototrophically under anaerobic conditions. In this latter respect, *Chloroflexus* resembles the Rhodospirillaceae (nonsulfur purple bacteria).

I have already discussed in Chapter 3 the existence of filamentous organisms in alkaline hot spring mats which had been classified erroneously as blue-green algae. During the work on these mats (Brock, 1968, 1969), I used fluorescence microscopy to determine whether these reputed blue-green algae possessed chlorophyll-a. Fluorescence microscopy is a rather sensitive method. Since chlorophyll-a fluoresces a brilliant red when excited by blue light, it is easy to determine at the microscopic level whether or not an organism possesses chlorophyll-a. I was able to show that none of the filamentous organisms found at higher temperatures in alkaline hot spring mats possess chlorophyll-a (Brock, 1968), and I con-cluded that they were thermophilic filamentous bacteria of uncertain taxo-nomic status (Brock, 1968). In another paper (Brock, 1969), I presented evidence that the filamentous bacteria in these mats provided the dominant structure of the mat, and postulated a crude type of symbiotic relationship between these bacteria and the unicellular blue-green alga, in which the latter organism synthesized organic matter that served as the nutrient source for the filaments. It was because of this hypothesis that Bauld began to work on the feeding of these filamentous bacteria by the blue-green alga (Bauld and Brock, 1974). In 1968–1969 I actually had cultures of these filamentous organisms growing heterotrophically in the dark on simple organic media (Brock, 1969). I had initially isolated these organisms by streaking on plates containing the blue-green alga, and then transferring filaments to organic-containing media. At that time I had no reason to believe that the organisms were photosynthetic, and was simply carrying them in the dark.

The discovery that this organism contained bacteriochlorophylls came about in an interesting way. Pierson was studying a rather unusual type of mat on the Warm Springs Indian Reservation in Oregon. In these mats there was a filamentous organism, designated by her as F-1, which formed prominent orange surface layers *on top of* a green layer of the blue-green alga *Synechococcus*. F-1 is present in only a very few hot springs, and is never seen in the majority of alkaline hot springs in Yellowstone Park. The original suggestion that F-1 was photosynthetic arose because F-1 mats were most extensively developed in the summer months and diminished in extent in the winter. It seemed logical to associate this seasonality with changes in light intensity. Pierson showed that if a mat in summer were covered with a dark shade, F-1 disappeared. She then showed that if a suspension of filaments is prepared from mat material, the filaments rapidly clump in the light, but if starved do not clump in the dark. This showed that

F-1 had some sort of light metabolism. Because of the large size of the F-1 filaments (1.5 μm diameter), it was relatively easy to study the metabolism of natural material using autoradiography, and Pierson was able to easily show that F-1 had a light-dependent uptake of ^{14}C-acetate and ^{14}C-glucose. She did a number of studies on the physiology of natural populations of F-1 using autoradiography as a tool (Pierson, 1973). Because F-1 formed virtually pure mats under some conditions, it was possible to obtain sufficient material to carry out chlorophyll analyses, and it could be shown that F-1 had bacteriochlorophyll-a. Despite numerous attempts, Pierson was unable to culture F-1, so that taxonomic and other types of studies could not be carried out.

As noted, F-1 mats were unusual, since the orange filamentous organism was *on top of* the alga. In the vast majority of mats in alkaline hot springs, the orange filamentous organisms are *underneath* the blue-green alga (Brock, 1969). Furthermore, the filamentous organism in these mats (which we now know is *Chloroflexus*) has a narrower diameter and differs morphologically in other ways from F-1. With the knowledge that F-1 contained bacteriochlorophyll, Castenholz suggested that perhaps the common orange filamentous organisms of the undermat might also be photosynthetic. Bacteriochlorophylls could be detected in these mats, and even more significantly, it was possible to isolate pure cultures of these filamentous organisms and characterize them. Thus, work on F-1 was not pursued further when pure cultures of *Chloroflexus* became available, and we really do not know to this day whether or not F-1 is closely related to *Chloroflexus*.

Isolation and Culture of *Chloroflexus*

I originally had cultures of *Chloroflexus* in 1968–1969 growing in the dark on mineral salts medium containing 0.1% tryptone and yeast extract, or on spring-water medium to which a lawn of the blue-green alga *Synechococcus lividus* had been added (Brock, 1969). Castenholz (see Pierson, 1973) used a similar method to isolate *Chloroflexus,* incubating in the light. Since the filaments glide away from the inoculum, it is possible to pick filaments and transfer them relatively free of other bacteria. By use of these methods, pure cultures of a number of strains of *Chloroflexus* were obtained by Castenholz (Pierson, 1973) and by Bauld (1973).

The purity of *Chloroflexus* cultures is of critical importance in the characterization, because otherwise it could not be absolutely concluded that the bacteriochlorophyll was in the filament rather than in a contaminating unicellular organism. Pierson (1973) described in detail her tests to ensure the purity of her cultures. First, careful microscopic examination revealed no unicellular organisms. Second, examination of old cultures often revealed cellular debris that resembled bacteria of small dimension, but when such cultures were transferred to fresh medium, either no growth was obtained or typical *Chloroflexus* filamentous growth was obtained.

Finally, inoculation of other media failed to result in growth of putative contaminants.

In our laboratory, cultures of *Chloroflexus* have been maintained routinely at 45–50°C in plant growth chambers under either pure tungsten light or a combination of tungsten and fluorescent light. In all of our later work (Madigan, 1976), cultures were maintained under photosynthetic conditions only, although dark heterotrophic cultures can be maintained if desired. The basal salts medium used was Medium D of Castenholz (see Chapter 4), supplemented with 0.2 g/l NH_4Cl, 0.05% yeast extract, and 0.05% glycyl glycine (added as a buffer). Sulfide was supplied as a solution of $Na_2S \cdot 9H_2O$ at a final concentration of 0.05% (w/v). Cultures were carried in screw-capped test tubes filled to the top, and rendered anaerobic by steaming or autoclaving (in the absence of sulfide), then cooled, supplemented with sulfide, and inoculated. For convenience in mixing the contents of the filled tubes, 2 or 3 small glass beads were added to each tube.

For autotrophic growth, inorganic carbon was added following the procedure of Pfennig (1965). One to two grams Na_2CO_3 were dissolved in 300–400 ml distilled water in a 1-liter bottle and autoclaved. The cooled solution was bubbled for 20 minutes with pure CO_2 using a sterile bubbling apparatus, which lowered the pH of the solution to about 7–7.5. The mineral salts-trace element mixture, the sulfide solution, and the carbonate solution were then mixed and the pH checked aseptically and adjusted to 7.6–8.0 with sterile NaOH if necessary. Any additional supplements such as vitamins or heat labile substrates were added at this time. Finally, the medium was dispensed in sterile containers (16 × 125-mm screw-capped tubes or 125–250-ml Vitro bottles), filling them completely to the top. All sulfide-containing media were stored in the dark for at least 24 hr before inoculating to ensure the removal of traces of oxygen. Light intensities during photosynthetic incubations were at about 50–200 foot candles. Stock cultures were maintained at low light (40 foot candles) at 48°C to keep transfers to a minimum.

Under these conditions, *Chloroflexus* grows in liquid medium as flocs, clumps, or small aggregates, depending on the strain and on the age of the culture. To obtain the largest amount of inoculum for transfers, it was found useful to homogenize the inoculum material gently with a sterile Teflon homogenizer, in order that there would be a larger number of inoculum points. The use of homogenized material was of considerable importance when carrying out physiological or nutritional experiments, except for culture J-10, which grew as a more homogeneous suspension.

Morphology of *Chloroflexus*

The morphological studies of Pierson and Castenholz (1974a) clearly showed that *Chloroflexus* was related to the green bacteria, in that it possessed typical "chlorobium vesicles." These structures are the site of the photosynthetic apparatus in the green bacteria.

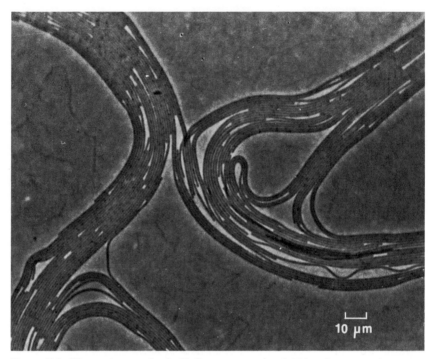

10 µm

Figure 7.1. Photomicrograph of a Yellowstone strain of *Chloroflexus* growing on an agar plate, showing the typical "flexibacterial" swirls.

On agar, *Chloroflexus* grows as a typical gliding bacterium, forming long swirls and flexuous strands (Figure 7.1). There is considerable variability from strain to strain in filament diameter, but there is also considerable variation within a single strain at different phases of growth. Filaments range in width from 0.45 to 1.0 μm, and they are of indeterminate length, ranging to much longer than 100 μm. In liquid medium the filaments often clump and form various-sized cell aggregates, although size of clumps varies from strain to strain, and with cultural conditions. Although septa are difficult to see in the light microscope, they are readily observed at low magnification in the electron microscope, by negative staining (Figure 7.2). Morphologically, *Chloroflexus* resembles fairly closely organisms of the genera *Flexibacter, Herpetosiphon,* and *Flexithrix* but the precise differentiation between these three genera is difficult to perceive from the published descriptions (see Reichenbach and Golecki, 1975).

In thin section, *Chloroflexus* appears to have a typical gram-negative cell wall (Figure 7.3), and there is nothing especially unusual about its fine structure. (This figure also reveals a fairly well-defined sheath, although insufficient work has been done to know if this is a constant characteristic of the organism.) Pierson and Castenholz (1974a) provided excellent electron micrographs of thin sections showing the "chlorobium vesicles" typical of the green sulfur bacteria. Madigan and Brock (1977c) studied the

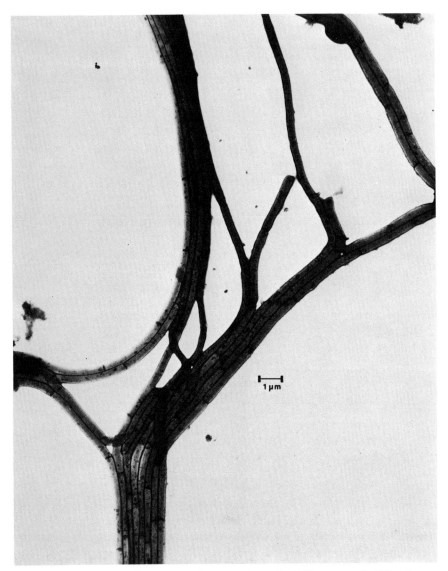

Figure 7.2. Electron micrograph of negatively stained filaments of *Chloroflexus* strain 254-2 (phosphotungstic acid stain). Septa and internal structures are readily apparent. (Courtesy of Michael Madigan.)

"chlorobium vesicles" using negative staining in the electron microscope, a technique that permits ready observation of these structures without fixation or alteration of cell structure (Figure 7.4). The vesicles in strain OK-70-f1 are ellipsoidal structures about 100–140 nm long and 45–70 nm wide. In general, strains which are narrower in filament diameter have smaller vesicles than strains with larger filament diameters. A high-magnifi-

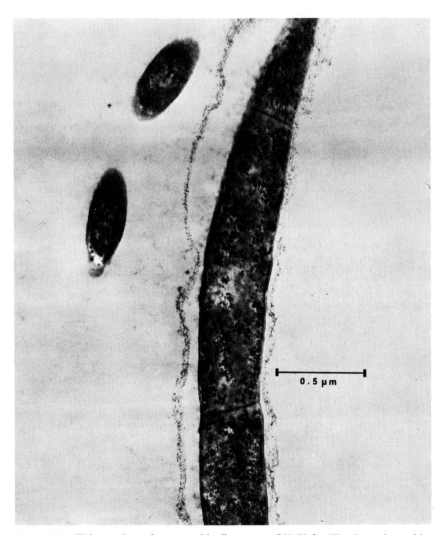

Figure 7.3. Thin section of autotrophically grown OK-70-f1. Fixation: glutaraldehyde followed by osmium. Doubly stained with uranyl acetate and lead citrate. Note septum. Typical gram-negative cell wall and the rather diffuse outer sheath are shown. (Courtesy of Michael Madigan.)

cation micrograph showing the angularity and characteristic structure of the vesicles from *Chloroflexus* is shown in Figure 7.5.

Oxygen represses bacteriochlorophyll synthesis as well as the synthesis of the vesicles. As seen in Figure 7.6, filaments from a culture grown aerobically in the dark show no evidence of "chlorobium vesicles." Photoautotrophically grown cells are noticeably greener than photoheterotrophically grown cells, and also contain considerably larger numbers of the photosynthetic vesicles.

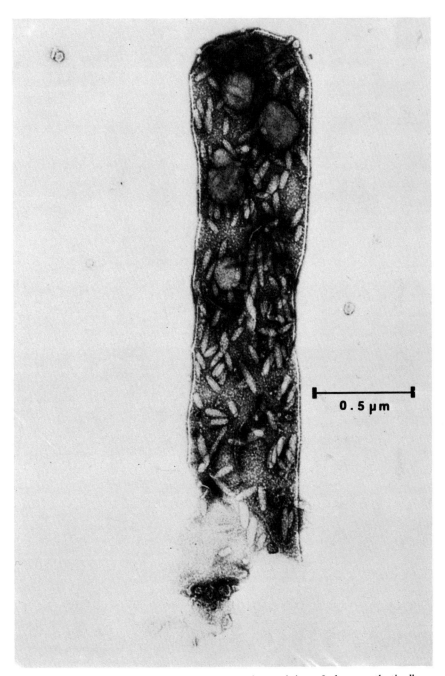

Figure 7.4. Electron micrograph by negative staining of photosynthetically grown strain 396-1. Note the large number of photosynthetic vesicles ("chlorobium vesicles"). The large structures are possibly storage granules. (Courtesy of Michael Madigan.)

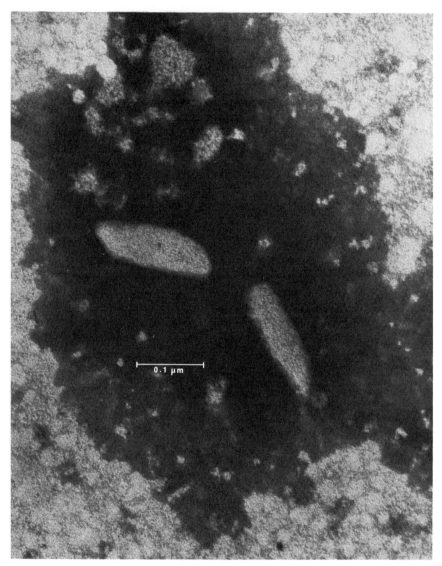

Figure 7.5. High-magnification electron micrograph by negative staining of the photosynthetic vesicles of strain Y-400-f1. Note the angular appearance. (Courtesy of Michael Madigan.)

Nutritional Studies on *Chloroflexus*

In her original work, Pierson (1973) was unsuccessful in culturing *Chloroflexus* photoautotrophically or on defined culture media. Madigan was first able to obtain good growth of strain OK-70-fl in defined media, and subsequently strains J-10-fl, 254-2, and 396-1 were also grown under defined conditions (Madigan et al., 1974). The secret here was apparently only to

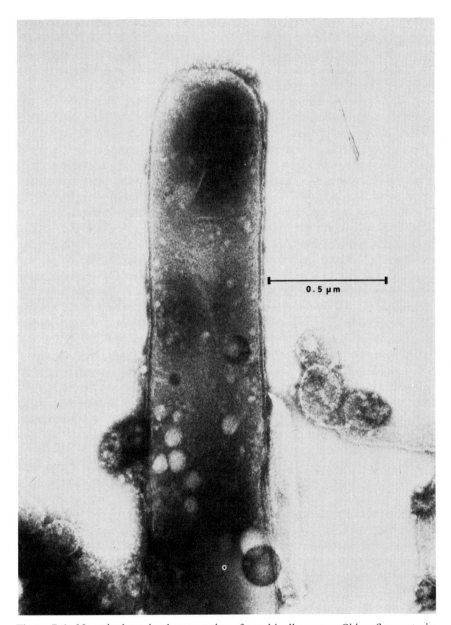

0.5 µm

Figure 7.6. Negatively stained preparation of aerobically grown *Chloroflexus* strain OK-70-f1. Note the absence of photosynthetic vesicles. (Courtesy of Michael Madigan.)

provide a source of inorganic carbon (as described above) and the necessary vitamins. In the original nutritional work, a mixture of vitamins was
used (Madigan et al., 1974), but Madigan (1976) subsequently showed that
two strains, OK-70 and J-10, required only folate and thiamine when
growing photoheterotrophically.

Chloroflexus is able to use a wide variety of organic compounds as
energy sources when growing aerobically in the dark. Table 7.1 lists the
results obtained with a number of strains, and it can be seen that tricarbox-
ylic acid cycle intermediates, short-chain alcohols, amino acids, hexoses,
and at least one pentose support the growth of *Chloroflexus* in the dark. In
many cases, there is some stimulation of growth in the light under these
conditions. Forced aeration was not used for the experiments reported in
Table 7.1, so that it is likely that conditions were only partially aerobic, so
that some photopigments were made. As also seen in Table 7.1, strictly
autotrophic growth does not occur under aerobic conditions, which would
be expected since no electron donor for photosynthesis would be present.

Table 7.2 gives the results of nutritional experiments carried out under
anaerobic conditions. In these experiments, sulfide was present to maintain
anaerobic conditions, and some growth of OK-70 occurred in the absence
of any organic compound. The other strains did not grow completely
photoautotrophically with sulfide (subsequently photoautotrophic growth
with these strains was obtained), but did show good growth with a variety
of organic compounds. The same range of compounds which served for
aerobic growth in the dark also permitted photoheterotrophic growth.
Sustained growth in the dark under anaerobic conditions was never
obtained with any organic compounds. Thus, in contrast to certain members of the Rhodospirillaceae (Uffen and Wolfe, 1970), fermentative growth
in *Chloroflexus* does not occur.

Madigan (1976) showed that ammonium ion and several amino acids
would serve as nitrogen sources for *Chloroflexus,* but growth did not occur
with nitrate or N_2. Since many photosynthetic bacteria do fix N_2, and since
N_2 fixation might appear to be of selective advantage in the low-nitrogen
hot springs in which *Chloroflexus* lives, Madigan made further attempts to
determine if *Chloroflexus* could use N_2. Sparse growth occurred during
initial transfer to a medium free of combined nitrogen, but repeated transfers from these initial N-free cultures did not lead to further growth.
Madigan also used the acetylene reduction technique to determine if the
enzyme nitrogenase were present. No significant acetylene reduction
occurred when the initial culture in nitrogen-free medium was used for the
assays. This experiment was done with both strains OK-70-f1 and J-10-f1.
The medium used was supplemented with sodium citrate as chelator (to
replace the nitrogen-containing chelator normally used), 0.2% sodium acetate, and 10 mM glucose, with 0.08–0.1% HCO_3^- and 0.05% sulfide. Growth
was in 500-ml bottles containing 150-ml medium, in an atomosphere of 99%
nitrogen–1% CO_2. Since not all photosynthetic bacteria fix N_2, the inability

Table 7.1. Utilization of Organic Compounds under Aerobic Conditions[a]

	Strain							
	OK-70-f1		J-10-f1		254-2		396-1	
Carbon Source	Light	Dark	Light	Dark	Light	Dark	Light	Dark
4% Formalin-killed control	−	−	−	−	−	−	−	−
Glutamate	+	+	++	+	++	+	+	+
Aspartate	++	+	+	+	+	+	+	+
Glycyl-glycine	−	−	−	−	−	−	−	−
Acetate	+	−	+	+	++	+	+	+
Pyruvate	++	+	+	+	+	+	++	+
Lactate	−	−	++	++	++	+	++	++
Succinate	+	+	+	−	+	+	++	+
Malate	+	+	++	++	++	+	+	+
Butyrate	++	−	++	+	++	+	+	+
Citrate	+	+	−	−	−	−	−	−
Ribose	++	+	+	+	+	+	+	+
Glucose	++	+	++	++	++	++	++	++
Galactose	++	+	++	++	++	+	++	++
Ethanol	+	+	+	−	+	−	+	−
Glycerol	+	−	++	++	+	+	++	++
Mannitol	++	++	+	+	+	+	+	+
Yeast extract	++	++	++	++	++	++	++	++
Casamino acids	++	++	++	+	++	++	++	+
HCO_3^-	−	−	−	−	−	−	−	−

[a]Relative growth values after 8 days incubation were assessed by determining whether the yield of cells was comparable to that of yeast extract (++), or whether the growth yield was significantly less than that derived from yeast extract-grown cells (+). A (−) value represents substrates or conditions not supporting growth of a particular strain of *Chloroflexus* when compared with a 4% formalin-killed control. (From Madigan et al., 1974.)

Table 7.2. Utilization of Organic Compounds under Anaerobic Conditions[a]

Carbon Source	OK-70-f1 Light	OK-70-f1 Dark	J-10-f1 Light	J-10-f1 Dark	254-2 Light	254-2 Dark	396-1 Light	396-1 Dark
4% Formalin-killed control	−	−	−	−	−	−	−	−
Glutamate	+	−	++	−	+	−	+	−
Aspartate	++	−	+	−	+	−	+	−
Glycyl-glycine	+	−	+	−	+	−	+	−
Acetate	+	−	+	−	+	−	+	−
Pyruvate	+	−	++	−	+	−	++	−
Lactate	−	−	+	−	+	−	−	−
Succinate	+	−	+	−	+	−	+	−
Malate	+	−	+	−	+	−	+	−
Butyrate	+	−	+	−	+	−	+	−
Citrate	+	−	+	−	+	−	−	−
Ribose	+	−	+	−	+	−	+	−
Glucose	++	±[b]	++	−	++	−	++	−
Galactose	+	−	+	−	+	−	+	−
Ethanol	+	−	+	−	+	−	−	−
Glycerol	++	−	+	−	+	−	+	−
Mannitol	+	−	+	−	+	−	+	−
Yeast extract	++	−	++	−	++	−	++	−
Yeast extract, no sulfide	++	−	++	−	++	−	++	−
Casamino acids	++	−	++	−	++	−	++	−
HCO_3^-	+	−	−[c]	−	−[c]	−	−[c]	−

[a]Relative growth values after 8 days incubation were assessed by determining whether the yield of cells was comparable to that of yeast extract (++), or whether the growth yield was significantly less than that derived from yeast extract-grown cells (+). A (−) value represents substrates or conditions not supporting growth of a particular strain of *Chloroflexus* when compared with a 4% formalin-killed control.
[b]Initial growth occurred, but sustained growth in the dark anaerobically was not achieved.
[c]Sustained autotrophic growth was not achieved in these strains by Madigan et al. (1974) but was achieved subsequently.

of *Chloroflexus* to fix N_2 is not surprising. It is always possible, of course, that other strains of *Chloroflexus* do fix N_2.

Sulfur Metabolism of *Chloroflexus*

Ecological studies by Bauld and Brock (1973) and by Castenholz (1973) had provided evidence that CO_2 fixation in natural material was markedly stimulated by sulfide. Figure 7.7 shows some data obtained by Bauld comparing $^{14}CO_2$ fixation in the presence and absence of sulfide. The adaptation by Madigan of *Chloroflexus* strain OK-70-f1 to completely autotrophic growth anaerobically using sulfide as electron donor made possible studies on the phototrophic sulfur metabolism of *Chloroflexus*. As seen in Figure 7.8, autotrophically grown cells oxidized sulfide during growth and produced elemental sulfur, but significant amounts of sulfate were not formed. In other experiments under photoheterotrophic conditions, similar results were obtained. Thus, elemental sulfur is a major

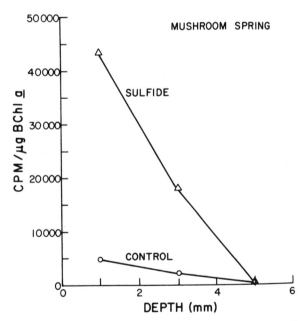

Figures 7.7. Effect of sulfide on photosynthetic $^{14}CO_2$ fixation by samples removed from different depths of a mat at Mushroom Spring. The algal layer was removed and discarded. Temperature of incubation 62–65.5°C. Light intensity 6000–7400 foot candles. $NaH^{14}CO_3$: 0.1 ml/vial, 20 μCi/ml (200 μg/ml). Incubation time: 2 hours. The top 6 mm of the bacterial portion of a core was sliced into three discs, each 2 mm thick. Each disc was then subdivided, one portion being incubated in the presence of sulfide and the other portion being used as a control. $Na_2S \cdot 9H_2O$ was added to give a final concentration of 0.01% (w/v). (From Bauld and Brock, 1973.)

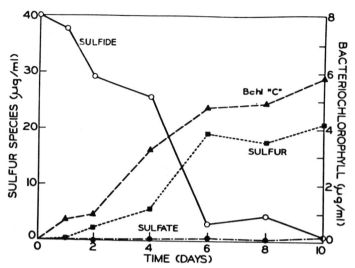

Figure 7.8. Oxidation of sulfide with the production of elemental sulfur during photoautotrophic growth of *Chloroflexus*. Sulfate expressed in terms of its sulfur content. (From Madigan and Brock, 1975.)

product of sulfide oxidation, but at most only 70% of the sulfide oxidized could be accounted for as elemental sulfur, so that some other reduced sulfur species such as thiosulfate or tetrathionate may also be formed.

Sulfur produced during sulfide oxidation is deposited in the medium as amorphous elemental sulfur (Figure 7.9). The typical green sulfur bacteria (Chlorobiaceae) produce elemental sulfur externally, thus further suggesting a relationship between *Chloroflexus* and the other green bacteria. Under some conditions the elemental sulfur is deposited externally along the long axis of the filament (Figure 7.10), and these elongated deposits are apparently quite loosely attached, since one frequently sees similar sulfur deposits floating free in the medium. This same phenomenon has also been observed in strains J-10-f1 and 244-3.

Several other reduced sulfur compounds were tested to see if they would support growth of *Chloroflexus*. Sulfite, thiosulfate, tetrathionate, thioglycolate, and methionine were unable to support phototrophic growth. Although crystalline elemental sulfur did not support growth, growth did occur when a colloidal suspension of sulfur was used. Growth also occurred when cysteine was used, but it is not certain whether cysteine was being used as an electron donor for photoautotrophic growth or as an organic compound for photoheterotrophic growth.

Thus, despite its filamentous shape, gliding nature, and its ability to grow aerobically in the dark, *Chloroflexus* resembles a typical green sulfur bacterium in its ability to use sulfide as a reductant with the deposition of elemental sulfur outside the cell.

The inability of *Chloroflexus* to oxidize sulfide all the way to sulfate is

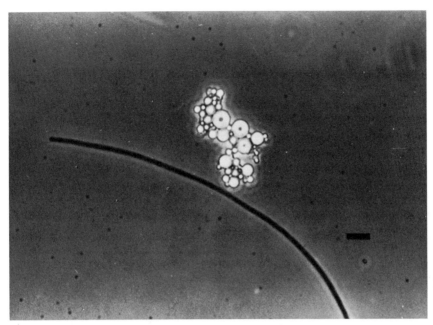

Figure 7.9. Phase contrast photomicrograph of elemental sulfur deposited in the medium during growth of *Chloroflexus* strain OK-70-f1. Marker bar represents 10 μm. (From Madigan and Brock, 1975.)

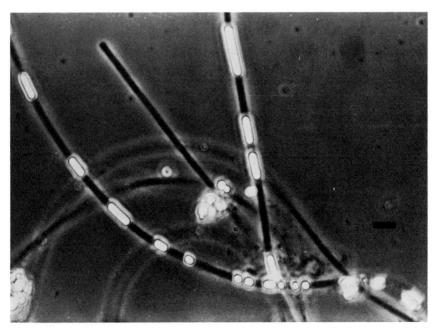

Figure 7.10. Phase contrast photomicrograph of filaments with deposited sulfur. Marker bar represents 4 μm (From Madigan and Brock, 1975.)

puzzling, especially in light of the fact that the organism can grow on elemental sulfur. It is possible that kinetic problems prevented further oxidation of the elemental sulfur formed from sulfide, perhaps due to insolubility of the sulfur and the lack of continuous mixing during the experiment.

Pigments of *Chloroflexus*

Pierson and Castenholz (1974b) separated the chlorophyll pigments of *Chloroflexus* by column chromatography on powdered sugar and identified them by spectrophotometry in various solvents. They showed that both bacteriochlorophyll-a and bacteriochlorophyll-c were formed (the latter is sometimes known as chlorobium chlorophyll 660). In natural mats, where *Chloroflexus* is almost always associated with blue-green algae, some method is necessary to distinguish between bacteriochlorophyll-c and algal chlorophyll-a, since these two chlorophylls absorb in organic solvents at almost the same wavelength. In earlier ecological work, Bauld and Brock (1973) did *in vivo* absorption spectra, since under these conditions bacteriochlorophyll-c and algal chlorophyll-a absorb at quite different wavelengths (740 and 680 nm, respectively). Later, Madigan and Brock (1976) developed a thin layer chromatographic method for separating bacteriochlorophyll-c from chlorophyll-a, and this technique was used to quantify *Chloroflexus* in mixed mats in various springs and under experimentally modified light intensities (see later section).

Pierson and Castenholz (1974b) showed that the ratio of BChl-c/BChl-a decreased in cultures with increasing light intensity. Furthermore, chlorophyll synthesis was suppressed under fully aerobic conditions in cells grown either in dark or light. In either case, upon transfer of pigmented cells to aerobic conditions, existing pigment was not destroyed but was diluted out by continued cell growth. A light-minus-dark difference spectrum of a cell-free extract of *Chloroflexus* provided evidence for a BChl-a-type reaction center similar to that occurring in purple sulfur bacteria.

Carotenoids are conspicuous in *Chloroflexus,* both in culture and in nature. Only under strictly photoautotrophic conditions at high sulfide levels or low light intensities are *Chloroflexus* cultures or natural populations visibly green (Castenholz, 1973). Under most conditions the organism is deep orange in color. Halfen et al. (1972) studied the carotenoids of *Chloroflexus* strain OK-70-f1 by thin layer chromatography and mass spectroscopy. Despite the similarities of *Chloroflexus* with the green sulfur bacteria mentioned earlier in this chapter, *Chloroflexus* differs from the green sulfur bacteria markedly in its carotenoids. In the green sulfur bacteria, the predominant carotenoids are of the chlorobactene series, but *Chloroflexus* completely lacks this type of carotenoid. However, gamma-carotene makes up 22% of the carotenoids of *Chloroflexus* and this carotenoid is found in small amounts in some green bacteria. In *Chloroflexus*, glycosides of gamma-carotenoid make up an additional 34% of the total carotenoid pigment. The other major carotenoid of *Chloroflexus* is beta-carotene, which makes up 28% of the total. This carotenoid is a minor constituent in some photosynthetic bacteria but is the primary carotenoid of blue-green algae. However, *Chloroflexus* lacks the oxygenated carotenoids characteristic of the blue-green algae, and does not contain any of the carotenoids found in the nonphotosynthetic flexibacteria.

Thus, some of the major carotenoids of *Chloroflexus* are found in both the green sulfur bacteria and the blue-green algae, but based on carotenoid characteristics there is no clear-cut relationship to any known microbial group.

Although there are no data on the function of the carotenoids of *Chloroflexus*, they may play some photoprotective role, because the organism is often exposed to intense solar radiation in the surface mats in hot springs. There may also be some distinct changes in carotenoids as a function of aerobiosis and light intensity; *Chloroflexus* nodes and other colonial structures found on the surface of mats, where O_2 and light intensity are high, are generally pink in color, whereas the organism (presumably the same strain) in the depths of the mat, in low light and anaerobiosis, is generally orange in color. Furthermore, the carotenoids seem to be more stable to decomposition that the chlorophylls of *Chloroflexus*, since in the deeper parts of mats, where decomposition is the primary process occurring, bacteriochlorophylls have disappeared (Bauld and Brock, 1973), whereas the carotenoids are still visible. Ultimately, the carotenoids also disappear, and the remaining mat becomes white or colorless.

Physiology of CO₂ Fixation

Since *Chloroflexus* will grow both photoautotrophically and photoheterotrophically, it was of interest to study the physiology of the CO_2 fixation process. As seen in Figure 7.11, $^{14}CO_2$ uptake in autotrophically grown cells is not only light-dependent but is also dependent on sulfide. Addition of acetate partially inhibits CO_2 fixation, probably because of competition between acetate and CO_2 for reducing power. It is likely that acetate is reduced to poly-β-hydroxybutyrate (PHB), since this polymer is known to occur in *Chloroflexus* (Pierson and Castenholz, 1974a) and Stanier et al. (1959) have shown a competition between acetate and CO_2 for reducing power in *Rhodospirillum*.

Cells grown photoheterotrophically with acetate fix comparable amounts of CO_2 whether or not sulfide is present during the isotope experiment, and acetate also has little effect on the process (Figure 7.12). This suggests that with acetate-grown cells, an endogenous reductant is used for CO_2 fixation, probably PHB, although as also seen in Figure 7.12, light is still required for

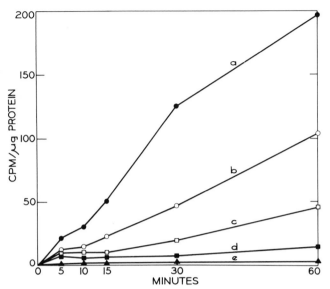

Figure 7.11. ¹⁴CO₂ incorporation in autotrophically grown *Chloroflexus* strain OK-70-f1. Cells were washed and resuspended in Medium D pH 8 and placed under anaerobic conditions in the light except where noted. Additions were made as follows: *a*, 1.25 mM sulfide; *b*, 1.25 mM sulfide and 15 mM acetate; *c*, no additions; *d*, no additions, aerobic incubation; *e*, 15 mM acetate. (From Madigan and Brock, 1977b.)

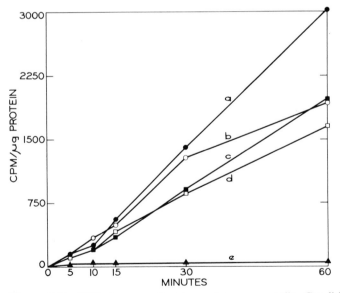

Figure 7.12. ¹⁴CO₂ incorporation in acetate-grown cells. Conditions anaerobic in the light except where noted. Additions were made as follows: *a*, 1.25 mM sulfide; *b*, 1.25 mM sulfide and 15 mM acetate; *c*, 15 mM acetate; *d*, no additions; *e*, no additions, dark. (From Madigan and Brock, 1977b.)

Table 7.3. CO$_2$-Fixation Enzymes in Crude Extracts of *Chloroflexus* Strain OK-70-f1, and *Phormidium* B-1

Growth condition	Activity (μmol/min/mg protein)[a]	
	PEP carboxylase	RuDP carboxylase
Autotrophic	15	<0.001
Acetate (15 mM)	12.4	0.012
Yeast Extract (0.05% w/v)	14.3	<0.001
Yeast Extract (0.001% w/v)	—[b]	0.023
Phormidium B-1 (used as positive control)	0.55	14.3

[a]All assays performed at 45°C.
[b]Assay not performed.
(From Madigan, 1976.)

CO$_2$ fixation, even with an endogenous reductant. Similar results were obtained with yeast extract-grown cells.

Madigan (1976) looked for the enzyme ribulosediphosphate (RuDP) carboxylase in *Chloroflexus*, in an attempt to obtain some idea about the enzymology of the CO$_2$ fixation process. The presence of this enzyme is essential if the Calvin cycle for CO$_2$ fixation is operable. The presence of this enzyme in other green photosynthetic bacteria is controversial. Sirevag (1974) and Buchanan et al. (1972) were unable to find the enzyme in one strain of *Chlorobium*, but Tabita et al. (1974) found small amounts of the enzyme in another strain. The enzyme from some sources is easily inhibited or inactivated, and the *Chlorobium* enzyme is known to be especially unstable (F. R. Tabita, personal communication to M. Madigan). Madigan assayed for the enzyme in four batches of *Chloroflexus* cells (strain OK-70-f1) grown under different conditions. Activity was either undetectable or extremely low in three batches of cells (Table 7.3), but in one batch of cells grown with 0.001% yeast extract, sufficient activity was obtained so that a few experiments could be done. As seen in Figure 7.13 the temperature optimum of the RuDP carboxylase activity was at 50°C, a temperature close to the optimum growth temperature.

The inability to find consistent RuDP carboxylase activity in *Chloroflexus* extracts may mean either that the enzyme is unstable, or that enzymatic activity is present only during certain phases of the growth cycle, and that the cells were harvested at the improper time. Because of the poor growth of *Chloroflexus* under photoautotrophic conditions, and the difficulty of obtaining large cell yields, there is a natural tendency to incubate until maximal growth has been obtained, at which time the enzyme may have become inactivated. Because at least two cell-free extracts did have detectable RuDP carboxylase activity, it seems reasonable to conclude that the enzyme is present, but it would be impossible to determine how significant this enzyme, or the Calvin cycle, is in autotrophic CO$_2$ fixation in *Chloroflexus*.

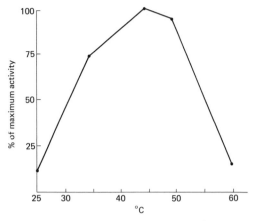

Figure 7.13. Temperature optimum of ribulosediphosphate carboxylase in crude extract of *Chloroflexus* OK-70-f1. Cells grown phototrophically with 2 mM sulfide, 0.1% HCO_3^- and 0.001% yeast extract. Reaction mixtures contained 60 mM tris-sulfate, 20 mM $MgCl_2 \cdot 6H_2O$, 4 mM EDTA, 25 mM HCO_3^-, 1 mM ribulosediphosphate, pH 7.9 in a volume of 0.6 ml. Atmosphere was rendered anaerobic by gassing with N_2. (Madigan, 1976).

Habitat of *Chloroflexus*

Chloroflexus has been primarily found living only in hot springs of neutral to alkaline pH, at temperatures up to 70°C. However, Gorlenko (1975) has reported the presence of filamentous phototrophic bacteria from freshwater lakes in Russia. These organisms have all of the characteristics of *Chloroflexus,* except thermophily, and hence it is reasonable to classify them in the same genus. Gorlenko isolated his strains from samples of bottom water and mud from stratified freshwater lakes that had been exposed to light for a few months, and in which a mat consisting of blue-green algae and filamentous flexibacteria had developed. He obtained pure cultures and characterized them physiologically and morphologically. The filaments were quite similar in morphology to *Chloroflexus,* and thin sections revealed typical "chlorobium vesicles." The main photopigment was bacteriochlorophyll-c. The organism also contained carotenoids but these were not characterized. The isolates grew anaerobically in the light, under which conditions the cell suspension was green, or aerobically in light or dark, under which conditions the cell suspension was pinkish orange. Photoheterotrophic growth occurred on complex organic substrates such as yeast extract or casein hydrolysate, and also on single organic compounds (not specified) when growing in a medium supplemented with vitamins. The optimum temperature for growth was 20–25°C, and the range of temperatures was 10–40°C. Gorlenko postulates that the freshwater *Chloroflexus* probably grows on the surface of the mud, where it forms a mat with other microorganisms. The discovery of this freshwater *Chloroflexus* considerably broadens the interest in this group of organisms, since it shows that the group is not restricted to the rather specialized thermal environment. Most enrichment culture techniques for photosynthetic bacteria are to some extent selective for rapidly growing unicellular organisms. It is conceivable that a wide range of filamentous photosynthetic bacteria exist, living primarily on mud surfaces at depth in the water column, and that they have been missed because of improper enrichment. Experience has shown (Brock, 1966) that specific isolation of filamentous bacteria from nature requires special attention to the physical characteristics of the medium. Specifically, liquid culture enrichments are not advisable, since they favor unicellular organisms. The best procedure is to go directly from the natural sample to agar surfaces, and examine inoculated surfaces periodically during incubation to detect filamentous organisms before they have been overgrown by unicellular (especially motile) bacteria. By enriching directly onto agar under anaerobic conditions in a sulfide-rich environment in the light, it may be possible to isolate *Chloroflexus*-like organisms from a wide variety of aquatic environments, both freshwater and marine.

Ecology of *Chloroflexus* in Hot Springs

Under most conditions, *Chloroflexus* is associated in hot springs with blue-green algae (Brock, 1968; 1969; Bauld and Brock, 1973), but in high-sulfide springs, blue-green algal development is suppressed so that at the higher temperatures pure stands of *Chloroflexus* can develop (Castenholz, 1973). Springs with sufficiently high sulfide to suppress blue-green algae are virtually absent from Yellowstone, although several such springs were found by Castenholz (1973) in New Zealand. Since hot spring waters are generally low in organic matter (see Chapter 2), it is likely that when *Chloroflexus* is growing alone in high-sulfide springs it is growing photoautotrophically using sulfide as electron donor (Castenholz, 1973), expecially since the organism will grow photoautotrophically with sulfide in culture. The fact that *Chloroflexus* requires certain vitamins for growth raises the question of the source of these vitamins in hot spring waters. Conceivably small amounts of vitamins are present in the natural waters.

The widespread distribution of *Chloroflexus* in hot springs is suggested by the data of Table 7.4, which shows the almost universal presence of bacteriochlorophylls (either a, c, or both) in Yellowstone hot springs. Castenholz has found *Chloroflexus* in New Zealand and Japanese and European hot springs (personal communication), and it is present throughout the western U.S. and in Mexico (Table 7.4).

Since *Chloroflexus* lives in nature primarily in association with blue-green algae, attention naturally turns to the relationship between these two organisms. The compact stratified mats formed by *Chloroflexus* in association with the unicellular blue-green alga *Synechococcus* are of interest

Table 7.4. Presence of Bacteriochlorophyll Pigments in Algal-Bacterial Mats Containing Filamentous Bacteria

Collection site	Temperature	pH	Methanolic extract		Chromatophore preparation[a]	
			BChl-c	BChl-a	BChl-c	BChl-a
Yellowstone						
Octopus	59	8.55	+[b]	+	+	+
Pool B	63	8.6	+	+	+	+
Five Sisters	65	8.9	ND	ND	+	+
Little Brother	63	8.9	+	+	+	+
Tri-spring	55	8.2	ND	ND	+	+
Bubbling Pool	63	7.4	ND	ND	+	+
Toadstool Geyser	63	8.7	+	+	+	+
Zomar Spring	63	8.05	?[c]	+	−	+
Grassland Spring Channel A	66	ND	?	+	ND	ND
Grassland Spring Channel B	61	ND	?	+	ND	ND
Mushroom Spring	63	8.6	+	+	+	+
Twin Butte Vista	64	8.8	+	+	+	+
Column Pool	30–40	ND	?	−	+	−
Sulfur Spring near Hot Lake	62	6.6	?	+	ND	ND
Others						
Steamboat Springs, Nevada	70.5	ND	+	−[d]	ND	ND
Araro thermal area, Mexico	62	ND	+	+	ND	ND

[a]Chromatophore preparations courtesy of R. W. Castenholz, University of Oregon.
[b]+, detectable peak; −, peak not detectable; ND, not done.
[c]Not certain whether BChl-c or Chl-a (methanol extract).
[d]BChl-a detectable in a strain isolated from this sample.
(From Bauld, 1973.)

because of their possible relation to Precambrian stromatolitic mats, and because they provide an interesting opportunity to study the interaction in nature between two microorganisms. Some detailed work on the structure, growth, and decomposition of these mats is discussed in Chapter 11. Here I concentrate primarily on the ecology of *Chloroflexus* itself.

The gross anatomy of a *Chloroflexus-Synechococcus* mat is illustrated in Figure 7.14. As seen, the blue-green alga is only found in a narrow surface zone, whereas the bacterium is present to considerably greater depths in the mat. A vertical profile of the distribution of photopigments through a mat of moderate thickness is shown in Figure 7.15 and it can be seen that the chlorophyll assays agree with the microscopic observations. Algal chlorophyll-a is found only in the narrow surface layer, whereas bacter-

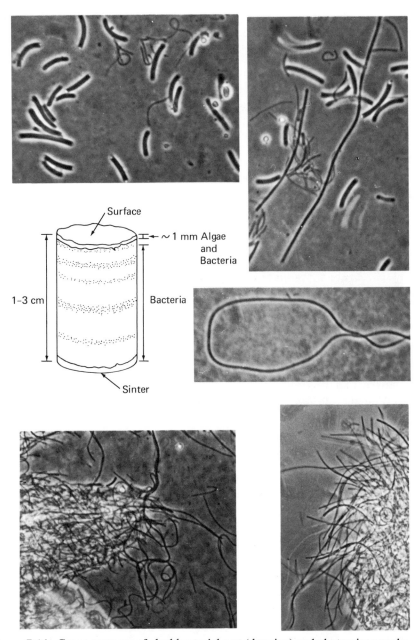

Figure 7.14. Gross anatomy of algal-bacterial mat (drawing) and photomicrographs of material taken from different levels. Top left: surface layer showing the blue-green alga *Synechococcus lividus* and a small unidentified flexibacterium; top right: *S. lividus* and *Chloroflexus;* center right: *Chloroflexus* filament from the top bacterial layer, about 2–3 mm from the surface of the mat; bottom left and right: filamentous bacteria from the bacterial layers. Note the general absence of *S. lividus* from the lower layers. Marker bars represent 10 μm. (From Bauld and Brock, 1973.)

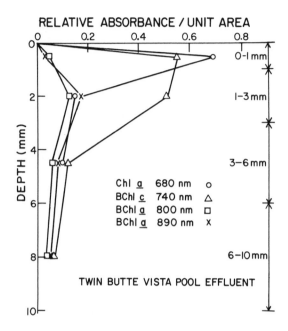

RELATIVE ABSORBANCE / UNIT AREA

TWIN BUTTE VISTA POOL EFFLUENT

Chl a   680 nm   o
BChl c 740 nm   △
BChl a 800 nm   □
BChl a 890 nm   x

Figure 7.15. Vertical profile of photosynthetic pigments from cores taken from the algal-bacterial mat at a 65°C site of Twin Butte Vista Pool effluent. The top 10 mm of each of the cores was cut into discs of the height shown by the scale on the right-hand side of the figure. Discs from a given mat depth were pooled and spectra were determined by making cell-free extracts and reading *in vivo* chlorophyll absorption maxima on the chromatophore preparations so made. (From Bauld and Brock, 1973.)

iochlorophylls-a and c are present not only in the surface but further down in the mat. Measurements of photosynthesis with $^{14}CO_2$, using the inhibitor DCMU to distinguish between bacterial and algal photosynthesis, also showed that *Chloroflexus* in the mats was photosynthetically active not only at the surface but down to about 3 mm (Bauld and Brock, 1973; Chapter 11). As will be shown in the following section, *Chloroflexus* is able to photosynthesize at considerably lower light intensities than *Synechococcus,* thus accounting for its ability to maintain active populations further down in the mats.

The distribution of *Chloroflexus* with respect to temperature was studied by Bauld and Brock (1973), using bacteriochlorophyll-c as a marker for the bacterium (this assumes that *Chloroflexus* is the only BChl-c-containing organism in the mats). As seen in Figure 7.16, along the thermal gradient of a single spring, there is a maximal amount of bacteriochlorophyll at around 55°C, and at temperatures of 45°C or less bacteriochlorophyll was virtually undetectable. Also, the standing crop of bacteriochlorophyll falls off sharply above 55°C, and is undetectable at 70°C. (At a temperature just above 70°C there is no visible mat.) The relationship between *Chloroflexus* biomass and temperature illustrated in Figure 7.16 is almost exactly the same as that found for algal chlorophyll (also illustrated in Figure 7.16), the only difference being that a small amount of algal chlorophyll is detected at the lower temperatures. As shown earlier (Brock, 1967; Brock et al., 1969), it is likely that the fall-off in standing crop at the lower temperatures is due to the introduction of grazing animals into the system. Although bacteriochlorophyll is difficult to detect in low-temperature mats, *Chloroflexus* is probably present in such mats in small amounts, since Bauld isolated a strain of *Chloroflexus* from a stromatolitic structure consisting predominantly of *Phormidium* that had developed in a hot spring at a temperature of 30–40°C.

Bauld and Brock (1973) determined the temperature optimum for photosynthesis for a series of natural samples taken from the thermal gradient of a single hot spring. As seen in Figure 7.17, bacterial photosynthesis was observed to be greatest at temperatures close to the environmental temperature for each sample, suggesting a series of temperature strains, each adapted optimally to the temperature of the habitat. So far, no attempt has been made to isolate *Chloroflexus* cultures with specific temperature optima, most strains having been isolated with enrichment temperatures around 50–55°C. As also seen in Figure 7.17, the population taken from a temperature of 72°C, just at the upper temperature limit at which visible mats are present, had a temperature optimum slightly lower than its habitat temperature. Similar results were obtained by Meeks and Castenholz (1971) for growth and photosynthesis of a high-temperature strain of *Synechococcus*. Thus, near the upper temperature limit, the habitat does not appear to be optimal for photosynthesis by natural populations of either the bacterium or the alga. This is probably why the mats existing at these temperatures are so thin.

Although little has been done on the effect of pH on *Chloroflexus*, extensive observations in acid hot springs (pH values 1–3) have never

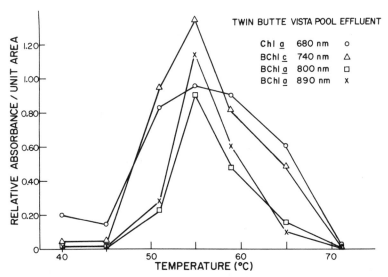

Figure 7.16. Relative concentration of photosynthetic pigments per unit area along the temperature gradient of Twin Butte Vista Pool effluent, as determined by *in vivo* chlorophyll assays on cell-free extracts of complete, intact algal-bacterial cores. (From Bauld and Brock, 1973.)

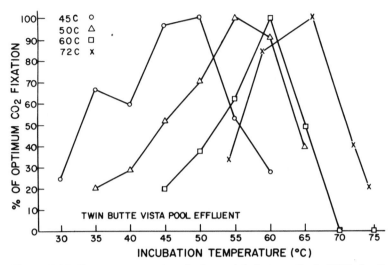

Figure 7.17. Temperature optima for bacterial photosynthetic $^{14}CO_2$ fixation in the presence of 10^{-5} M DCMU, for samples from sites at temperatures of 72°C, 60°C, 50°C, and 45°C in the effluent of Twin Butte Vista Pool. (See Bauld and Brock, 1973, for details.)

revealed organisms morphologically resembling *Chloroflexus*. So far, all of the habitats in which *Chloroflexus* has either been seen or cultured have had pH values on the neutral or alkaline side. (It is difficult to find hot springs with pH values around 5–6.)

Bauld showed in his studies that the blue-green alga of the mats excreted a significant amount of its photosynthate, and that *Chloroflexus* was able to assimilate this organic matter in a light-stimulated fashion (Bauld and Brock, 1974). As shown in Figure 7.18, there is a linear rate of excretion of labeled material when intact cores from mats are incubated with $^{14}CO_2$, and more material is excreted in the light than in the dark. Since in intact cores an unknown proportion of the excreted material may be assimilated by *Chloroflexus* (or other bacteria) immediately, and thus would not be assayed as excretion products, an experiment was done in which natural populations from a mat were homogenized and the homogenate diluted to various cell densities before the excretion experiment was performed. As seen in Table 7.5, as the cell density was decreased, there was a marked increase in the percent of the photosynthate that was excreted, so that at a cell density of 10^6 cells of *Synechococcus* per ml, over 50% of the photosynthate was excreted. Although this experiment could be interpreted in several ways, one reasonable interpretation is that in intact mats considerably more of the photosynthate is excreted than is measured in the excretion assay, and that as soon as it is excreted it is assimilated by the bacteria present, so that it is not assayable as soluble materials. This would suggest that there is an extensive and very efficient transfer of organic matter from the alga to the bacterium.

Table 7.5. The Effect of Cell Density on Excretion by Natural Populations of *Synechococcus lividus*[a]

Cells/ml	Relative cell density	Chl-a, µg/ml	cpm/µg Chl-a					
			Total		Excreted		Percent excreted	
			Light	Dark	Light	Dark	Light	Dark
1.2×10^8	1	4.73	14,713	614	1,156	431	7.9	70.2
6.0×10^7	1/2	2.31	18,332	740	1,416	459	7.7	62.0
2.4×10^7	1/5	0.97	13,624	706	2,233	398	16.4	56.4
1.2×10^7	1/10	0.40	11,773	530	2,433	365	20.7	68.9
1.2×10^6	1/100	0.05	3,240	1,260	1,700	1,260	52.5	100

[a]Sampling and incubation site, Octopus Spring. Temperature of incubation site 56–58°C. Light intensity 1750–6000 foot candles. The algal layer was removed, homogenized, and diluted to give the cell densities designated. (From Bauld and Brock, 1974.)

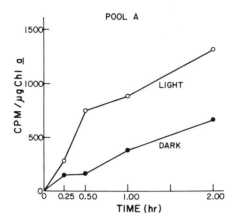

Figure 7.18. Time course for excretion of ^{14}C-labeled compounds by intact cores of the algal layer in Octopus Spring (Pool A). Temperature of the incubation site 55–56°C. Light intensity during the experiment varied from 740 to 6700 foot candles. (From Bauld and Brock, 1974.)

That *Chloroflexus* is at least one of the organisms in the mat assimilating the excreted material is shown by the experiment illustrated in Figure 7.19. Labeled materials which had been previously excreted by the blue-green alga were assimilated by natural populations of bacteria taken from the algal-free zone immediately beneath the top layer of mat. As seen, assimilation of this material was stimulated by light, suggesting that it was indeed *Chloroflexus,* rather than common heterotrophic bacteria, that was assimilating the excreted products.

On the basis of these experiments, it can be concluded that in hot springs *Chloroflexus* is probably growing primarily photoheterotrophically, and that it obtains its organic compounds for growth from the blue-green alga with which it is intimately associated. Only under the rare situation in high-sulfide springs, where blue-green algae are absent and an inorganic electron donor for photosynthesis is present, is it likely that *Chloroflexus* grows photoautotrophically.

It is also possible that strictly heterotrophic growth of *Chloroflexus* occurs, but since heterotrophic growth only occurs aerobically, and most of the mat below the top 1–2 mm is probably anaerobic (Doemel and Brock, 1976), it seems likely that heterotrophic growth would only occur with the *Chloroflexus* filaments present immediately at the surface of the mat. However, these surface filaments would be exposed to light, and $^{14}CO_2$ experiments (Chapter 11) show that photosynthesis by *Chloroflexus* does occur in surface filaments. Thus, the only time that aerobic heterotrophic growth would seem to be significant would be at night, when light is absent. As shown by Doemel and Brock (1974), *Chloroflexus* does migrate to the surface of the mat at night, possibly because of a positive aerotaxis. Thus, it seems reasonable that populations of *Chloroflexus* grow at the surface of the mat heterotrophically at night, and photoautotrophically in the daytime. Populations from the lower part of the mat probably grow photoheterotrophically during the daytime, and do not grow at all at night. Although it is well established that O_2 represses bacteriochlorophyll synthesis in *Chloro-*

Figure 7.19. Time course for the bacterial assimilation of algal extracellular products. Temperature of the incubation site 56–57°C. Light intensity varied from 1250 to 6000 foot candles. Radioactivity of the solution of algal extracellular products was 16,000 cpm/ml. (See Bauld and Brock, 1974, for details.)

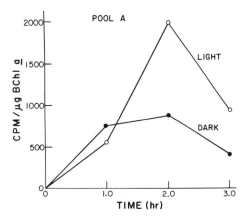

flexus (Pierson and Castenholz, 1974b), it seems unlikely that at the slow growth rates of the natural populations bacteriochlorophyll content would be completely diluted out during the night period, so that the transition from heterotrophic to photoheterotrophic growth could readily occur upon daybreak.

Adaptation of *Chloroflexus* to Various Light Intensities

The effect of light intensity on photosynthesis of *Chloroflexus* was studied initially by Bauld (1973) and in a more detailed way by Madigan (Madigan, 1976; Madigan and Brock, 1977a). Experiments were done with $^{14}CO_2$ using neutral density filters to determine the optimum light intensity for natural populations, and then other neutral density filters were placed over portions of mats to experimentally reduce the habitat light intensity. Changes in the populations were then measured as a function of time (Figures 7.21 and 7.22). As shown in Figure 7.20, the optimum light intensity of the natural (unadapted) populations of *Chloroflexus* and *Synechococcus* was fairly high, around 4000–6000 foot candles, but *Chloroflexus* was able to photosynthesize at light intensities considerably lower than those utilizable by the alga. This experiment confirms previous observations (see above) that *Chloroflexus* can live farther down in the depths of the mat because it is able to photosynthesize at lower light intensities than the alga.

The changes in algal and bacterial standing crop with time after neutral density filters were placed over a mat are illustrated in Figure 7.23. When the neutral density filter reduces the light 73%, little change in standing crop of either organism is seen, but when a filter reducing light intensity 98% is used, the alga disappears (probably because of washout and lysis). Interestingly, no change in standing crop of *Chloroflexus* is observed over the 10-

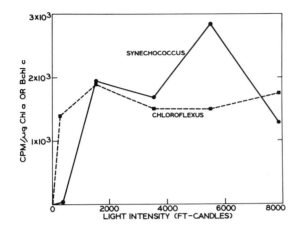

Figure 7.20. Photosynthesis at various light intensities by unadapted natural populations of *Synechococcus* and *Chloroflexus* at Ravine Spring, White Creek area. (From Madigan and Brock, 1977a.)

Figure 7.21. Ravine Spring, White Creek area, Lower Geyser Basin, illustrating the manner in which light intensity over the mat was experimentally reduced. The source is in the background, and the algal-bacterial mat has developed in the outflow channel. The white filter reduced light 73% and the black filter 98%. Photographed in summer 1975.

Figure 7.22. The mat at Octopus Spring, White Creek area, illustrating the manner by which light intensity was experimentally reduced. Left to right: dark cover; uncovered site; 73% reduction; 93% reduction; 98% reduction. Photographed in summer 1975. See Figure 11.11 for the appearance of the mat when uncovered.

day period when the light intensity is reduced 98%, showing that the bacterium can survive quite low light intensities. Over the next 30 days, however, the *Chloroflexus* population slowly decreased and with time completely disappeared, leaving only fragmented filaments embedded in a matrix of siliceous sinter. This result clearly shows that *Chloroflexus* is ultimately dependent on the blue-green alga for survival. It can maintain itself for a while after the alga disappears, probably because there is considerable organic material in the lower part of the mat that can be mobilized for photoassimilation, but once the residual organic matter is gone, the *Chloroflexus* population disappears. (An alternative explanation is that *Chloroflexus* is less subject to washout and lysis than *Synechococcus*.)

The ability of *Chloroflexus* to adapt to low-light intensity is shown by the experiment illustrated in Figure 7.24. This figure shows the effect of light intensity on photosynthesis by *Chloroflexus* populations adapted to different light intensities for 16 days. As seen, there is not only a lowering of the optimum light intensity, but considerable increase in sensitivity to high-light intensity. In the population that developed under the 98% light-

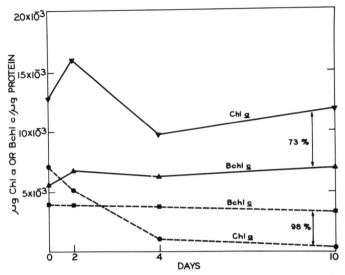

Figure 7.23. Changes with time in chlorophyll-a or bacteriochlorophyll-c for the mat at Ravine Spring (White Creek area), after installation of neutral density filters. The percent reduction in light by the filters is given on the figure. Note the washout of chlorophyll-a at the 98% reduction site. The values for photopigments are normalized to the protein contents of the samples to reduce sampling variability. (From Madigan and Brock, 1977a.)

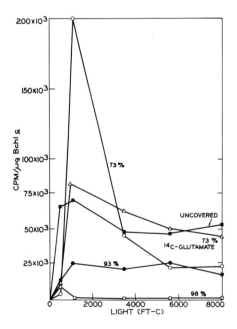

Figure 7.24. Photosynthesis measurements at Octopus Spring (Pool A), 16 days after installation of neutral density filters over the mat. Note the clear adaptation to lower light in populations subjected to light reductions of 73% and 98%. Photosynthesis was measured by $^{14}CO_2$ uptake. Uptake of ^{14}C glutamate was also measured and is predominately in the *Chloroflexus*. (From Madigan and Brock, 1977a.)

Figure 7.25. Replot of data from Figure 7.24 for the *Chloroflexus* population subjected to light reduction of 98%, to illustrate the marked low-light optimum for photosynthesis and high-light inhibition. (From Madigan and Brock, 1977a.)

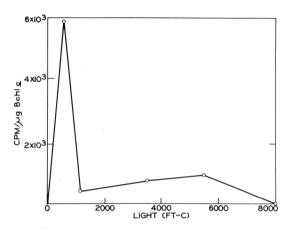

reduction filter, strong high-light inhibition was seen (Figure 7.25), and a sharp optimum light intensity for photosynthesis at around 400 foot candles was found. No obvious increase in bacteriochlorophyll content per cell was observed during the adaptation period, and the residual population after 16 days adaptation (rather moribund because of the absence of the alga), looked salmon pink rather than green. It would be of considerable interest to repeat these experiments in a high-sulfide spring, where *Chloroflexus* is able to grow photoautotrophically, since under these conditions the adaptation to reduced light intensity could be observed independently of any dependence on photosynthate from the alga. (As will be discussed in Chapter 8, the blue-green alga in these mats also shows marked adaptation to lowered light intensities, including significant pigment changes.)

Evolutionary Significance of *Chloroflexus*

Although some case might be made that *Chloroflexus* is a missing link between the blue-green algae and the photosynthetic bacteria, this case would be rather tenuous. *Chloroflexus* is clearly a typical photosynthetic bacterium, and is obviously related to the green sulfur bacteria (Chlorobiaceae) in a variety of ways. It is more versatile than the other green sulfur bacteria, since it can grow aerobically in the dark and as well as phototrophically. But in this respect, its nutritional versatility is similar to that of the "nonsulfur" purple bacteria (Rhodospirillaceae), some of which are now known to use sulfide as an electron donor. Perhaps the only notable difference nutritionally between *Chloroflexus* and the sulfide-utilizing Rhodospirillaceae is that *Chloroflexus* is more tolerant to high-sulfide concentrations.

The morphological relationship between *Chloroflexus* and the blue-green algae is no greater than that between the nonphotosynthetic flexibacteria and the blue-green algae. Although some case can be made that a number of nonphotosynthetic flexibacteria are really apochlorotic cyano-

Order:		*Rhodospirillales*		
Suborders:		*Rhodospirillineae*	*Chlorobiineae*	
Families:	*Rhodospirillaceae*	*Chromatiaceae*	*Chlorobiaceae*	*Chloroflexaceae*
Type Genera:	*Rhodospirillum*	*Chromatium*	*Chlorobium*	*Chloroflexus*
Type species:	*R. rubrum*	*C. okenii*	*C. limicola*	*C. aurantiacus*

Figure 7.26. Proposed taxonomic scheme for the photosynthetic bacteria to accomodate *Chloroflexus*. (From Trüper, 1976.)

phytes (Pringsheim, 1949), there is now such a diversity of filamentous gliding bacteria that it becomes moot which are and which are not derived from blue-green algae.

Trüper's suggestion (Trüper, 1976) that *Chloroflexus* should be placed in a new family, the Chloroflexaceae, under the suborder Chlorobiineae, provides a reasonable taxonomic solution, since it points out the obvious relationship between *Chloroflexus* and the other green sulfur bacteria, but signals the considerable differences in structure and nutrition between these two groups. The position of *Chloroflexus* in the overall taxonomy of photosynthetic bacteria is indicated in Figure 7.26.

From my point of view, the most important evolutionary aspect of the discovery of *Chloroflexus* resides in the light it sheds on the interpretation of Precambrian microfossils and stromatolites. It has been traditional to assume that laminated rocks of obvious biogenic origin were formed by lithification of blue-green algal mats. It has also been traditionally assumed that filamentous microfossils seen in thin sections of Precambrian rocks were the remains of blue-green algae. Because of the major biogeochemical significance of blue-green algae in the formation of O_2 in the atmosphere, it is especially important to be certain at which time in geological history the blue-green algae arose. Because *Chloroflexus* can form laminated mats such as the blue-green algae (Doemel and Brock, 1974; also Chapter 11), and because *Chloroflexus* resembles microscopically small-diameter blue-green algae (as noted earlier, it was in fact classified as a blue-green algal by phycologists), two commonly accepted criteria for the presence in the fossil record of blue-green algae (stromatolites, filamentous microfossils) can no longer be used. Thus, the existence of *Chloroflexus* calls attention to the critical importance of interpreting the fossil record in terms of modern organisms. Further study on *Chloroflexus,* as well as a continued search for other photosynthetic organisms, is desirable, since it will considerably broaden our understanding of the origin and evolution of life.

References

Bauld, J. 1973. Algal-bacterial interactions in alkaline hot spring effluents. Ph.D. thesis, University of Wisconsin-Madison, 115 pp.

Bauld, J. and T. D. Brock. 1973. Ecological studies of *Chloroflexis,* a gliding photosynthetic bacterium. *Arch. Mikrobiol.* **92**, 267–284.

Bauld, J. and T. D. Brock. 1974. Algal excretion and bacterial assimilation in hot spring algal mats. *J. Phycol.* **10**, 101–106.

Brock, T. D. 1966. The habitat of *Leucothrix mucor,* a widespread marine microorganism. *Limnol. Oceanogr.* **11**, 303–307.

Brock, T. D. 1967. Relationship between standing crop and primary productivity along a hot spring thermal gradient. *Ecology* **48**, 566–571.

Brock, T. D. 1968. Taxonomic confusion concerning certain filamentous blue-green algae. *J. Phycol.* **4**, 178–179.

Brock, T. D. 1969. Vertical zonation in hot spring algal mats. *Phycologia* **8**, 201–205.

Brock, M. L., R. G. Wiegert, and T. D. Brock. 1969. Feeding by *Paracoenia* and *Ephydra* (Diptera: Ephydridae) on the microorganisms of hot springs. *Ecology* **50**, 192–200.

Buchanan, B. B., P. Schurmann, and K. T. Shanmugam. 1972. Role of the reductive carboxylic acid cycle in a photosynthetic bacterium lacking ribulose 1,5-diphosphate carboxylase. *Biochim. Biophys. Acta* **283**, 136–145.

Castenholz, R. W. 1973. The possible photosynthetic use of sulfide by the filamentous phototrophic bacteria of hot springs. *Limnol. Oceanogr.* **18**, 863–876.

Doemel, W. N. and T. D. Brock. 1974. Bacterial stromatolites: origin of laminations. *Science* **184**, 1083–1085.

Doemel, W. N. and T. D. Brock. 1976. Vertical distribution of sulfur species in benthic algal mats. *Limnol. Oceanogr.* **21**, 237–244.

Gorlenko, V. M. 1975. Characteristics of filamentous phototrophic bacteria from freshwater lakes. *Microbiology* **44**, 682–684.

Halfen, L. N., B. K. Pierson, and G. W. Francis. 1972. Carotenoids of a gliding organism containing bacteriochlorophylls. *Arch. Mikrobiol.* **82**, 240–246.

Madigan, M. 1976. Studies on the physiological ecology of *Chloroflexus aurantiacus,* a filamentous photosynthetic bacterium. Ph.D. dissertation, University of Wisconsin-Madison, 239 pp.

Madigan, M. T. and T. D. Brock. 1975. Photosynthetic sulfide oxidation by *Chloroflexus aurantiacus,* a filamentous, photosynthetic, gliding bacterium. *J. Bacteriol.* **122**, 782–784.

Madigan, M. T. and T. D. Brock. 1976. Quantitative estimation of bacteriochlorophyll *c* in the presence of chlorophyll *a* in aquatic environments. *Limnol. Oceanogr.* **21**, 462–467.

Madigan, M. T. and T. D. Brock. 1977a. Adaptation by hot spring phototrophs to reduced light intensities. *Arch. Microbiol.* **113**, 111–120.

Madigan, M. T. and T. D. Brock. 1977b. CO_2 fixation in photosynthetically-grown Chloroflexus aurantiacus. *FEMS Microbiol. Lett.* **1**, 301–304.

Madigan, M. T. and T. D. Brock. 1977c. 'Chlorobium-type' vesicles of photosynthetically-grown *Chloroflexus aurantiacus* observed using negative staining techniques. *J. Gen. Microbiol.* **102**, 279–285.

Madigan, M. T., S. R. Petersen, and T. D. Brock. 1974. Nutritional studies on *Chloroflexus,* a filamentous photosynthetic, gliding bacterium. *Arch. Microbiol.* **100**, 97–103.

Meeks, J. C. and R. W. Castenholz. 1971. Growth and photosynthesis in an extreme thermophile *Synechococcus lividus* (Cyanophyta). *Arch. Mikrobiol.* **78**, 25–41.

Pfennig, N. 1965. Anreicherungskulturen für rote und grüne Schwefelbakterien. *Zentralbl. Bakter., I. Abt. Suppl.* **1**, 179–189.

Pierson, B. K. 1973. The characterization of gliding filamentous phototrophic bacteria. Ph.D. dissertation, University of Oregon, Eugene, 240 pp.

Pierson, B. K. and R. W. Castenholz. 1974a. A phototrophic gliding filamentous bacterium of hot springs, *Chloroflexus aurantiacus* gen. and sp. nov. *Arch. Microbiol.* **100**, 5–24.

Pierson, B. K. and R. W. Castenholz. 1974b. Studies of pigments and growth in *Chloroflexus aurantiacus*, a phototrophic filamentous bacterium. *Arch. Microbiol.* **100**, 283–305.

Pringsheim, E. 1949. The relationship between bacteria and Myxophyceae. *Bacteriol. Rev.* **13**, 47–98.

Reichenbach, H. and J. R. Golecki. 1975. The fine structure of *Herpetosiphon,* and a note on the taxonomy of the genus. *Arch. Microbiol.* **102**, 281–291.

Sirevag, R. 1974. Further studies on carbon dioxide fixation in *Chlorobium. Arch. Microbiol.* **98**, 3–18.

Stanier, R. Y., M. Doudoroff, R. Kunisawa, and R. Contopoulou. 1959. The role of organic substrates in bacterial photosynthesis. *Proc. Natl. Acad. Sci.* **45**, 1246–1260.

Tabita, F. R., B. A. McFadden, and N. Pfennig. 1974. D-Ribulose-1,5-bisphosphate carboxylase in *Chlorobium thiosulfatophilum* Tassajara. *Biochim. Biophys. Acta* **341**, 187–194.

Trüper, H. G. 1976. Higher taxa of the phototrophic bacteria: *Chloroflexaceae* fam. nov., a family for the gliding, filamentous, phototrophic "green" bacteria. *Int. J. System. Bacteriol.* **26**, 74–75.

Uffen, R. L. and R. S. Wolfe. 1970. Anaerobic growth of purple nonsulfur bacteria under dark conditions. *J. Bacteriol.* **104**, 462–472.

Chapter 8
The Thermophilic Blue-green Algae

I have discussed in Chapter 3 in a general way the thermophilic blue-green algae. Here I propose to discuss the biology and ecology of those blue-green algae that live around the upper temperature limits. An excellent review on the whole group of thermophilic blue-green algae has been provided by Castenholz (1969a), but a few key concepts have arisen from work done since his review was published. As seen in Table 3.3, when discussing the thermal limit species, only two blue-green algae need be considered, the unicellular *Synechococcus,* a member of the Chroococcales, and the filamentous, heterocystous *Mastigocladus* (Figure 8.1), a member of the Nostocales. Although the genus *Synechococcus* covers a variety of unicellular forms, the thermal limit species is *S. lividus,* an elongate rod, rather narrow in diameter (Figure 8.2). *Synechococcus* is present at higher temperatures than *Mastigocladus* (70–72°C as opposed to 60–64°C), and is the thermal limit species in Western United States and Japan (Molisch, 1926), but seems to be absent from New Zealand (Brock and Brock, 1971) and Iceland (Castenholz, 1969b), where the thermal limit species is *Mastigocladus. Mastigocladus* is the most cosmopolitan thermophilic blue-green alga, being found in thermal areas in every continent and in both hemispheres (Castenholz, 1969a). Even where the thermal limit *Synechococcus* is present, *Mastigocladus* is also almost always present, although generally greatly reduced in amount at temperatures near its thermal limit, probably due to competition from *Synechococcus.*

As I have discussed in Chapter 3 (see also Castenholz, 1969a), a variety of blue-green algae have been reported at temperatures above 70–72°C, but it seems likely that most of these reports have been erroneous. Either temperatures have not been measured precisely at the site or the temperature of the site has increased since growth occurred and the now scalded

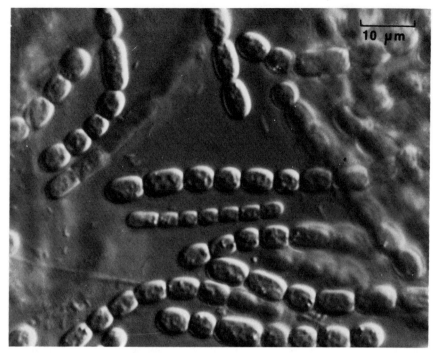

Figure 8.1. Photomicrograph of *Mastigocladus laminosus,* from a culture isolated from effluent of Basin Geyser, Norris Geyser Basin. Nomarski interference contrast.

blue-greens had not been completely eliminated at the time of sampling. Where careful attention has been paid to the temperature of the habitat, and using isotope techniques to prove that the algae were active, it can be readily shown that 70–72°C is the upper temperature limit (Brock, 1967a)

Cultivation of Thermophilic Blue-green Algae

The work on cultivation has been done primarily by Castenholz, and has been well reviewed (Castenholz, 1969a, 1970; see also Castenholz, 1972). It is relatively easy to culture the blue-green algae that grow at 45–50°C, using general blue-green algal culture media. Medium D (see Chapter 4), first designed by R. P. Sheridan when a student of Castenholz, supports the growth of most thermal blue-greens. Either liquid or solid medium can be used, although the former is more convenient for most purposes. Medium D has a salinity of about 1200 mg/l, which is similar to the salinity of many hot spring waters, but the concentration of nitrate and phosphate are considerably higher than hot spring waters. The criterion for the development of Medium D was that it supported the growth of one strain of *S. lividus* at high rate and to a high cell yield.

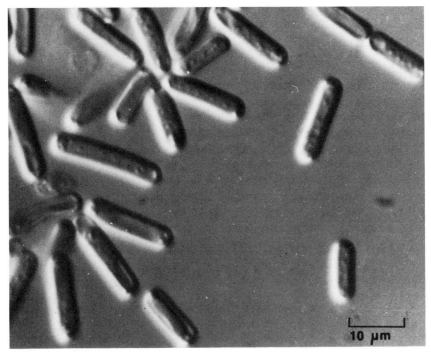

Figure 8.2. Photomicrograph of *Synechococcus lividus* cells from a culture isolated from effluent of Basin Geyser, Norris Geyser Basin. Nomarski interference contrast.

For routine cultivation, Castenholz (1970) used a temperature of 45°C, even if the alga had been isolated from a habitat of other temperature. A light intensity of 500 to 1500 lux is used, being provided continuously from fluorescent lamps. When temperatures above 55°C are used, it is essential to use stainless steel closures to reduce evaporation. For several years, I cultured a strain of *S. lividus* in a New Brunswick water bath shaker with a lucite lid to prevent evaporation and permit lighting. Better growth yields can be obtained by enriching the atmosphere with CO_2, although since Medium D is not very strongly buffered, it is important that the CO_2 concentration in the gas phase not be too high, or otherwise the pH will drop. A mixture of 1% CO_2 in air is preferable.

Castenholz (1970) has been able to preserve many thermophilic blue-green algae by freezing in liquid nitrogen or by lyophilization, although not all species will survive these treatments. *Mastigocladus* is more tolerant than *Synechococcus,* and can even survive drying at room temperature. For those cultures that cannot be frozen or lyophilized, it is necessary to make transfers every 8–9 weeks.

Jackson and Castenholz (1975) have shown that a temperature of 45°C in the initial enrichment of blue-green algae from nature is selective for thermophilic forms. Most blue-green algae isolated at 45°C will grow at

50°C and 55°C. The thermophilic blue-green algae that have a growth temperature optimum of over 45°C occur exclusively in heated waters (either hot springs or artificially heated waters), and are not found in temperate waters of normal temperature. However, studies (Jackson and Castenholz, 1975) carried out in Everglades National Park, Florida, showed that blue-green algae able to grow at temperatures as high as 45°C could be isolated from sun-heated waters that normally only reach temperatures of 35–40°C.

The Genus *Mastigocladus*

This member of the Nostocales is a cosmopolitan thermophile, being found world-wide. This alga was first described from Karlsbad hot springs (now Karlovy Vary in Czechoslovakia) by Ferdinand Cohn (1862), who pointed out the evolutionary and biogeochemical significance of organisms living in hot springs. This alga is one of the easiest to recognize in natural material, because of its morphological complexity, although at temperatures near its thermal limit the filaments may become very short, lacking the characteristic branching morphology.

Many authors even in recent times have used the genus name *Hapalosiphon* as a synonym for *Mastigocladus,* but Geitler (1932) uses the name *Hapalosiphon* for another group of algae. As described by Geitler (1932), *Mastigocladus* forms single filaments with long, narrow side branches, the branches generally arising from one side. Heterocysts are formed intercalarly, rather than at the base of the branches, and hormogonia are not known. Geitler only recognizes a single species, *M. laminosus.* The organ-

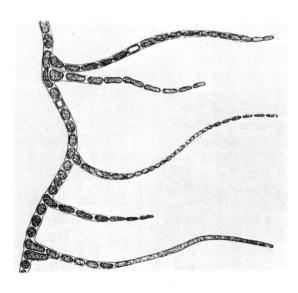

Figure 8.3. Geitler's drawing of *Mastigocladus laminosus* f. *typica,* illustrating the characteristic branches and position of the heterocysts. (From Geitler, 1932.)

Figure 8.4. Geitler's drawings of the three forms of *Mastigocladus*, (a) f. *typica*, (b) f. *anabaenoides*, and (c) f. *phormidioides*. (From Geitler, 1932.)

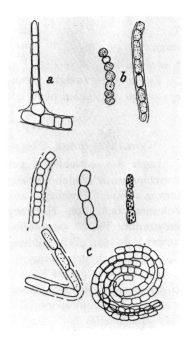

ism forms tough mats. The filaments are 4–6 μm wide, and the side branches are around 3 μm wide, and oriented at right angles to the main filaments. The cells of the main filaments are barrel-shaped or short cylinders, and the cells of the side branches are long cylinders. The heterocysts are up to 6.5 μm wide, and occur either singly or doubly. Geitler (1932) describes three forms of *M. laminosus: M. laminosus* f. *typica* (Figure 8.3), *M. laminosus* f. *anabaenoides,* in which the filaments are not branched but the cells are barrel-shaped or spheres (Figure 8.4); and *M. laminosus* f. *phormidioides* (Figure 8.4); in which the filaments are not branched, and heterocysts are not formed. I personally have never been able to figure out how f. *anabaenoides* could be distinguished from the genus *Anabaena,* or f. *phormidioides* from the genus *Phormidium.* Castenholz, who has studied *Mastigocladus* in considerable detail, can recognize a *Mastigocladus* filament in the field in natural material even if the filament is only a few cells long (personal communication). As he notes (Castenholz, 1969a), the variability in *Mastigocladus* may simply be a developmental or temperature-controlled phenomenon, and can only be deciphered by careful cultural studies. Since all of Geitler's taxonomy is based on field material, it is unlikely to be reliable. Castenholz (1969b) has found in preliminary studies with cultures that there are at least two genetic "races" with unique morphologies and temperature tolerances. Clones isolated from Iceland material at temperatures near the upper temperature limit (60–62°C) grew primarily as unicells and 2–10-celled filaments, 2.5–6.0 μm broad, without branching, whether grown at 45°C or 62°C. Strains of *Mastigocladus* isolated from habitats of 45°C and cultured at 45°C grew in the "normal" *Mastigocladus* morphology.

The tolerance of *Mastigocladus* to nonthermal conditions, and its rapid rate of dispersal, are illustrated by the report of Castenholz (1972) that this alga colonized fumaroles on the Icelandic island of Surtsey by 1970, only a few years after the island was formed. Both the short-form and the typical form of *Mastigocladus* were isolated. There are no thermal waters in the vicinity of Surtsey, the closest hot springs being approximately 75–90 km north. *Mastigocladus* also colonizes the man-made thermal effluents of the Savannah River Nuclear Plant, near Aiken, South Carolina (C. G. Flier-mans, personal communication). Large thermal streams are created by the effluents from the reactors on this plant (Figure 8.5), and *Mastigocladus* has successfully colonized them. These thermal streams are unique in that they do not flow constantly, and may be completely shut down for periods of a few weeks or months. Within a week or so after the flow of hot water resumes, visible mats of *Mastigocladus* can be seen. Although there are no extensive thermal areas for several thousand kilometers from the Savannah River Plant, there are warm springs in northern Georgia and northern Virginia, which might contain *Mastigocladus,* and it is also possible that this alga lives in waters subject to solar heating (see Jackson and Casten-holz, 1975).

Mastigocladus is interesting because it fixes nitrogen (Fogg, 1951), and is the most thermophilic nitrogen-fixing blue-green alga (Stewart, 1970). As with other heterocystous blue-green algae, the formation of heterocysts is repressed by combined nitrogen. Stewart (1970) and Stewart, Brock, and Burris (unpublished) carried out nitrogen-fixation studies on algal mats in Yellowstone National Park using $^{15}N_2$ and acetylene reduction to measure nitrogen fixation. No evidence of nitrogen fixation was obtained for algal mat samples collected at temperatures above 60°C, although active fixation was obtained in most samples collected from temperatures below 60°C. The rate of nitrogen fixation varied markedly from habitat to habitat, and was greatest at temperatures around 45°C.

As described by Holton et al. (1968), the fatty acid composition of *Mastigocladus* differs from that of the other blue-green algae. It lacks polyunsaturated fatty acids, the dominant fatty acids being 16:0, 16:1, 18:0, and 18:1. It is of interest that this suite of fatty acids is quite similar to that shown by the eucaryotic alga *Cyanidium caldarium* (see Chapter 9), which also grows in the same temperature range as *Mastigocladus,* but at acid pH.

As Castenholz (1969a) has pointed out, in areas such as Yellowstone Park, where both *Mastigocladus* and *Synechococcus* coexist, some springs are dominated by *Mastigocladus,* whereas others (superficially similar) are dominated by *Synechococcus*. Since *Mastigocladus* fixes nitrogen and *Synechococcus* does not, a logical hypothesis might be that in springs low in combined nitrogen, *Mastigocladus* is present, whereas if

Figure 8.5. The thermal stream derived from the cooling water of a nuclear reactor on the Savannah River Plant, near Aiken, South Carolina. An example of a man-made habitat for *Mastigocladus*.

there is sufficient combined nitrogen, the more rapidly growing *Synechococcus* could dominate. So far, there is no evidence to support this hypothesis, and it might be difficult to test, because it is difficult to be certain of the limiting concentration of a nutrient such as nitrogen in the flowing stream situation. [As shown by Sperling and Hale (1973), the photosynthetic activity of *Mastigocladus* is considerably stimulated by water movement.]

It is surprising that despite the cosmopolitan distribution of *Mastigocladus,* the ease of cultivation, and the interesting morphological and biochemical features, more extensive work has not been done on the physiological ecology of this interesting organism. Our own work concentrated on *Synechococcus* because this was the dominant alga of the springs we studied. In other parts of the world, where *Mastigocladus* is dominant, ecological studies on this alga would be quite easy to do, and could be readily coupled to sophisticated laboratory experiments.

The Genus *Synechococcus*

Although the taxonomy of the unicellular blue-green algae has been in a chaotic state, it has been greatly clarified by the careful cultural and biochemical studies of Stanier et al. (1971). There is a variety of unicellular blue-green algae with cylindrical, ellipsoidal, or spherical cells which divide by repeated binary fission in a single plane, frequently forming short chains of cells. Stanier et al. (1971) classify such organisms in an arbitrary typological group, group IA, and note that if the taxonomic system of Geitler is used, these organisms would be classified in the genus *Synechococcus.* The thermophilic strains referable to this group do not differ greatly from the nonthermophilic strains, except in temperature limits for growth, and in the fact that polyunsaturated fatty acids are low in amount or absent. A similar impoverishment in polyunsaturated fatty acids was also noted for *Mastigocladus* (see above), and probably for the same reason, the necessity of saturated fatty acids for the construction of a thermostable membrane (see Chapter 4).

It is thus of considerable evolutionary interest that the blue-green alga capable of growing and photosynthesizing at the highest temperature of any photosynthetic organism cannot be distinguished morphologically or biochemically from similar nonthermal blue-green algae.

The ubiquity of *Synechococcus* in Yellowstone hot springs was noted by Copeland (1936), who described a number of new species. It is extremely unlikely that all of Copeland's species are valid, and it is probably best to lump all of the high-temperature *Synechococcus* of Yellowstone hot springs into a single species, *S. lividus* [a name first used in the modern literature by Dyer and Gafford (1961)]. For the present purposes, this name can be taken to comprise those narrow-form, elongately unicellular blue-green algae living at temperatures of 50°C and above in hot springs in Yellowstone and other parts of Western United States (Castenholz, 1969a).

Morphology of *Synechococcus lividus*

Cells of *S. lividus* are generally about 1.5 μm wide by 8–10 μm long (Figure 8.2). In culture, chains may often form, but in nature the cells appear mostly singly. There has been some suggestion (Brock, unpublished) that size of the cells found near the upper temperature (68–72°C) limit is larger than size of cells at lower temperatures (50–65°C), but no precise data have been obtained. Some strains are motile, moving by a twitching or jerky gliding motion, but most strains are immotile.

The fine structure of *S. lividus* has been studied by Edwards et al. (1968) and by Holt and Edwards (1972). The ultrastructural features of *S. lividus* (Figures 8.6 and 8.7) are virtually identical to those nonthermal blue-greens of the same Stanier typological group. It was not possible by either thin sectioning or freeze-etching to discern any ultrastructural features that might relate to the high-temperature habitat of this organism (Holt and Edwards, 1972). As noted in Chapter 4, electron microscopy also revealed nothing unusual about the fine structure of *Thermus aquaticus*.

S. lividus contains phycocyanin as its sole phycobilin, and Edwards and Gantt (1971) have shown that both cultures and natural samples of this alga (collected from Yellowstone hot springs) contain structures called *phycobilisomes,* which are aggregates of phycobiliproteins. The phycobilisomes of *S. lividus* did not differ significantly from those previously described in nonthermal blue-green and red algae. Furthermore, the phycocyanin from *S. lividus* has the same amino acid composition, molecular weight, sedimentation, and antigenic properties as the phycocyanin from a nonthermal blue-green alga (Edwards and Gantt, 1971), although it does show a thermal difference, requiring a temperature of 49°C to form large aggregates whereas the optimum temperature for aggregate formation in mesophilic blue-green algae is 25°C (Edwards and Gantt, 1971).

Temperature Strains of *Synechococcus lividus* and the Upper Temperature Limit

Peary and Castenholz (1964) first recognized the possibility that temperature strains of *S. lividus* might exist at different temperatures along hot spring effluent channels. Their work was done at Hunter's Hot Springs, a small spa in the desert of southeastern Oregon. In this location, a number of hot springs rise in a gravel flat and flow gently across the gravel to form a marsh. Striking thermal gradients are seen, traced by the development of blue-green algae, perhaps made more dramatic by the fact that the gentle water flow has little erosional force. In the temperature range from the upper limit (around 75°C) down to 50°C, *S. lividus* is the only blue-green alga present, and the populations at the various temperatures appear to be quite similar.

Peary and Castenholz (1964) reported the culture of *S. lividus* from various temperatures in the Hunter's Hot Springs. Enrichment cultures developed at different temperatures were maintained at those temperatures and then clones isolated at the same temperatures. Initially, eight clones were isolated, but when the temperature optima were determined, they grouped themselves into four groups, each representing a single temperature strain, as shown in Figure 8.8. Strains from 45°C, 48°C, and 53°C

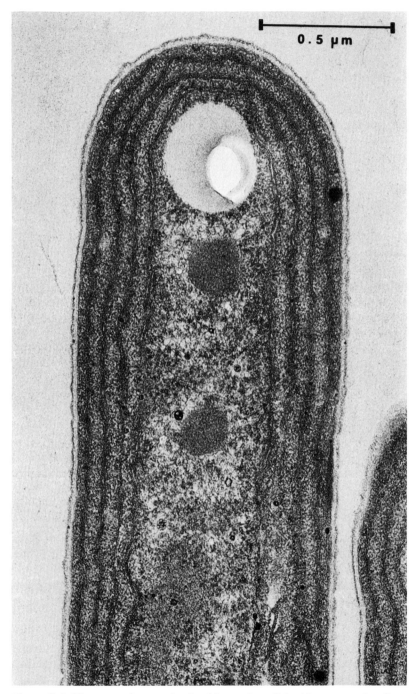

Figure 8.6. Electron micrograph of a thin section of the blue-green alga *Synecho-coccus lividus*. (From Edwards et al., 1968.)

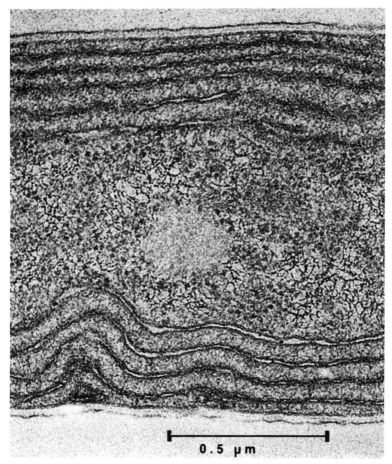

Figure 8.7. Higher-power electron micrograph of *S. lividus,* showing details of the photosynthetic membranes and cell wall. (Courtesy of Mercedes Edwards.)

represent one group, with a temperature optimum of 45°C, strains from 55°C and 60°C represent a second group, with an optimum at 50°C, strains from 66°C (and possibly 71°C) represent a third group, with an optimum of 55°C, and one strain from 75°C represents the fourth group, with an optimum around 65°C.

Most striking in the data given in Figure 8.8 is the wide variation in growth rate shown by the several groups, even at their optimum temperature. The low-temperature strains, with optima of 45°C, show exceedingly rapid growth rates, as many as 10 doublings a day, a rate that is matched by few other algae. On the other hand, the strain from 75°C had a low growth rate under all conditions, and seems to be an "injured" strain. Further work on the high-temperature *Synechococcus* is discussed later in this section.

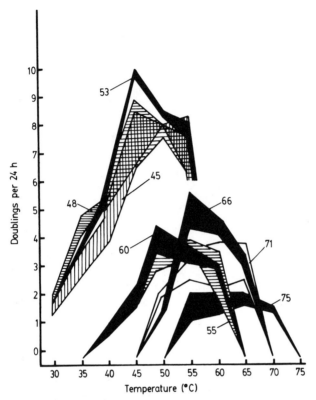

Figure 8.8. Growth rates of eight clones of *Synechococcus lividus* at 5-degree Celsius intervals. The clones can be grouped into four temperature strains. Clones 45, 48, and 53: strain I; 55 and 60: II; 66 and (?) 71: III; 75: IV. (From Peary and Castenholz, 1964.)

In other work, Peary (1964) attempted to change the temperature optima of some of the isolates, by growing them for a number of generations at different temperatures. He was unable to change the temperature optimum in this way, indicating that the temperature optimum is genetically fixed.

In our early work at Yellowstone, we used a ^{14}C technique (Brock and Brock, 1967) to determine the temperature optimum for photosynthesis of natural populations of *Synechococcus*. The approach here (Brock, 1967a) was to select populations that had developed at different temperatures along the effluent channel, remove a series of cores of mat from each location and place in vials with spring water, equilibrate several vials at each of a variety of temperatures for a few minutes, and then inject $^{14}CO_2$ and measure the rate of photosynthesis over the next hour of incubation. Generally, a characteristic temperature–rate curve was obtained, with a well-defined optimum (Figure 8.9). When the optimum temperature for each series was plotted against the temperature of the habitat, a good

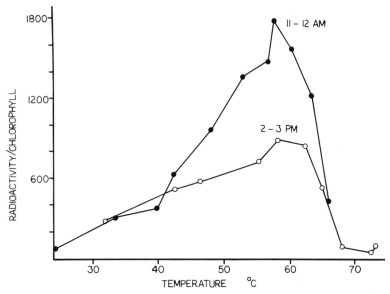

Figure 8.9. Photosynthesis of algal cores taken from a location at 58.5°C in Mushroom Spring and incubated at various temperatures. Curves are for separate experiments on 29 August 1966 (11–12 A.M.) and 21 August 1966 (2–3 P.M.). Values on the ordinate are cpm/μg chlorophyll. (From Brock, 1967a.)

correlation was obtained (Figure 8.10). The line in this figure is the theoretical line that would be obtained if the optimum temperature and the habitat temperature were identical. The agreement between the theoretical line and the experimental points is remarkably close.

These results strengthen to some extent the conclusion of Peary and Castenholz (1964), since they involve natural populations that have not

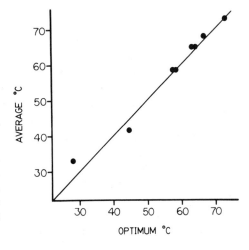

Figure 8.10. Temperature optimum for photosynthesis and habitat temperature of each station. The ordinate was determined from data of temperature measurements made frequently over a 1-month period during the time that the experiments were being carried out. The abscissa was determined from optima for separate experiments such as shown in Figure 8.9. (From Brock, 1967a.)

been forced to adapt to culture conditions. There is some suggestion that perhaps each population is adapted to its own habitat temperature, and if there are only four temperature strains, as found by Peary and Castenholz (1964), then there may be some additional physiological adaptation that is superimposed upon the genetic adaptation. The existence of physiological adaptation was later shown in cultural studies by Meeks and Castenholz (see below).

The striking conclusion from both our results and those of Peary and Castenholz (1964) is that even the blue-green algae which are living at a temperature close to the upper temperature limit for photosynthetic life are optimally adapted to their environmental temperature. Thus, it is not the case that these high-temperature populations consist of more normal forms that have merely extended their range to higher temperatures, because of the lack of competition, as has been occasionally suggested, but they are forms that are specifically adapted to these environments and can grow nowhere else.

In subsequent work, Meeks and Castenholz (1971) carried out some detailed studies on the high-temperature strain of *S. lividus*. Although they found it difficult to isolate axenic cultures, they obtained unialgal cultures from several enrichments and measured the effect of temperature on growth and photosynthesis. The high-temperature *S. lividus* has a minimum temperature for growth of 54°C, a maximum of 72°C, and an optimum between 63°C and 67°C (Figure 8.11). The high minimum temperature is noteworthy, as several of the most widely studied thermophilic strains of *S. lividus* have *optima* at the temperature at which the high-temperature form is no longer able to grow. As had also been shown earlier by Peary and Castenholz, even at its optimum temperature, the high-temperature form grew slowly, with a minimum generation time of about 11 hours.

As also seen in Figure 8.11, the minimal temperature for photosynthesis is considerably lower than the minimal temperature for growth, about 33°C, but the maximal temperature is near 75°C. However, although some photosynthesis could occur at temperatures up to 75°C in short-term experiments, when a population grown at 70°C was incubated at slightly higher temperatures for short periods of time and then tested for photosynthesis at 70°C, it was found that the ability to photosynthesize at 74°C was rapidly lost, suggesting thermal destruction of some part of the photosynthetic apparatus, so that the "stable" upper limit for photosynthesis was about 73°C (see Figure 8.12). As also shown in Figure 8.12, there was some physiological adaptation to high temperature, as 57°C-grown cells preincubated at 71°C rapidly lost ability to photosynthesize, whereas 70°C-grown cells photosynthesized well at 71°C. Evidence for physiological adaptation to supraoptimal temperatures has also been shown for a lower temperature strain of *S. lividus* by Sheridan and Ulik (1976). Meeks also showed that there was an interrelationship between light intensity and temperature. When cultured at 55°C, the saturating light intensity for growth was 400

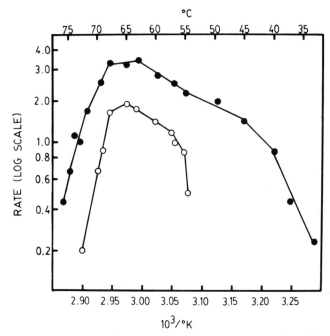

Figure 8.11. Growth rate and photosynthesis of a high-temperature strain of *S. lividus* (unialgal culture) as a function of temperature. Open circles: growth rate, each point represents doublings per 24 hr; solid circles: photosynthetic rate, each point × 10³ cpm/g chlorophyll-a. (From Meeks and Castenholz, 1971.)

footcandles, whereas at 65°C the saturating intensity was 550 footcandles and at 68°C the saturating intensity was 700 footcandles.

The lower growth rate of the high-temperature *S. lividus,* even at its optimum temperature, may indicate that this strain has had to sacrifice growth efficiency even to be able to grow at all. However, since this is the only photoautotroph able to grow in the temperature range of 65 to 73°C, it is also conceivable that it grows poorly because it has not been faced with competition from other organisms. In the absence of competitive pressure, all that would be necessary would be a growth rate sufficiently rapid to form new cells to replace those lost by washout, lysis, etc. (see next section).

Meeks and Castenholz (1971) conclude: "The 'stable' upper temperature limit for photosynthesis in the laboratory fits closely to the upper limit for growth in the laboratory and field (Brock and Brock, 1968) as well as to the photo-dependent incorporation in the field of $^{14}CO_2$ (Brock, 1967a) and $^{32}PO_4$ (Kempner, 1963). This may imply that some aspect of the photosynthetic system determines the upper temperature limit for growth, although it would seem an unnecessary or unusual adaptation for any of the metabolic systems to be potentially operative at temperatures much above the

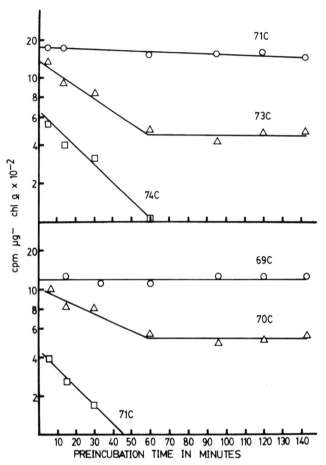

Figure 8.12. Time course of photosynthesis at supraoptimal temperatures. The points correspond to the time of preincubation at the desired temperature, not including a 5-minute preincubation and 15-minute incubation in the presence of radioisotope. Top: culture grown at 70°C; bottom: culture grown at 57°C. (From Meeks and Castenholz, 1971.)

upper limit. No single system necessarily determines the upper temperature limit of growth of this obligate thermophile. However, there are strong indications that derangements of the photosynthetic system in this organism are occurring over the entire upper (supraoptimal) portion of its growth range (68 to 72°C). The tendency to require a greater light intensity for growth saturation with increasing temperatures may suggest that light is not being efficiently utilized at these supraoptimal temperatures. This is emphasized by the observation that under conditions of adequate nutrients and CO_2, growth at 72°C occurred only at intensities greater than 1000 footcandles.''

Returning to the ecological observations, the striking fact is that there is a clear-cut upper temperature limit for photosynthetic life at about 73°C. The fact that nonphotosynthetic bacteria and chemolithotrophs are found at higher temperatures (see Chapter 3) suggests that the 73°C limit has something to do with the sensitivity of the photosynthetic apparatus. It seems as if it has not been possible for organisms to construct a photosynthetic apparatus that is both functional and stable at very high temperatures. Possibly the photosynthetic membranes must retain a degree of fluidity that is incompatible with thermostability at very high temperatures.

Growth Rates of *Synechococcus* in Nature

One of the simplest experiments we did was to darken the channel of a flowing spring and observe what happened. Within a few days, the color of the mat had changed from yellowish green to pink, as the *Synechococcus* washed out. At the end of the experiment, only the undermat of *Chloroflexus* (see Chapter 7) was left. This simple experiment immediately gave me the idea for a technique for measuring the growth rate of *Synechococcus* directly in its natural habitat. The technique is possible because the hot spring mat is in a steady state, analogous to a chemostat (Brock and Brock, 1968). In a chemostat, the generation time of the culture is a function of the replacement rate of the medium, which is determined by the volume of fluid in the vessel and the flow rate. As cells are washed out of the vessel, they are replaced by new cells resulting from cell division. Given a chemostat in operation under steady-state conditions but with an unknown flow rate (and hence an unknown generation time), in principle, we can still calculate the generation time if we decrease to zero the concentration of the limiting nutrient in the inflow. Further growth cannot occur because of the lack of nutrient and only washout occurs. We can measure the rate of washout by obtaining cell counts on samples at periodic intervals. The rate of loss of cells will be exponential with time because this rate follows the same kinetics as any dilution process. The time required for the population to be reduced to half its initial level is then equivalent to the washout rate or replacement rate. Because in a chemostat the generation time is proportional to flow rate, the generation time can thus be calculated. Although in this example growth was stopped by eliminating an essential nutrient, other methods of stopping growth could be used. In the present case, darkening of the mat stops growth because *Synechococcus* is an obligate autotroph.

In the studies we carried out, we selected effluent channels with good constant flow rates and relatively constant temperatures. The water in all areas of the channels used had approximately laminar flow characteristics, resulting in relatively flat homogenous algal mats over regions of at least 100 cm². Permanent stations were established, and preliminary cell counts were obtained to confirm that the mats were in steady state. Cores were taken at intervals with a cork borer, the material homogenized in spring water, and the cells counted with a Petroff-Hausser counting chamber. The alga was easily recognized and occurred almost exclusively as unicells. Chains of cells were never seen, although occasionally two cells remained attached after cell division.

After samples had been taken for several days to establish the steady-state levels, the areas were darkened in a way that did not impede water flow (Figure 8.13). The covers used were constructed from opaque black plastic sheeting wrapped around a metal frame and suspended from a wooded support over the spring channel. Cores were removed only from the central region under the plastic where total darkness existed. The total darkness of this central area was verified by measuring $^{14}CO_2$ incorporation into the algal material; the rate of incorporation was quite low and was the same in transparent vials as in vials wrapped with opaque covers.

The initial darkening was done just before sundown, so that the initial dark period corresponded with normal night. Figure 8.14 summarizes the data obtained for the next 2 weeks at a station at 69.7°C in Mushroom Spring. The rate of loss of cells was exponential with time, as would be predicted. The half-time of the loss rate was 40 hours, and the generation time was 40 hours, if we assume that cell division can occur in both light and darkness. Within 1 week, no more algae were visible and only bacteria

Figure 8.13. The darkening experiment at Mushroom Spring, summer 1968. The black cover is constructed of thick plastic and is wrapped around a metal frame to give it strength. It is suspended from a wooded frame so that it is just above the water level. Water flow and habitat temperature are unaffected. The frame in the background contains a neutral density glass for light-adaptation experiments.

Figure 8.14. Loss rate of *S. lividus* cells at Station II, Mushroom Spring, after darkening. Day zero, 15 August 1967. (From Brock and Brock, 1968.)

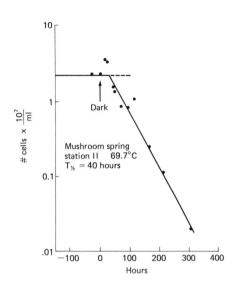

remained. Figure 8.15 depicts the appearance of this area after 2 weeks of darkness (the difference in appearance is much more dramatic in color photographs).

Similar data were obtained at a 72°C station at Grassland Spring and are presented in Figure 8.16. Again, an exponential rate of cell loss was observed, with a half-time of 22 hours. For two stations, after the loss of algal numbers appeared complete, the covers were removed and the rate of recovery was measured. After a delay, there was an exponential increase in cell numbers, followed by a continued slower increase. In neither case did the population density return to the original steady-state level during the time of the observations. Presumably, the later stages of the recovery phase involved a gradual slow return to the original population density, because even during the recovery period there was probably some washout, and thus the recovery rate can never represent the actual growth rate in the steady state (see Figure 7.23 for an analogous study in a nonflowing spring). As shown in another study (Brock and Brock, 1969a), when the *Synechococcus* populations were wiped out from a spring by a catastrophe (a violent hailstorm), complete recovery of the mat had occurred within 6 months. However, again the doubling times were much slower than the washout times. However, if it is assumed that washout occurs at the same rate during the whole recovery process, then the slow doubling times seen during the recovery represent growth rates that were even more rapid than

Figure 8.15. Appearance of the mat at Station II after 13 days of darkening. Note the light color of the mat, due to virtually complete disappearance of the *S. lividus* cells. (cf. Figure 8.13).

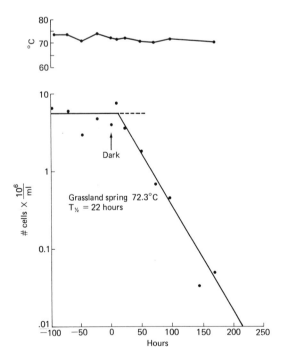

Figure 8.16. Loss rate at a 72.3°C station at Grassland Spring. Day zero, 19 July 1967. (From Brock and Brock, 1968.)

the steady-state rates. (Put another way, if the growth rate during recovery were identical to the steady-state growth rate, then recovery could never occur.)

In the summer of 1968, we used this technique in a number of additional studies which never got published. We determined the washout rate for four stations along the thermal gradient of Mushroom Spring to determine whether there was any relationship between growth rate and temperature (Table 8.1). A very striking pattern emerged: the growth rate was the most rapid at the *high temperature,* and decreased markedly as the temperature was *lowered* about 10°C. At all four stations, the only organism present was *Synechococcus,* and there were no significant morphological differences. As also shown in Table 8.1, the standing crop of *Synechococcus* at all four stations was just about the same. These results suggest that growth and productivity were highest at the upper temperature. If standing crop is multiplied by washout rate, the productivity of the population can be calculated, and it can be determined that productivity was the highest at the high temperature, dropping as the temperature was reduced. This result is at variance with the cultural studies of Peary and Castenholz (1964) and Meeks and Castenholz (1971). These workers found that the growth rate of the high-temperature strain of *Synechococcus* was much lower than that of the lower-temperature strains. (I present an explanation for this discrepancy at the end of this section.)

During the 1968 study presented in Table 8.1 we also did several additional studies in the nature of controls. I was concerned that the washout technique might be biased if there were considerable settling of algae from upstream onto the mat farther down, thus reducing the apparent washout rate. Consequently, we placed microscope slides on top of the mat to measure the rate of settling. The slides used were Shoemaker Fungus Microculture slides, from Clay-Adam, which were 3×1 inch by 3 mm thick, and had a channel in the center that was of unpolished glass, $\frac{1}{2}$ inch wide and about 1 mm deep. The slides were tied so that the open ends of the channel were at right angles to the current flow. These slides were removed at intervals and the number of cells per unit area determined microscopically. The results are given in Table 8.1. As would be predicted, the settling rate is lower at the hottest station, since this station is nearest the source and hence has less algal biomass upstream of it. (Since the source of Mushroom Spring had a temperature of about 72°C during the time that these experiments were done, and because of the large size of the Mushroom pool, there was a fair bit of algal productivity upstream even of Station I.) However, at all stations, the settling rate was many orders of magnitude lower than the standing crop, and since the washout rate as measured from the darkening experiments was of the order of a few days, the growth rates were thus many times faster than the settling rates. Therefore, it seems unlikely that settling would invalidate the values for growth rates calculated from the darkening experiments.

Table 8.1. Steady-State Growth Rates at Several Stations of Mushroom Spring, as Determined by Measuring the Washout Rate after Darkening[a]

Station	Temperature (°C)	Standing crop before darkening (cells/core)[b]	Cells/m²	Drift: cell counts in water collected at each station (cells/1000 ml)[c]	Washout rate, determined from graph of exponential loss rate (days)	Settling rate (cells/m²/day)[d]
I	68.0	1.7×10^8	6.1×10^{12}	8.1×10^6	3.8	3.3×10^6
II	65.2	1.4×10^8	5.0×10^{12}	1.32×10^7	4.1	2.3×10^7
V	59.1	1.6×10^8	5.8×10^{12}	2.31×10^7	6.8	9.7×10^6
VI	57.2	1.6×10^8	5.8×10^{12}	1.88×10^7	10.2	2.2×10^7

Productivity of Stations at Mushroom Spring, Based on the Data above (Cells/m²/day)

Station	Productivity, based on washout rate and standing crop	Productivity, based on cell counts in water and flow rate data for Mushroom Spring
I	1.6×10^{12}	8.1×10^{12}
II	1.2×10^{12}	
V	8.5×10^{11}	
VI	5.7×10^{11}	

[a]Mushroom Spring flow rate was 70 liter/min or 10^6 liter/day. Productivity from drift was only calculated at the upper station, since water at the lower stations contains cells derived from all of the mats above and cannot reflect productivity at the lower stations.
[b]Core size, 0.23 cm².
[c]Cell counts in water over the mat done by fluorescence microscopy on membrane filters (see text).
[d]Settling rates determined from counts on special slides immersed in the channel (see text).

At the same time that the washout and settling rates were measured, drift was measured by collecting 1-liter water samples just below each station, filtering 100-ml amounts on membrane filters, and doing algal cell counts by fluorescence microscopy. The values obtained are also given in Table 8.1. For the calculation of productivity based on these measurements, only the uppermost station was used, since this is the only site where cells in the water could come only from the vicinity of the mat. All that is necessary is to multiply the flow rate by the cell concentration in the water, and convert to a standard time period. The calculations are given in the lower part of Table 8.1, adjacent to the productivity values calculated from the washout data. As seen, the two calculations for Station I are of the same order of magnitude, and considering the errors involved, and the various assumptions that must be made (see below), the agreement is surprisingly good. For the two estimates of productivity to be equated, it would of course be essential that all cell loss in the darkening experiment were due to a washout, and losses due to lysis and predation would have to be minimal. From the response of the *Synechococcus* population to reduced light intensity, there is reason to believe that lysis does occur (see next section). (However, if significant lysis occurred, then rate of cell disappearance on darkening should be higher than the rate calculated from cell counts of the water, whereas as seen in Table 8.1 the rate is actually lower.) On the other hand, estimating productivity by counting cell numbers in water samples requires the assumption that drift rate is constant throughout the 24-hour period, whereas in fact it may be much higher at one time than another. It must also be assumed that flow rate remains constant, and that the estimate of flow rate, done rather crudely by measuring the rate at which a large plastic bag is filled, is accurate.

Since all the controls just described do suggest that the darkening experiments indeed provide a reasonable measure of growth rate, it must be explained why growth rate is more rapid at the higher temperatures, even though cultural data (Meeks and Castenholz, 1971; Peary and Castenholz, 1964) suggest that the reverse should be the case. One likely possibility is that conditions are considerably more favorable for photosynthesis at the high-temperature station than at the lower, due to availability of light. As noted earlier in this chapter, near the upper temperature limit the mat is quite thin, whereas as the temperature drops, the mat becomes considerably thicker. This thickness may be due primarily to organisms other than *Synechococcus,* since the cell count per unit area is about the same throughout the temperature gradient (Table 8.1). One significant organism in the mats is *Chloroflexus,* which was discussed in detail in Chapter 7. This filamentous photosynthetic bacterium forms thick mats and provides a matrix within which the *Synechococcus* cells become embedded, which may help to some extent to provide stability to the *Synechococcus* cells, but at the same time should cause additional shading of the *Synechococcus* cells from light. *Chloroflexus* lives at the high-temperature station, but since its upper temperature limit is similar to that of *Synechococcus,* it is present only as a thin film and thus causes considerably less shading than at the lower temperatures. I thus hypothesize that the reason *Synechococcus* grows better at the high-temperature station in nature, even though the high-temperature strains grow poorly in culture, is that reduction in growth rate due to light limitation does not occur. I discuss in more detail the relationship of *Synechococcus* to light intensity in the next section. Finally, it should be noted that the growth rate at the high-temperature station is lower than the growth rate obtained by Meeks and Castenholz (1971) in culture.

Light Responses and Adaptation of Thermophilic Blue-green Algae

Synechococcus in Yellowstone

During the discussion of *Chloroflexus* (see Chapter 7), I discussed the vertical distribution of chlorophyll pigments within the mats, and showed (Figures 7.14 and 7.15) that virtually all of the chlorophyll-a and *Synechococcus* cells were in the top 1–2 mm of the mat. Autoradiographic studies using $^{14}CO_2$ permitted a study of the distribution of photosynthetic activity within this narrow region (Brock and Brock, 1969b). Cores were removed from the mat at a station at 57°C (where the mats are thickest) and incubated intact in spring water containing $^{14}CO_2$, then frozen, and thin slices used to prepare autoradiograms. Quantitation was done by counting the number of silver grains per *Synechococcus* cell, and the data are presented in Table 8.2. The advantage of autoradiography is not only that it permits analysis of photosynthesis in layers too thin to provide sufficient radioactivity for conventional isotope counting, but it ensures that measurements are being made in the organism of interest. As seen in Table 8.2,

Table 8.2. Radioactivity of *Synechococcus* Cells at Different Levels through an Algal Core[a] as determined by Autoradiography

Layer	Labeled cells (%)	No. silver grains per labeled cell
Surface	95	4.8
Next to surface	57	3.5
Next to bottom	20	1.8
Bottom	35	1.3

[a]Thickness of core, about 5 mm; each layer analyzed, about 1.2–1.3 mm. Core 0.75-cm diameter. Source of core, Mushroom Spring, Station VI, 57°C. Incubation 4 hr at temperature of origin in 1 μCi/ml NaH^{14}CO$_3$ in 5-ml volume. (From Brock and Brock 1969b.)

photosynthesis is maximal in the surface layer and falls off progressively in the deeper portions of the core, but some photosynthesis still occurs in the lower portions of the core.

The results in Table 8.2 would suggest that self-shading occurs in the mat, so that the cells near the bottom of the mat receive insufficient sunlight for optimal photosynthesis. Also, the maximum thickness of the *Synechococcus* layer is probably determined by self-shading phenomena.

These results raise the question of the minimum light intensity required for growth in nature of *Synechococcus,* and how extensively the alga can adapt to changes in sunlight intensity. Studies done in 1968 (Brock and Brock, 1969b) provided some evidence of adaptation, but they were done before the existence of the photosynthetic bacterium *Chloroflexus* was known to be present in mats, and hence may not be too clear-cut. Doemel and Brock (1974) showed the ability of *Chloroflexus* to adapt to extremely low-light intensities, and Bauld and Brock (1973) had used DCMU to distinguish bacterial from algal photosynthesis. Then Madigan and Brock (1976) developed a thin layer chromatographic technique for quantifying algal chlorophyll-a and bacteriochlorophyll-c in mats. It thus became possible to restudy the question of light adaptation of *Synechococcus,* being certain that results obtained were due to this organism and not to *Chloroflexus*. These studies were carried out in the summer of 1975 by Madigan, as part of his study on light adaptation by *Chloroflexus,* and are presented in detail in Madigan's thesis (Madigan, 1976). The springs studied and the technique for experimental reduction of incident light are described in Chapter 7.

The relationship of photosynthesis to light intensity for natural (unadapted) populations of *Synechococcus* at Ravine Spring was shown in Figure 7.20. As seen, the *Synechococcus* population shows a relatively broad optimum light intensity for photosynthesis (between 2000 and 8000 footcandles), but does not photosynthesize at a low-light intensity (560 footcandles) at which *Chloroflexus* photosynthesizes well. This result confirms the less detailed data of Doemel and Brock (1974) regarding the ability of *Chloroflexus* to develop at lower light intensities than *Synechococcus*. Two neutral density filters were placed over the mat in this spring effluent, reducing the light 73% and 98%. The changes in chlorophyll with time after installation of the filters were shown in Figure 7.23. As seen, with light reduced 73%, the *Synechococcus* population remains unchanged in amount (as indicated by chlorophyll assay) over the next 10 days, whereas if the light intensity is reduced 98% (to a midday intensity of about 160 footcandles), the *Synechococcus* population washes out exactly as it does when completely darkened, but the *Chloroflexus* population remains intact. Thus, after 10 days under a neutral density filter reducing light intensity 98%, a population is obtained which consists of *Chloroflexus* as the sole photosynthetic component.

The alga shows a marked ability to adapt to reduced light intensity.

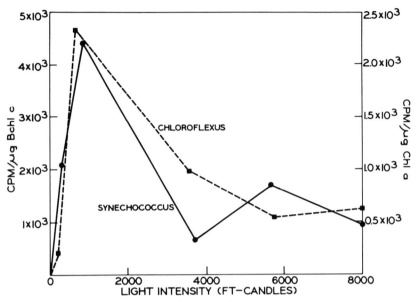

Figure 8.17. Photosynthesis at various light intensities in the *Synechococcus* population that developed under the filter reducing light intensity 73%; 14 days after installation of the filter. *Chloroflexus* data also shown for comparison. (From Madigan and Brock, 1977.)

Figure 8.17 gives the response to varying light intensity of the *Synechococcus* population that had developed for 14 days under a filter that reduced the light intensity 73%. This figure also shows that the adapted population showed marked high-light inhibition of photosynthesis, as had been reported earlier from Mushroom Spring (Brock and Brock, 1969b).

Similar experiments were done in Octopus Spring (Pool A), but with a mat that developed in a nonflowing situation (Figure 7.22). The data showed that during a 14-day period of adaptation, the *Synechococcus* cells gradually shifted their optimum light intensity for photosynthesis to a lower value, but that the population under the 98% filter became moribund and disappeared. Figure 8.18 shows the results obtained after the 14-day adaptation period. Note the marked adaptation to lower light intensity of the population that had developed under the 73% filter, and the lack of adaptation of the population that had developed under the 93% filter. Also, the population under the 98% filter photosynthesized very poorly, suggesting that it was quite moribund (note that since the ordinate has values normalized to chlorophyll, the *quality* of the population was affected as well as the *quantity*). One possible explanation for the lack of adaptation of the population under the 93% filter is that there was insufficient light for cell growth and development of a new population, but still sufficient light for maintenance of the population that was already present, and this population remained adapted to the full light intensity. In an attempt to determine whether this explanation was correct, Madigan did an experiment in which he allowed a new population of *Synechococcus* cells to develop under filters. This was done by laying down layers of silicon carbide (Carborundum powder), and overlaying with neutral density filters, and then allowing new populations to develop. It had previously been shown (Doemel and Brock, 1974) that new populations developed rapidly on top of silicon carbide powder in full sunlight. Growth was rapid over the silicon carbide marker at the uncovered and 73% reduction sites, but much slower at the 93% reduction site, and absent at the 98% and total darkness sites. Seventeen days after these new populations had developed, the optimal intensity for photosynthesis was determined. As seen in Figure 8.19 the population that had developed under the 73% reduction filter was highly adapted to low-light intensity, showing optimal photosynthesis at about 1500 footcandles, and showed strong inhibition at high-light intensities. Photosynthesis was completely undetectable at light intensities of 3000 footcandles or higher.

The population which developed over silicon carbide powder under the 93% filter did not show low-light adaptation (Figure 8.19), as would have been expected. Thus, the suggestion that the lack of adaptation in natural populations when the light was reduced 93% was due to insufficient light for the development of a new population cannot be maintained. A new population *did* develop under a neutral filter reducing light intensity 93%, but this

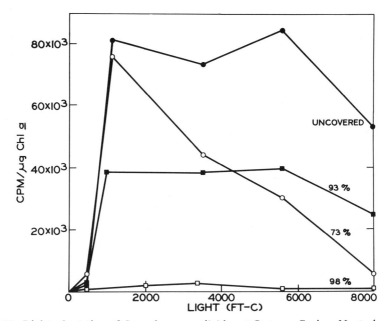

Figure 8.18. Light adaptation of *Synechococcus lividus* at Octopus Spring. Neutral density filters in place for 16 days. (From Madigan and Brock, 1977.)

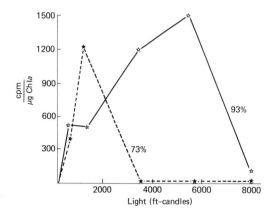

Figure 8.19. Photosynthesis versus light intensity for the *Synechococcus* population which had grown for 17 days on top of a silicon carbide marker at Octopus Spring. Note clear adaptation by the population under the 73% filter and the lack of adaptation of the population under the 93% filter. (From Madigan, 1976.)

population still showed optimal photosynthesis at high intensity. To date, no explanation for this interesting phenomenon is available.

It is striking that in both Madigan's experiments and in the experiments of Brock and Brock (1969b), there were dramatic color changes when neutral density filters were placed over the mat, but chlorophyll assays showed that these color differences were not due to the synthesis of increased amounts of this photopigment. In the experiments of Brock and Brock (1969b), after almost a year of adaptation to low-light intensity, the chlorophyll content had increased only about twofold (cell number had not increased at all) and during the first 2 weeks of adaptation, there was no chlorophyll increase at all. Yet, visible changes in color can be observed as soon as 6 hours after neutral density filters are put in place, and within a week the mats under the filters are very dark green. Since chlorophyll does not change this rapidly, it seems likely that the color changes are due to changes in phycocyanin content. Indeed, in some preliminary assays, Mercedes Edwards (personal communication) showed increases in phycocyanin content under filters. Also, she found considerable increases in number of thylakoid membranes (as shown by electron microscopy) of populations that had developed for several weeks under neutral density filters.

One point should be emphasized regarding the high-light inhibition that Madigan found even in natural populations that had developed in full sunlight. Madigan did all his photosynthesis experiments using cell suspensions prepared by homogenizing in spring water the thin top layers of mat. In the experiments of Brock and Brock (1969b), which did not show significant high-light inhibition, the layers removed from the mat were not homogenized before incubation with $^{14}CO_2$, but were used intact. It seems likely that in intact mats, self-shading results in adaptation of a portion of the population to light intensities below maximal, and as long as the mat is intact these populations remain protected from high light. When a suspension is made, the cells from the interior of the mat are brought out into bright light, and high-light inhibition occurs.

In both the experiments of Madigan and in those of Brock and Brock, when photosynthetic efficiency was calculated (photosynthesis per unit of light received), the maximal efficiency was at very low light, even in unadapted populations, although the total amount of photosynthesis (not considering the amount of light received) occurred at high light. This suggests that some of the light received at high-light intensities is "wasted," in the sense that it is not photosynthetically active. This is not especially surprising, but does show that the organism has somewhat of a cushion of available light, so that when light intensities decrease, it need not adapt instantly in order to maintain populations.

Light intensities vary markedly at Yellowstone throughout the year. In the summer, climatic conditions are often warm and dry for long periods of time, and intense solar radiation impinges on the mats. In the winter, on the other hand, days are much shorter, and cloudy weather is common during the long period of deep snow. It was of interest to know if the algae could maintain populations throughout the year, especially near the upper temperature limit. Observations of the mats in winter have shown that the upper temperature limit is similar to that in summer (although more extensive cooling means that the thermal gradient moves closer to the spring source). Thus, there is sufficient light to maintain populations all winter, and the ability to adapt to changing light regimes probably has played a significant role.

The situation in Yellowstone should be compared with that in Iceland, which because of its higher latitude has a greatly reduced day length in winter. Tuxen (1944) calculated that in Iceland the hot spring algae receive 50 times more light in June than in December and has estimated that for at least 2 months around the winter solstice, there will not be sufficient light for photosynthesis to exceed day and night respiration. As shown below, this calculation was probably erroneous.

Mastigocladus in Iceland

Mastigocladus is the dominant thermal alga in Iceland, and during 2 months in the winter there appears to be insufficient light for net photosynthesis (Tuxen, 1944). Schwabe (1936) had presented some observations in Iceland which indicated that *Mastigocladus* did not maintain winter populations, but died out and had to be replaced each summer. Sperling (1975) carried out a more detailed study in Iceland in 1968–1969, surveying hot spring habitats for presence of algae and carrying out $^{14}CO_2$ photosynthesis measurements. Sperling confirmed Schwabe's observations of a marked decrease in algal standing crop in winter [in contrast to the situation in Yellowstone (see above)], but *Mastigocladus* persisted in still-water zones at temperatures up to 55°C in late December. *Mastigocladus* encased in siliceous sinter was found abundantly in rapid flow in many streams descending a south face (thus with good incidence to the direction of the sun), although yellow to pale green in color. Following a brief period of sunny weather in early January, the streams became bright green. Some *Mastigocladus* was also seen in a north-facing effluent in moderate to rapid flow at a temperature between 40°C and 50°C. Sperling also found *Mastigocladus* living subaerially along the margins of thermal effluents, and there was little change in appearance of these populations between summer and winter. Either this subaerial population or the thin coating remaining in the streams could serve as an inoculum for the renewed growth of the mat during the following summer.

Sperling measured radiation in Iceland and concluded that previous estimates of radiation available in winter may have been on the low side, since they neglected diffuse radiation reaching the algae as a result of scattering from the sky. He presents measurements made over a number of years by Icelandic meteorologists which show considerable year-to-year variation due to the weather, with a range in December from 1.7 $g \cdot cal \cdot cm^{-2} \cdot day^{-1}$ (year 1962) to 4.6 $g \cdot cal \cdot cm^{-2} \cdot day^{-1}$ (year 1968). Thus, it is possible that the algae are more severly light-limited in some years than in others. Sperling shows, however, that it is quite difficult to measure precisely the radiation impinging on a natural algal mat, due to the difficulty of arranging the geometry of the sensor to mimic the geometry of the mat.

Sperling measured both growth and photosynthesis in the winter. Photosynthesis measurements with $^{14}CO_2$ were actually a better measure of radiation available to the algae than physical measurements. Primary productivity calculations were made based on the ^{14}C data, and the values for southerly or horizontally exposed cores were close to the values reported by Brock (1970) for Yellowstone hot springs in summer. Artificial substrates (polyvinylchloride sheets) placed in the springs during the winter showed some signs of colonization in 8 of 14 sites studied. All the sites showing evidence of colonization were either south or horizontal facing. The fact that colonization occurred (almost certainly not due to drift, since the standing crop was low) showed that the 24-hour compensation point must have been exceeded. Thus, *Mastigocladus* can maintain populations even under the low-light intensities of Iceland in the winter. These data, taken together with the experimental data reported above for *Synechococcus,* show that populations of hot spring algae can adapt to low-light intensities, and maintain populations, at least to some extent, even where light intensities are very low.

The Effect of Wide Temperature Fluctuations on Blue-green Algae

Although many springs have remarkably constant flow, other springs fluctuate widely in flow. Also, the outflow channels of geysers show wide fluctuations in flow during eruption cycles. In such situations, temperature of the channel bottom varies with flow rate. Blue-green algae develop in these habitats of widely fluctuating temperature, generally to much less extent than in habitats of constant temperature, and it was of interest to determine how these algae have managed to cope with wide temperature fluctuations. A knowledge of this topic has certain practical implications in management of thermal pollution. In a study done in 1970, Dr. Jerry Mosser studied the blue-green algae in the outflow channel of Bead Geyser in the Lower Geyser Basin and in an unnamed fluctuating spring in the River Group of the Lower Geyser Basin (Mosser and Brock, 1971).

Bead Geyser (Figure 8.20) was chosen because this geyser had a highly predictable eruption cycle. On the numerous occasions when it had been observed over the previous 5 years, it had erupted every 23–25 minutes for 2.5 minutes. In the effluent channel, a thin algal mat developed, composed primarily of *Synechococcus*. Between eruptions the water drained from the effluent channel and the mat approached ambient temperature, but the period between eruptions was too short to permit desiccation of the mat. The effluent had formed two shallow channels, one flowing west to join the effluent of Shelf Spring, and the other flowing east. In the latter channel, which was the site of the study, the effluent produced peak temperatures from about 70°C to less than 40°C at different locations along the drainway. Stations I and II were established in this channel at sites reaching maximum temperatures of 65°C and 58°C, respectively, during eruption. Temperature data for the two stations during two eruption cycles are shown in Figure 8.21. Shortly after eruption the temperature rises sharply and then decreases exponentially to a base value. At Station I, the average temperature was 36°C and the ranges were from about 30°C to 65°C. At Station II, the average was 34°C and the range 30–58°C. As indicated in Figure 8.21, the mean temperatures of the two stations were fairly similar, 34°C and 36°C, considerably lower than the mean temperature at which *Synechococcus* develops in springs of constant flow.

The algal mat in this channel was quite thin (0.1–1 mm), bright orange, and occurred on the smooth white surface of siliceous sinter. At Station I, the mat contained only *Synechococcus* and flexibacteria (not identified but probably *Chloroflexus*), and the mat at Station II contained a sheathed, filamentous blue-green of the *Phormidium* type, the individual cells of which were microscopically very similar to *Synechococcus* rods. *Synechococcus* therefore might have been present, but indistinguishable.

Table 8.3 gives the estimated biomass of the mat at the two stations, and compares it with the biomass of a habitat of constant temperature at Mushroom Spring. Both the standing crop of chlorophyll and the number of cells per unit area were much lower than at Mushroom Spring. (It really did not matter which temperature at Mushroom Spring was used as a control, since the standing crop at *all* the Mushroom stations was markedly higher than at either of the Bead Geyser stations.) From the data in Table 8.3, it is also possible to calculate the chlorophyll content per cell, and it can be determined that chlorophyll/cell is much lower at Bead Geyser than at Mushroom Spring. Analyses for the common chlorophyll degradation prod-

a

Figure 8.20. (a) Bead Geyser, Lower Geyser Basin. The effluent channel is toward the back of the photograph. Photographed summer 1975. The eruption cycle was unchanged from 1970, when the experiments were carried out. (b) Close-up of the geyser throat between eruptions.

Table 8.3. Biomass of Bead Geyser Algal Mat and Comparisons with That of Mushroom Spring, a Habitat of Constant Temperature[a]

	Algal cells/m²			Chlorophyll-a (mg/m²)		
	Mean	Standard deviation	No. of determinations	Mean	Standard deviation	No. of determinations
Station I	4.0×10^9	1.7×10^9	19	1.28	0.12	15
Station II	1.17×10^{12}	0.19×10^{12}	19	7.33	0.78	15
Mushroom Spring	1.03×10^{13}		—	203	—	—

[a]From Mosser and Brock (1971).

b

Figure 8.20b. Legend see opposite page.

uct, pheophytin, were also carried out, but pheophytin levels were below the limits of detection. The data in Table 8.3 thus show that the Bead Geyser algae are not developing well, and it is reasonable to assume that it is because of the wide temperature variation.

To study the adaptation of the Bead Geyser algae to the fluctuating temperatures, photosynthesis was measured with the ^{14}C technique at a series of incubation temperatures. The results are given in Figure 8.22. It can be seen that algae from Station I photosynthesize at a maximum rate at a temperature of about 46–50°C, while those from Station II have a relatively broad optimum with a midpoint at about 41–42°C. Some cultural studies carried out with the Bead Geyser algae (Mosser and Brock, 1971) showed that the optimal temperatures for growth were similar to the optimal temperatures as measured by ^{14}C. In view of the 15–20 degree Celsius difference between the optimum temperature and the temperatures to which the algae were exposed during eruption, it was of interest to examine the heat resistance of these organisms. Samples were placed in

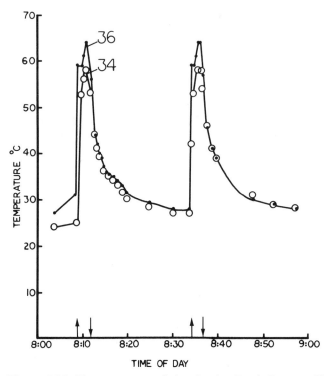

Figure 8.21. Temperature variation in the Bead Geyser effluent channel. Solid circles: Station I; open circles: Station II. The numerical labels are the mean temperatures, computed by averaging the temperatures for each 1-minute interval. Upward arrows indicate the beginning of eruption; downward arrows, the end. (From Mosser and Brock, 1971.)

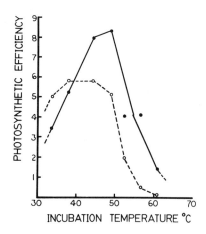

Figure 8.22. Photosynthetic efficiency of Bead Geyser algae at different temperatures. Solid circles: Station I; open circles: Station II. The ordinate is expressed as cpm/mg chlorophyll-a/hour $\times$ 10^{-6}. (From Mosser and Brock, 1971.)

vials and exposed to 55°C and 65°C for varying lengths of time and then photosynthesis determined by ^{14}C at the optimum temperature of 45°C. For comparison, the experiment was also performed with algae taken from a 45°C site at Grassland Spring. As seen in Figure 8.23, the Bead Geyser algae were not inhibited or killed by heat treatment at temperatures and of duration similar to the conditions they experienced at eruption. In contrast, the algae from Grassland Spring lost their photosynthetic ability more rapidly during the treatment, although a residual portion of the photosynthetic activity, amounting to about 15–20%, appeared to be relatively resistant to the treatment. This residual activity may be related to the fact that the Grassland Spring samples contained small amounts of other blue-green algal genera in addition to *Synechococcus,* although the latter alga was dominant.

From these studies, it can be concluded that the major factor contributing to the sparsity of the Bead Geyser algal mat was the fact that the alga

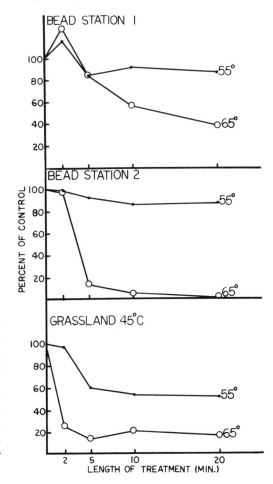

Figure 8.23. The effect of heat treatment on the photosynthesis of algae from Bead Geyser and Grassland Spring. Samples were heated at 55°C or 65°C and then transferred to 45°C for determination of photosynthesis. (From Mosser and Brock, 1971.)

experiences temperatures optimal for photosynthesis and growth only
during a small fraction of the time, namely at the beginning and end of the
eruption. Damage or death caused by the periodic high temperatures could
also contribute to the sparsity, but the heating experiment and the absence
of pheophytin in the mats indicate that this is probably not a major factor.
In view of the low mean temperature of the habitats at Bead Geyser (34°C
and 36°C, for the two stations), the temperature optimum of 45°C and the
tolerance to temperatures of 55°C are of interest. There was no *a priori*
reason to expect that these algae would exhibit optimal temperatures
markedly higher than the mean environmental temperatures. During an
eruption cycle, the algae experience their optimum temperature for less
than 5% of the time. It would appear that an inherent limitation has forced
the algae to forego photosynthesis and growth during a large fraction of the
time in order to acquire resistance to the high temperatures to which they
are exposed only 10% of the time. (Over the flow cycle, the temperature
fluctuated 45 degrees Celsius.)

In contrast to the situation at Bead Geyser, in the other fluctuating spring
studied (Mosser and Brock, 1971), water flowed about 90% of the time, and
then stopped completely. The algal mats in the outflow channel of this
spring were more extensive, and they showed a temperature optimum close
to the computed mean environmental temperature. Here, despite the wide
fluctuation, the temperature conditions appeared not to differ sufficiently
from those of a constantly flowing spring to prevent the algae from adapting
maximally to their environment. Thus, it seems that not only the wide
fluctuation, but the manner in which the temperature varies with time, is
important in determining how thermal algae adapt to a thermally fluctuating
environment.

This work has relevance to problems of thermal pollution, since in steam
power plants, cooling water is often subjected to sudden marked increases
in temperature. Our data suggest that organisms may adapt to such sudden
temperature shocks, although they cannot at the same time evolve the
ability to grow well throughout most of the temperature cycle. However,
the geyser runoff studied varied in a highly predictable manner, whereas
power plant effluents may fluctuate erratically. Thus, not only the range of
temperature is critical, but the time constant of the system and the predictability of the fluctuation are also important. It is interesting that the thermal
algae have adapted as well as they have to the kinds of fluctuations seen at
Bead Geyser, but erratic power plant variations would probably be too
difficult for algal evolution to overcome.

References

Bauld, J. and T. D. Brock. 1973. Ecological studies of *Chloroflexus,* a gliding
photosynthetic bacterium. *Arch. Mikrobiol.* **92**, 267–284.
Brock, T. D. 1967a. Microorganisms adapted to high temperatures. *Nature* **214**,
882–885.

Brock, T. D. 1967b. Life at high temperatures. *Science* **158**, 1012–1019.

Brock, T. D. 1970. High temperature systems. *Ann. Rev. Ecol. System.* **1**, 191–220.

Brock, T. D. and M. L. Brock. 1967. The measurement of chlorophyll, primary productivity, photophosphorylation, and macromolecules in benthic algal mats. *Limnol. Oceanogr.* **12**, 600–605.

Brock, T. D. and M. L. Brock. 1968. Measurement of steady-state growth rates of a thermophilic alga directly in nature. *J. Bacteriol.* **95**, 811–815.

Brock, T. D. and M. L. Brock. 1969a. Recovery of a hot spring community from a catastrophe. *J. Phycol.* **5**, 75–77.

Brock, T. D. and M. L. Brock. 1969b. Effect of light intensity on photosynthesis by thermal algae adapted to natural and reduced sunlight. *Limnol. Oceanogr.* **14**, 334–341.

Brock, T. D. and M. L. Brock. 1971. Microbiological studies of thermal habitats of the central volcanic region, North Island, New Zealand. *N.Z. J. Mar. Freshwater Res.* **5**, 233–257.

Castenholz, R. W. 1969a. Thermophilic blue-green algae and the thermal environment. *Bacteriol. Rev.* **33**, 476–504.

Castenholz, R. W. 1969b. The thermophilic cyanophytes of Iceland and the upper temperature limit. *J. Phycol.* **5**, 360–368.

Castenholz, R. W. 1970. Laboratory culture of thermophilic cyanophytes. *Schweiz. Z. Hydrol.* **32**, 538–551.

Castenholz, R. W. 1972. The Occurrence of the Thermophilic Blue-Green Alga, *Mastigocladus laminosus* on Surtsey in 1970. The Surtsey Progress Report VI, pp. 14–19.

Cohn, F. 1862. Üeber die Algen des Karlsbader Sprudels, mit Rücksicht auf die Bildung des Sprudelsinters. *Abhandlungen der Schlesischen Gesellschaft für vaterländische Cultur. Abt. Naturwiss. Med.* **II**, 35–55.

Copeland, J. J. 1936. Yellowstone thermal Myxophyceae. *Ann. N.Y. Acad. Sci.* **36**, 1–229.

Doemel, W. N. and T. D. Brock. 1974. Bacterial stromatolites: origin of laminations. *Science* **184**, 1083–1085.

Dyer, D. L. and R. D. Gafford. 1961. Some characteristics of a thermophilic blue-green alga. *Science* **134**, 616–617.

Edwards, M. R., D. S. Berns, W. C. Ghiorse, and S. C. Holt. 1968. Ultrastructure of the thermophilic blue-green alga, *Synechococcus lividus* Copeland. *J. Phycol.* **4**, 283–298.

Edwards, M. R. and E. Gantt. 1971. Phycobilisomes of the thermophilic blue-green alga *Synechococcus lividus*. *J. Cell Biol.* **50**, 896–900.

Fogg, G. E. 1951. Studies on nitrogen fixation by blue-green algae. II. Nitrogen fixation by *Mastigocladus laminosus* Cohn. *J. Exp. Bot.* **2**, 117–120.

Geitler, L. 1932. Cyanophyceae. Volume 14 in Rabenhorst's *Kryptogamen-Flora*. Akademische Verlagsgesellschaft, Leipzig (reprinted, Johnson Reprint Corp., New York, 1971).

Holt, S. C. and M. R. Edwards. 1972. Fine structure of the thermophilic blue-green alga *Synechococcus lividus* Copeland. A study of frozen-fractured-etched cells. *Can. J. Microbiol.* **18**, 175–181.

Holton, R. W., H. H. Blecker, and T. S. Stevens. 1968. Fatty acids in blue-green algae: possible relation to phylogenetic position. *Science* **160**, 545–547.

Jackson, J. E., Jr. and R. W. Castenholz. 1975. Fidelity of thermophilic blue-green algae to hot spring habitats. *Limnol. Oceanogr.* **20**, 305–322.

Kempner, E. S. 1963. Upper temperature limit of life. *Science* **142**, 1318–1319.

Madigan, M. T. 1976. Studies on the physiological ecology of *Chloroflexus auran-tiacus,* a filamentous photosynthetic bacterium. Ph.D. dissertation, University of Wisconsin, Madison.

Madigan, M. T. and T. D. Brock. 1976. Quantitative estimation of bacteriochloro-phyll *c* in the presence of chlorophyll *a* in aquatic environments. *Limnol. Oceanogr.* **21**, 462–467.

Madigan, M. T. and T. D. Brock. 1977. Adaptation by hot spring phototrophs to reduced light intensities. *Arch. Microbiol.* **113**, 111–120.

Meeks, J. C. and R. W. Castenholz. 1971. Growth and photosynthesis in an extreme thermophile, *Synechococcus lividus* (Cyanophyta). *Arch. Mikrobiol.* **78**, 25–41.

Molisch, M. 1926. *Pflanzenbiologie in Japan.* Gustav Fischer, Jena.

Mosser, J. L. and T. D. Brock. 1971. Effect of wide temperature fluctuation on the blue-green algae of Bead Geyser, Yellowstone National Park. *Limnol. Oceanogr.* **16**, 640–645.

Peary, J. A. 1964. Ecology and growth studies of thermophilic blue-green algae. Ph.D. Thesis, University of Oregon, Eugene.

Peary, J. and R. W. Castenholz. 1964. Temperature strains of a thermophilic blue-green alga. *Nature* **202**, 720–721.

Schwabe, G. H. 1936. Beiträge zur Kenntnis islandischer Thermalbiotope. *Arch. Hydrobiol.* **6**, (Suppl.), 161–352.

Sheridan, R. P. and T. Ulik. 1976. Adaptive photosynthesis responses to tempera-ture extremes by the thermophilic cyanophyte *Synechococcus lividus. J. Phycol.* **12**, 255–261.

Sperling, J. A. 1975. Algal ecology of southern Icelandic hot springs in winter. *Ecology* **56**, 183–190.

Sperling, J. A. and G. M. Hale. 1973. Patterns of radiocarbon uptake by a thermo-philic blue-green alga under varying conditions of incubation. *Limnol. Oceangr.* **18**, 658–662.

Stanier, R. Y., R. Kunisawa, M. Mandel, and G. Cohen-Bazire. 1971. Purification and properties of unicellular blue-green algae (Order *Chroococcales*). *Bacteriol. Rev.* **35**, 171–205.

Stewart, W. D. P. 1970. Nitrogen fixation by blue-green algae in Yellowstone thermal areas. *Phycologia* **9**, 261–268.

Tuxen, S. L. 1944. The hot springs of Iceland, their animal communities and their zoogeographical significance. *In The Zoology of Iceland,* Vol. 1. Munksgaard, Copenhagen, pp. 1–206.

Chapter 9
The Genus *Cyanidium*

One of the strange features of life in the Yellowstone hot springs is that during summer the microbial mats in those springs rich in blue-green algae are usually orange or yellow in color, whereas the springs that are strikingly blue-green in color do not contain blue-green algae, but instead the eucaryotic alga *Cyanidium caldarium*. Hot springs containing *Cyanidium* are always acidic, with pH values less than 4, and at temperatures above 40°C this alga is the sole photosynthetic component, since no photosynthetic bacteria such as *Chloroflexus* live in such acidic habitats. Because of the striking appearance of *Cyanidium* mats, the alga has been observed for a long time, and has been variously called *Chroococcus varius* (Tilden, 1898), *Protococcus botryoides* f. *caldarium* (Tilden, 1898), *Pleurocapsa caldaria* (Collins et al., 1901), *Palmellococcus thermalis* (West, 1904), *Pluto caldarius* (Copeland, 1936), *Dermocarpa caldaria* (Drouet, 1943), and *Rhodococcus caldarium* (Hirose, 1958). The name *Cyanidium caldarium* was first used by Geitler and Ruttner (1936) the same year that Copeland described the organism as *Pluto,* but despite the euphony of the latter name, the name *Cyanidium* has taken precedent. A photomicrograph of typical cells of *C. caldarium* is shown in Figure 9.1.

The modern history of *Cyanidium* begins with the late Mary Belle Allen, who developed methods for cultivating blue-green algae (Allen, 1952) and isolated an unidentified unicellular alga from Lemonade Spring, an acid spring at The Geysers, Sonoma County, California. As she noted: "... it is extremely interesting as providing an example of a blue-green alga which presents a glaring exception to the rule that organisms of this class thrive only in alkaline environments. It will tolerate and grow well in extremely acid solutions (up to 1 N sulfuric acid); it grows on organic media in the dark at a rate comparable to its rate of growth in the light; and when grown

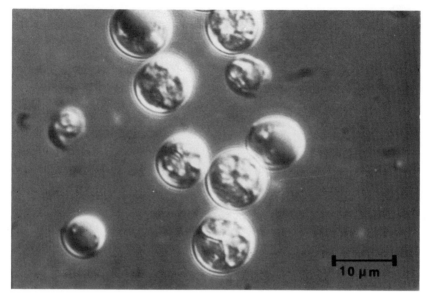

Figure 9.1. Nomarski interference contrast photomicrograph of *Cyanidium caldarium* cells. The larger cells have already divided.

in the dark it loses its chlorophyll and phycocyanin and becomes a light straw yellow in color. It will be referred to as the 'acid alga' in this paper.'' She also noted that the morphology and manner of cell division differed from that of other unicellular blue-green algae. Two years later (Allen, 1954) she published further description of this alga, now calling it a blue-green *Chlorella,* because its cell cycle is quite similar to that of this green alga. Although she did not specifically note the presence of nuclei or chloroplasts, she reported that the alga was devoid of diaminopimelic acid, an acid present in the cell walls of blue-green algae and most bacteria.

Independently of Allen, Hirose (1950) isolated the alga from an acid hot spring at Noboribetsu in Hokkaido, Japan, and showed that it possessed chloroplasts. He used the name *Cyanidium* for the following reason: "Because both the genus *Cyanidium* Geitler and *Pluto* Copeland were established in the same year (1936) and moreover the exact dates of their publications were unknown to him, the author could not assure himself which of the two has the priority. As F. E. Fritsch (1945) in his treatise . . . used *Cyanidium caldarium* as the name of the present alga, the author also adopted '*Cyanidium*' after Fritsch's opinion as the genus name.'' Hirose also proposed that *Cyanidium* be placed as a member of the Chlorophyta, because of its morphological resemblance to *Chlorella.* When Allen (1959) carried out more detailed work on this organism than she had done earlier, she could readily perceive that her isolate was similar to Hirose's, and because Hirose had already used the name *Cyanidium,* she used it also.

Although in her 1959 paper, Allen still classified *Cyanidium* with the Chlorophyta, she noted that the alga was anomolous, since none of the other chlorophytes possess phycobilin pigments. "Two possibilities which have been advanced to resolve this anomaly are (1) that *Cyanidium* is a colorless chlorophyte with a symbiotic blue-green alga, and (2) that it is not a chlorophyte at all, but a coccoid cryptomonad." She rejects the interesting first suggestion primarily because when grown in the dark *Cyanidium* loses its pigmentation, whereas blue-green algae grown in the dark retain their pigmentation. However, the possibility that the *Cyanidium* chloroplast is a symbiotic blue-green alga cannot be dismissed as lightly as Allen did (see later). The second possibility, that *Cyanidium* is a cryptomonad, Allen rejects because although cryptomonads do contain phycobilin-type pigments, they are distinctly different from the phycobilin of *Cyanidium*, which is a C-phycocyanin indistinguishable from the blue-green algae. She concludes: "In the absence of any evidence which would permit placing *Cyanidium* in some other group of algae, it appears necessary to consider it as a chlorophyte. It may be noted that other examples of chlorophytes with anomalous pigmentation are known."

More recently, immunological and other evidence has suggested a relationship between *Cyanidium* and the red algae (Rhodophyta). The possibility that *Cyanidium* was a red alga was suggested by Hirose (1958), who showed by Feulgen staining that the alga had a nucleus. Hirose (1958) proposed that the name be changed to *Rhodococcus*. Although Geitler (1958) did not object to the classification of *Cyanidium* with the red algae, he pointed out that it would be taxonomically incorrect to change the name, since the name *Cyanidium caldarium* clearly had priority.

A detailed comparison of the characteristics of *Cyanidium* with that of the other algae with which it has been classified is given in Table 9.1. The relationship with the *Rhodophyta* seems the strongest, especially since there are already well-established unicellular red algae (e.g., *Porphyridium*, *Phragmonema*). There is also some support for this idea from immunodiffusion studies on glucose phosphorylase (Fredrick, 1976) and from ultrastructural studies (Edwards and Mainwaring, 1973), although ultrastructural studies can also support the case that the chloroplast of *Cyanidium* may have originated from an endosymbiosis between a procaryotic blue-green alga and a nonphotosynthetic eucaryotic organism (Edwards and Mainwaring, 1973).

Our own work on *Cyanidium* began early in our Yellowstone project (Brock and Brock, 1966), but only got serious when W. N. Doemel joined my laboratory as a graduate student. I was looking for a project that he could do in Yellowstone which would be different from the other work we were carrying out, but yet would use similar techniques and facilities. At about this time, I spent the month of May 1967 in Italy and met Dr. Carmelo Rigano, who showed me the extensive development of *Cyanidium* in hot

Table 9.1. Characteristics of *Cyanidium caldarium*[a]

Characteristic	*Cyanidium caldarium*	Rhodophyta	Chlorophyta	Cyanophyta	Cryptophyta
Morphology					
Spherical cells that divide by formation of daughter cells	+[b]	+/−	+/−	+/−	+/−
Sexual cycle	−	+/−	+/−	−	+/−
Internal structure					
True chloroplast	+	+	+	−	+
True nucleus	+	+	+	−	+
True mitochondrion	+	+	+	−	+
Golgi bodies	−	+/−	+	−	?
Pyrenoid	−	+/−	+	−	−
Storage granules	+	+	+	+	+
Vacuoles	+	+	+	−	+
Chloroplast lamellae stacked	−	−	+	−	+
Chloroplast lamellae parallel	+	+	−	+	−
Pigments					
Chlorophyll-a	+	+	+	+	+
Chlorophyll-b	−	−	+	−	−
Chlorophyll-c	−	−	−	−	+
Chlorophyll-d	−	+/−	−	−	+
Phycocyanin	+	+	−	+	+
Phycoerythrin	−	+	−	+	+
α carotene	−	+	−	−	−
β carotene	+	+	+	+	+

ψ carotene	−	−	+	−	−
ε carotene	−	+	?	−	−
Flavacene	−	+	−	−	−
Lutein	−	?	+	+	+
Zeaxanthin	+	−	−	+	+
Violoxanthin	−	−	+	+	+
Flavoxanthin	−	−	−	−	−
Neoxanthin	−	−	+	−	−
Myxoxanthophyll	−	+	−	−	−
Oscillaxanthin	−	+/−	−	−	−
Cryptoxanthin	+	−	−	−	+
Cell wall					
High protein	?	−	−	+	+
Hemicellulose	?	−	+	+	+
Low cellulose	?	+	−	+/−	+
DAP	−	+	−	−	−
Amino sugars	−	+	+	−	−
Muramic acid	−	+	−	−	−
Storage product					
Polyglucoside with many 1,6-glucosyl linkages	?	+	−	+	+
Branching enzyme active on amylose and amylopectin	?	+	−	+	+

[a] From Doemel, (1970).

[b] + character present; − character absent; ± character variable in the group; ? presence of character uncertain.

acid soils in Agnano crater, near Naples. When I arrived in Yellowstone in June 1967, about a few days before Doemel, I observed that an excellent *Cyanidium* spring in the Norris Geyser Basin had just become available for detailed study, because the road that had until that season run immediately next to the spring had been moved. Because the old road had been taken out, an area of Norris that had previously been one of the most public in the Park had now become a very private area in the far reaches of the Park. On such trivia are research projects built! Doemel called the spring and its effluent Cyanidium Creek and proceeded to carry out the first ecological studies on *Cyanidium*. His work soon ramified in a number of different directions. I have already discussed some of the work on thermophilic fungi and acidophilic bacteria that live in *Cyanidium* mats in Chapter 3.

Culture, Isolation and Structure of *Cyanidium*

If one has spent much time culturing and isolating blue-green algae, turning to *Cyanidium* is a refreshing experience. Here is an alga that lives in nature as a unialgal culture and that develops rapidly in enrichment cultures. Simple inorganic media provide for luxuriant growth, and because of the low pH, the problem of providing iron as a nutrient is greatly simplified, as no chelators are necessary. Because of the low pH and lack of alkalinity in the medium, growth is considerably increased if the cultures are bubbled with CO_2, or if in closed containers the gas phase is replaced with CO_2. Oxygen is not necessary for growth (Seckbach et al., 1970), and an atmosphere of 100% CO_2 in a closed container provides sufficient carbon for the development of high cell densities.

After initial enrichment in Allen's salts (Allen, 1959) at pH 2 and 45°C, cultures can be purified by streaking onto Allen's 1.5% agar with 1% glucose. Incubation in the light or dark leads to the development of small yellow colonies which can be picked and reinoculated into liquid medium. Alternatively, streaking can be done on autotrophic agar with incubation in the light. Cultures can be maintained in screw-capped tubes and transferred bimonthly (or less frequently). A source of carbon can be provided by briefly gassing the tubes with 100% CO_2 before screwing the caps on tightly. Although the optimum temperature for most *Cyanidium* cultures is 45°C (Doemel and Brock, 1971b), stock cultures can be most conveniently kept in a 30°C or 37°C incubator, using a small fluorescent light. I have kept cultures in sealed tubes in this manner for over a year without transferring. *Cyanidium* cultures and natural samples also survive transport well. Doemel recovered cultures from every sample I sent back from New Zealand by air post, and I have obtained cultures from material from other parts of the world that have sat in my suitcase for days.

Figure 9.2 Cell division cycle in *Cyanidium caldarium*.

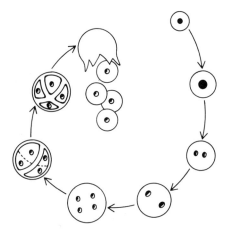

The life cycle of *Cyanidium* is diagrammed in Figure 9.2. This cycle is identical to that of *Chlorella,* and in fact *Cyanidium* and *Chlorella* cannot be told apart under the light microscope (see later). Electron microscopic studies on *Cyanidium* were first done by Mercer et al. (1962) and Rosen and Siegesmund (1961), but both studies used heterotrophically grown cells. Since glucose represses photopigment synthesis, even in the light, the chloroplasts of the cells were quite small and probably atypical. However, in both of these studies, the eucaryotic nature of *Cyanidium* was clearly evident. Mercer et al. (1962) made a comparison of the fine structure of *Cyanidium* with *Chlorella* and *Nostoc.* They noted that *Cyanidium* was distinctly different from the blue-green alga, but that it also had notable differences from *Chlorella.* The chloroplast of *Cyanidium* was not so elaborate or distinct as that in *Chlorella,* and the photosynthetic lamellae of the *Cyanidium* chloroplast are frequently indistinguishable from the lamellae and adjacent cytoplasmic membranes. They also noted that *Cyanidium* differed from *Chlorella* in other structural features: absence of a pyrenoid, absence of starch grains, indistinctness of the chloroplast membrane, fewer discs in the lamellae bands, absence of Golgi bodies, poor development of endoplasmic reticulum, and structure of the cell wall.

In later work, Seckbach (1972) and Edwards and Mainwaring (1973) carried out electron microscopic studies on autotrophically grown *Cyanidium.* Autotrophically grown cells possess much more extensive chloroplast development, and under many conditions virtually the whole cell may be filled with chloroplast (Figure 9.3). As noted by Edwards and Mainwaring (1973), the chloroplast contains nearly parallel, concentric, nonstacked thylakoids, and on the outer faces of both thylakoid membranes there are rows of phycobilisomes. The phycobilisomes of *Cyanidium* are similar to those of other algae where phycocyanin is present (Edwards and Gantt, 1971). Edwards and Mainwaring (1973) also noted a striking similarity

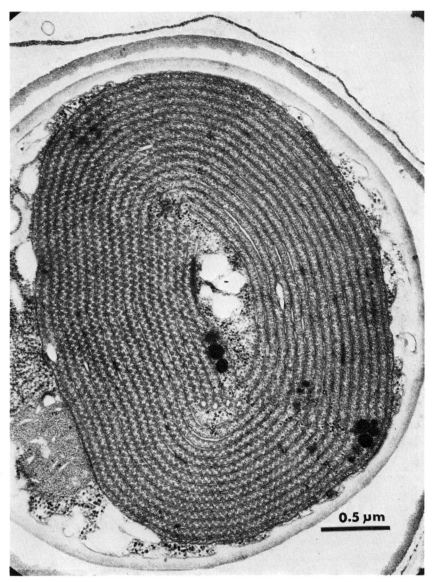

Figure 9.3. Electron micrograph of a thin section of a young cell of *Cyanidium caldarium* showing typical large chloroplast. (Courtesy of Mercedes Edwards.)

between the *Cyanidium* chloroplast and certain unicellular blue-green algae in the manner in which the phycobilisomes associated with the chloroplast thylakoid membrane.

   Staehelin (1968) has studied the ultrastructure of the membrane and cell wall of *Cyanidium* by the freeze-etching technique. An overall view of *Cyanidium* as revealed by freeze-etching is shown in Figure 9.4.

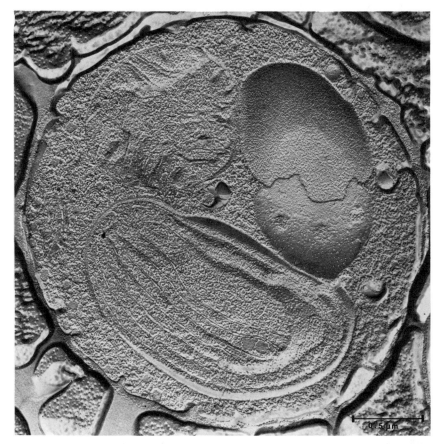

Figure 9.4. A *Cyanidium caldarium* cell as revealed by the freeze-etching technique. The bar represents 0.5 μm. (Courtesy of Andrew Staehelin.)

Lipids of *Cyanidium*

Perhaps because of its unusual environmental requirements, a surprisingly large number of investigations have been carried out on the lipids of *Cyanidium*. Initial work was done by Allen et al. (1970), and Kleinschmidt and McMahon (1970a,b), and additional studies have been carried out by Adams et al. (1971) and Ikan and Seckbach (1972). In general, the results of these various investigators agree. Allen et al. (1970) found the major lipids in *Cyanidium* cells cultured at optimal temperature of 45°C to be monogalactosyl diglyceride, digalactosyl diglyceride, phosphatidyl glycerol, plant sulfolipid, phosphatidyl ethanolamine, phosphatidyl choline, and phosphatidyl inositol. Among the lipids, there are some that are typically present in eucaryotic algae and higher plants as well as in blue-green algae: mono- and digalactosyl diglyceride, phosphatidyl glycerol, and sulfolipid. Also present are lipids not found in blue-green algae but present in more highly evolved green plants: phosphatidyl choline, phosphatidyl inositol, and phosphatidyl ethanolamine. Kleinschmidt and McMahon (1970b) found essentially the same lipid pattern as Allen et al. (1970) and additionally showed that cells grown at 20°C contained significantly larger quantities of both glycolipids and phospholipids than cells grown at 55°C.

The fatty acids of *Cyanidium* contained predominantly palmitic (16:0), stearic (18:0), oleic (18:1), and linoleic acids (18:2), a series quite similar to that of many other eucaryotic algae as well as some blue-green algae. *Cyanidium* contains lesser amounts of unsaturated fatty acids than higher plants, a likely consequence of the higher temperature for growth, since when cultured at 20°C, a marked increase in amount of unsaturated fatty acids occurred, including the synthesis of large amounts of linolenic acid (18:3), a fatty acid not seen at all in cultures grown at 55°C, and seen only in traces in cultures grown at 45°C (Kleinschmidt and McMahon, 1970b). Almost certainly, this increase in unsaturated fatty acids in organisms grown at low temperature reflects the common phenomenon of increased membrane fluidity, necessary to permit low-temperature growth. Kleinschmidt and McMahon (1970b) showed that the membranes of cells grown at low temperature lysed more readily upon heating than the membranes of cells grown at high temperature.

Seckbach and Ikan (1972) have shown that the predominant sterols of *Cyanidium* were ergosterol, beta-sitosterol, and campesterol, with traces of cholesterol and 7-dehydrositosterol. Seckbach and Ikan (1972) attempted to use sterol composition to relate *Cyanidium* to the red algae, and discuss the significance of sterols and chloroplast structure in the phylogeny of this organism.

Pigments and Photosynthesis of *Cyanidium*

As discussed earlier in this chapter, *Cyanidium* possesses only one chlorophyll, chlorophyll-a, and the accessory pigment C-phycocyanin. The phycocyanin has been purified and characterized by Kao et al. (1975) to determine whether there was anything unique about this biliprotein in view of the extreme environment under which *Cyanidium* lives. Like other biliproteins, the *Cyanidium* phycocyanin forms aggregates of various molecular sizes; the monomer molecular weight is 30,000. The protein consists of two nonidentical polypeptide chains, similar to those reported for other C-phycocyanins. The absorption and fluorescence properties of the *Cyanidium* phycocyanin are also similar to those of other C-phycocyanins. The amino acid content is very similar to the C-phycocyanin from the thermophilic blue-green alga *Synechococcus lividus* and the halophilic blue-green alga *Cocchochloris elabens*. The isoelectric point of the *Cyanidium* phycocyanin is 5.11, an unusually high value. Kao et al. (1975) believe that the molecular structure of the *Cyanidium* phycocyanin relates to an increase in protein stability at higher temperatures, possibly explaining the mode of adaptation of this protein to the environmental stress of high temperature. Since the internal pH of *Cyanidium* is close to neutrality (Allen, 1959), there is no reason to expect that the phycocyanin possesses any unusual characteristics involving adaptation to acidic conditions.

Troxler and Lester (1968) have shown that the phycocyanobilin chromophore of *Cyanidium* constitutes about 3.6% of the phycobilin, a value similar to that for other phycocyanin-containing organisms. Because *Cyanidium* does not synthesize pigment in the dark, yet grows well heterotrophically, it is possible to label phycobilin specifically by transferring dark-grown cells to the light in medium containing radiolabeled precursors (Troxler and Lester, 1968). Delta-aminolevulinic acid is a direct precursor of the phycocyanobilin, as it is in other organisms (Troxler and Lester, 1967). Phycocyanobilin is a bile pigment, and studies have been carried out in *Cyanidium* to determine whether the mechanism of bile pigment formation is similar to that in mammals, where large amounts of bile pigment are formed. In man, bile pigments are formed during the breakdown of hemoglobin, by cleavage and oxidation to carbon monoxide of the alpha-methyne bridge carbon of the porphyrin ring (Figure 9.5). Troxler et al. (1970) showed that in *Cyanidium* carbon monoxide and phycocyanobilin were produced in stoichiometric amounts at comparable rates, indicating that the

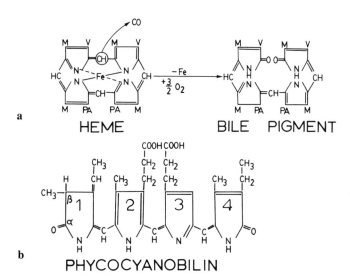

Figure 9.5. (a) Manner in which bile pigments are formed from heme via oxygenase attack on the methyne carbon with liberation of carbon monoxide. (b) The structure of phycocyanobilin. M, methyl; V, vinyl; PA, propionate.

pathway of bile pigment formation parallels that in mammals. When grown in the dark, *Cyanidium* incorporates ^{14}C-delta-aminolevulinic acid into phycocyanobilin and carbon monoxide, and the protein-free phycocyanobilin is excreted into the culture medium (Troxler, 1972). In the light, phycocyanin apoprotein and phycocyanobilin are synthesized simultaneously in stoichiometric amounts (Troxler and Brown, 1970), suggesting that the pathways for protein and chromophore are coordinately regulated. However, since the chromophore is synthesized in the dark (and excreted), it must be concluded that light affects primarily the synthesis of the apoprotein.

Nichols and Bogorad (1960) first isolated mutants of *Cyanidium* deficient in photopigments. *Cyanidium* is an excellent organism for the isolation of mutants, since dark growth on glucose occurs at rapid rates, so that pigmentless mutations are not lethal. Four general types of mutants were found: (1) Dark-greening mutants, which produce chlorophyll in significant amounts in darkness (wild-type cells produce only small amounts of chlorophyll in the dark) and also produce phycocyanin in the dark (wild-type cells produce no phycocyanin in the dark). Light-grown cells of these mutants also produce more photopigments than do comparable wild-type cells. (2) Phycobilin-less mutants. These mutants produce chlorophyll in the light, but no phycobilin. (3) A chlorophyll-less mutant. An obligately heterotrophic mutant was isolated that completely lacked chlorophyll, although it produced phycocyanin when grown in the light. (4) A chlorophyll-less, phycocyanin-less mutant was derived from the chlorophyll-less mutant by ultraviolet irradiation.

Nichols and Bogorad (1962) used the chlorophyll-less mutant to measure the action spectrum for phycocyanin formation. The action spectrum showed maxima at 420 nm and 550–600 nm, suggesting that the primary photoreceptor may be an unidentified heme compound. Nichols and Bogorad (1962) also made the intriguing observation, apparently not followed up, that *Cyanidium* would form phycocyanin in darkness if it were cultured on medium solidified with Fisher agar, U.S.P., instead of Difco Bacto-agar. This suggests that an unidentified factor might permit phycocyanin synthesis in the dark.

One of the significant results of the study of the *Cyanidium* pigment mutants was the clear-cut demonstration of the essentiality of chlorophyll for photosynthesis, since the mutant containing phycocyanin but not chlorophyll would only grow heterotrophically. Another interesting study using these *Cyanidium* mutants was that of Volk and Bishop (1968) on the role of phycocyanin in photosynthesis. These workers used the *Cyanidium* mutant devoid of phycocyanin and studied photosynthetic efficiency, the Emerson enhancement phenomenon, and the action spectrum for the quantum yield. Volk and Bishop showed that in the phycocyanin-less mutant there was a somewhat lower rate of oxygen evolution at saturating light intensity when compared to the wild type, but that the saturation intensity was the same for both phenotypes and there was no difference in rates of oxygen evolution at low intensities. Hill-reaction activity showed a similar picture. The action spectrum for photosynthesis in the mutant followed the absorption spectrum, showing the absence of a peak at the wavelength at which phycocyanin absorbed, but a broad plateau extending across the area where chlorophyll absorbed. The action spectrum for the quantum yield for photosynthesis of the mutant showed a decreased efficiency at shorter wavelengths and a shift in peak efficiency toward longer wavelengths. The red drop was also shifted toward the far red. Emerson enhancement experiments were done, measuring the effect of 620-nm light (absorbed by phycocyanin) on oxygen evolution when 750-nm light (absorbed by chlorophyll) was used. A 30% enhancement by 620-nm light was observed in the wild type, but none in the mutant. They conclude:

> It appears that in an organism possessing the complete pigment system and in which the chlorophylls of Systems I and II are absorbing as well as the phycocyanin pigment, a careful balance between these systems must be maintained to allow both systems to function efficiently. In the red region where phycocyanin stops absorbing, System II cannot provide the reducing power required by System I and efficiency falls off. . . . In an organism, however, in which System II depends solely upon its chlorophyll *a* component for light absorption, efficiency is high throughout the region of chlorophyll absorption. There is no sudden 'shut-down' of a large percentage of the light-gathering machinery resulting in an imbalance between System I and System II. The R-II mutant has adjusted to the loss of phycocyanin and the concomitant decrease in reducing potential of System II by decreasing the oxidizing capacity of System I, i.e., by a decrease in chlorophyll *a* content. The chlorophyll content of R-II per unit cell volume is approximately one-half that of wild type. This decrease in pigment content has the advantage of maintaining high efficiency in a system in which normal light-gathering capacity has been impaired. . . . It appears, therefore, that phycocyanin acts solely as an energy-gathering system broadening the wavelength range over which an organism may absorb energy for photosynthesis. A loss of this 'gathering' potential is not detrimental to the organism but does result in an adjustment in System I to compensate for a decreased electron flow from System II.

Möller and Senger (1972) were able to synchronize growth in *Cyanidium* cultures by use of a method that had previously been used with *Chlorella*. This involves alternating light-dark and temperature cycles. During the light at 35°C, there was marked increase in cell size but little increase in cell number, and during the subsequent dark period (at 40°C) there was a rapid increase in cell number. An examination of the life cycle of *Cyanidium* illustrated in Figure 9.2 helps to explain this synchronization: light apparently inhibits cell division but is essential for cell growth, so that in the light the cells increase in size, and division into autospores follows as soon as darkness ensues. Discussion of ecological aspects of photosynthesis in *Cyanidium* is given in a subsequent section of this chapter.

Habitat of *Cyanidium*

As noted in the introduction to this chapter, *Cyanidium* has been frequently described in acid thermal areas, from a wide variety of places. In our detailed studies of the ecology of *Cyanidium,* Doemel and I studied more than 150 thermal areas from throughout the world. Virtually any habitat of appropriate temperature and pH had *Cyanidium.* The organism exists in both terrestrial and aquatic habitats. Large free-flowing acid hot springs are actually fairly rare on earth, although small springs, seepages, wet soils, and fumarole condensates containing *Cyanidium* are much more common. A few dramatically large flowing acid hot springs exist in Yellowstone Park, mostly outside the main Geyser Basins. The characteristics of three acid springs, which we studied in some detail, are given in Table 9.2. One of these springs, dubbed by us "Cyanidium Creek," is in the Norris Geyser Basin (Figure 9.6). Nymph Greek, which has the best *Cyanidium* mat in the Park, is a cluster of small springs that join together to make the creek that flows into Nymph Lake. This creek is very accessible, as it is immediately adjacent to the Norris-Mammoth Road, but because of the particular layout of the road and a dense stand of trees, the creek is not very public and thus

Table 9.2. The Chemistry of Acid Effluents

	Cyanidium Creek	Nymph Creek	Roaring Mountain
pH [a]	2.95–3.10	2.70–2.80	2.0–2.3
Particulate carbon (mg/l) [b]	4.50	0	0
Soluble carbon (mg/l)	9.20	15.13	7.22
PO_4^{-3} (mg/l)	0.705–1.14	0.00–0.707	0.379–0.788
$SO_4^=$ (mg/l)	113	256	528
Silica (mg/l) [c]	271	220	N.D.
Conductivity (mhos)	1.59–1.64×10^{-3}	1.26–1.28×10^{-3}	2.85×10^{-3}
NH_4^+ (mg/l) [d]	1.71	2.32	3.08
Cl^- (mg/l)	387	36	41
NO_2^- (mg/l)	0	0.01	0
NO_3^- (mg/l)	0.41	0.98	0.70
H_2S (mg/l)	0	0.7	0
CO_2 (mg/l)	0	—	8
Acidity (mEq/l)	4.1	—	16.2
As (mg/l)	1.21	0.065	0.10
Fe (mg/l)	0.91	—	3.60
Hg (mg/l)	0.40	—	0.07

[a] Determinations of pH, $PO_4^=$, $SO_4^=$ and conductivity made by Doemel, Indiana University, 1970.

[b] Determinations of particulate carbon and soluble carbon made by Wilson, Montana State University, 1969.

[c] Determinations of silica made by Coller, Indiana University, 1968.

[d] Determinations of NH_4^+, Cl^-, NO_2^-, NO_3^-, H_2S, CO_2, acidity, As, Fe, and Hg made by Noguchi, Tokyo, 1971.

Figure 9.6. Cyanidium Creek, a small acid creek fed by three pools. Situated in the southcentral portion of the Norris Geyser Basin (see location on the map, Figure 2.3). The experimental channels and neutral density filters are in evidence. (W. N. Doemel sampling.)

can be easily studied without much concern about vandalism. The southern effluent of Roaring Mountain (Figure 2.6) is also quite accessible by road, yet invisible to the public, and has the additional feature that it has the lowest pH of a flowing stream, with a value of 2.0. Obsidian Creek (Figure 9.7), which is fed by seepage from Roaring Mountain, warms up during the summer as snow melt finishes and its algal population converts from diatoms to *Cyanidium* (see later in this chapter).

Figure 9.7. Obsidian Creek receives much of its water by seepage from Roaring Mountain. It is not thermal in early summer, due to snow melt runoff, but warms up by July and converts to *Cyanidium*.

Another excellent *Cyanidium* habitat in Yellowstone is Ebro Springs, which is similar to Nymph Creek in character. Ebro Springs is in the Pelican Creek area, near the Fishing Bridge tourist area, and is only accessible by foot. Another location, Lemonade Creek, is fed by springs in the Amphitheater Springs area, and has a large outflow channel (Figure 6.7), although it suffers from considerable variation due to rainfall and dilution with surface run-off.

Since most acid springs are relatively small, and depend markedly for their water on seepage from shallow near-surface soils which themselves are dependent for water on precipitation, acid springs can change markedly in flow rate throughout the year. In Yellowstone, where most precipitation in winter is tied up as snow, acid springs flow the least in winter, and the peak flow is during snow melt and the rainy season in May and June. Then flow decreases again during the late summer, especially in years when summer rainfall is low. Thus, the marked constancy observed in the large springs of neutral to alkaline pH (Figure 2.1) is considerably less in evidence in the acid springs, although during the time that Doemel studied them, Cyanidium Creek and Nymph Creek showed reasonably constant temperatures throughout the thermal gradients (Figures 9.8 and 9.9). The chemistry does not change nearly as markedly as the flow rate, so that pH values are fairly constant throughout the year.

Although *Cyanidium* is most conspicuous in hot spring effluents, almost

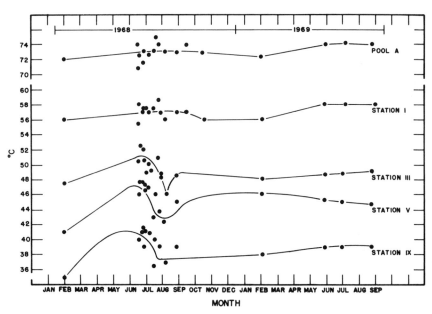

Figure 9.8. Temperatures at Cyanidium Creek, February 1968 through August 1969. The stations were numbered in series downstream from the source. For simplicity, only a few stations are shown. (From Doemel, 1970.)

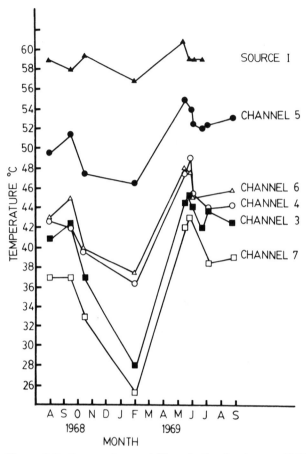

Figure 9.9. Temperatures at Nymph Creek, August 1968 through August 1969. Note the marked drop in winter. This is due to the very low flow at this time of year due to the fact that the ground water which feeds the springs is depleted (all precipitation tied up as snow on the ground).

certainly its most common habitat is in hot acid soils. Acidity develops in soils as a result of oxidation of sulfide and elemental sulfur (discussed in Chapters 6 and 12), and acid soils can develop even in habitats where acid springs and other manifestations of solfataras do not occur. All that is necessary is the emanation to the surface of the soil of steam enriched in H_2S. We found suitable acid soils in areas such as Steamboat Springs, Nevada, where all of the springs were neutral to alkaline pH. Temperatures in acid soils of course vary much more than do temperatures of springs, so that it is much more difficult determining the temperature of the habitat. As Smith and Brock (1973) showed, an even more important environmental factor in acid soils is water availability, and this may be the prime factor controlling the density of *Cyanidium* populations in acid soils. Of considerable interest from the viewpoint of the pH of the *Cyanidium* habitat, acid soils may have much lower pH values than aquatic environments. Although springs with pH values of 2–3 are common, and a few springs with pH values as low as 0.9–1.0 exist, soils with pH values even lower than this can be found, since given sufficient evaporation, the sulfuric acid present becomes greatly concentrated. We measured a soil pH in one location where *Cyanidium* was present (Steamboat Springs, Nevada) at 0.05, which is almost 1 N sulfuric acid! [The measurement of pH in such acidic soils is not straightforward. For a discussion of this question, and a method for measuring such pH values, see Doemel and Brock (1971a).] Allen (1959) actually grew *Cyanidium* in culture in 1 N sulfuric acid.

Temperature Limits of *Cyanidium*

When Doemel began his work on *Cyanidium,* he was faced with the fact that several workers had reported the existence of *Cyanidium* at temperatures of 75–80°C (Allen, 1959; Copeland, 1936). This high temperature, if true, would make *Cyanidium* the most thermophilic alga known, since not even blue-green algae are able to grow at such temperatures. The reports of *Cyanidium* living at high temperatures, however, did not actually demonstrate that growth was taking place, or that the organism from such temperatures was actually viable, and when the temperature variability of the acid environment is considered, it obviously becomes of considerable importance to ascertain that the organism is really viable. Furthermore, it has not always been clear in the reports of earlier workers that they took care to measure the temperature exactly at the site where the organism was seen. Many reports in the earlier literature reported the temperature of the *spring,* even if the organism was not present in the spring but in the *outflow channel.* In Doemel's work, he was careful to measure the temperature precisely at the site where the organism was seen and also to ascertain viability.

In the initial work (Doemel and Brock, 1970), an extensive survey was made by Doemel and myself in more than 150 thermal areas in Yellowstone National Park; The Geysers, California; Steamboat Springs, Nevada; Beowawe, Nevada; Solfatara and Agnano Crater, Italy; Iceland (many thermal areas); North Island, New Zealand [many thermal areas, see Brock and Brock (1971)]; Gunma Prefecture, Japan; Ahuachipan, El Salvador; Dominica, West Indies. Natural populations were identified tentatively in the field by their characteristic blue-green color, and the temperatures of the populations were determined with a small diameter "baby banjo" thermistor probe. Samples were then collected for pH determination, microscopic examination, and, in some instances, culture.

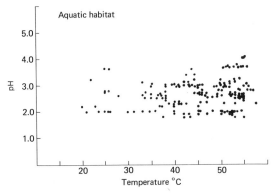

Figure 9.10. Distribution of *Cyanidium* in aquatic habitats with respect to temperature and pH. Each dot represents a pH and temperature at which *Cyanidium* has been observed in an aquatic habitat. (From Doemel, 1970. See Appendix 2 of this reference for details.)

Figure 9.10 presents the data obtained in aquatic environments and Figure 9.11 gives the data from soils. In no location was *Cyanidium* found at a temperature above 57°C.

In addition to these random observations, the thermal gradients of a number of acid springs were thoroughly examined to determine the upper temperature at which visible *Cyanidium* was present. The results of these observations are given in Table 9.3. In these thermal gradients there is usually a well-defined demarcation between high-temperature algal-free areas and cooler regions where *Cyanidium* is present. Temperatures and samples were taken of both the obvious *Cyanidium* populations and of the sediment material where *Cyanidium* appeared to be absent. Table 9.3 shows that the upper temperature limit for the existence of *Cyanidium* in nature is 55–57°C.

Copeland (1936) reported *Cyanidium* in Yellowstone at 75–80°C, but mentions only one specific location where the organism was found at such a high temperature: Great Sulphur Spring in the Crater Hills area. In the summer of 1967 we visited this area and found that although the characteristics of this spring were essentially the same as those reported by Allen and Day (1935) in 1933 (about the time Copeland was working in Yellowstone), no evidence of *Cyanidium* at a temperature greater than 41°C was found. We did find that Great Sulphur Spring fluctuates widely in flow (as also reported by Allen and Day), and Copeland may have observed an unusually large flow that was in the process of killing a *Cyanidium* population. At The Geysers, California, Allen (1959) reported *Cyanidium* from Lemonade Spring and although the spring had a temperature of 70–75°C, she did not specifically state the precise temperature at which the alga was taken. We examined carefully all possible *Cyanidium* habitats at The Geysers but

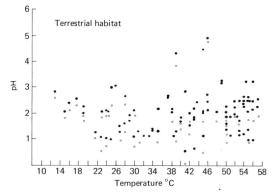

Figure 9.11. Distribution of *Cyanidium* in terrestrial habitats (soils, fumarole condensates) with respect to temperature and pH. Each dot represents the pH and temperature at which *Cyanidium* was observed. Solid circles, soil pH. Open circles, soil water pH. (From Doemel, 1970. See Appendix 2 of this reference for details.)

Table 9.3. The Upper Temperature Limits of *Cyanidium caldarium* in Aquatic Thermal Gradients[a]

Location	pH	Maximum temperature (°C)
Norris Geyser Basin, Yellowstone Park		
Blue Geyser effluent	3.8	55.5
Cyanidium Creek	3.0	54–55
Norris Annex effluent	3.0	55
Pinwheel Geyser effluent	3.1	53
Primrose Spring effluent	3.6	55
Gibbon Geyser Basin, Yellowstone Park		
Paint Pot Hill effluent	3.0	51.0
Sylvan Springs, main effluent	2.7	53–54
Sylvan Springs, secondary effluent	2.3	50.8
Miscellaneous effluents, Yellowstone Park		
Nymph Creek	2.6–3.0	55–57
Roaring Mountain, south effluent	2.0–2.3	54.5
Lemonade Creek	2.0	55
Mt. Lassen Volcanic Park		
Sulphur Works effluent	2.6	53–54
North Island, New Zealand		
Waimangu Caldron outlet	3.8	54
Gunma Prefecture, Japan		
Kusatsu (Yubatake) Spring	1.85	52.5

[a]From Doemel and Brock (1970).

could find no location where the alga was present at temperatures above 55°C.

Although it appeared that the maximum temperature of *Cyanidium* in nature was 55–57°C, we felt it important to establish the optimum and maximum temperature for photosynthesis by natural populations. Photosynthesis was measured with $^{14}CO_2$, using methods described in detail elsewhere (Doemel and Brock, 1970; Doemel, 1970; Doemel and Brock, 1971b). The results of four experiments with populations collected from different temperatures are shown in Figure 9.12. As seen, there is a sharp temperature optimum for photosynthesis at 45°C and no photosynthesis at temperatures above 60°C. The same results were obtained whether the natural populations used came from habitats at 30°C, 39°C, 49°C, or 56°C. Cultural studies (see next section) also supported this conclusion.

Doemel then tested the thermostability of the photosynthetic system by preincubating samples at 60°C or 70°C for various times before measuring photosynthesis at the optimal temperature of 45°C. As seen in Figure 9.13, the ability of *Cyanidium* populations to photosynthesize is rapidly destroyed at temperatures of 60°C and 70°C.

From these results, it seems evident that *Cyanidium* is not an extreme thermophile, and it is likely that the results of other workers are erroneous. The establishment of the upper temperature limit for *Cyanidium* was of considerable fundamental importance, since this alga is the most thermophilic eucaryotic alga, and is living at temperatures almost approaching the upper temperature limit for eucaryotic life (see Chapter 3).

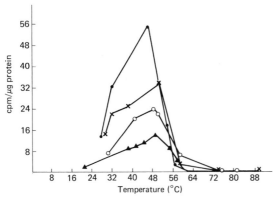

Figure 9.12. Temperature optimum for photosynthesis by natural populations of *Cyanidium* from habitats of various temperatures: 56.0°C (triangles), 48.8°C (X), 38.8°C (open circles) at Cyanidium Creek, and from a Yellowstone terrestrial habitat having a temperature of 30°C (closed circles). Samples were suspended in spring water, transferred to vials, and incubated at various temperatures. $^{14}CO_2$ was added and the rate of photosynthesis measured. Only light-stimulated uptake is shown, and incorporation is normalized to the amount of protein present in each suspension. (From Doemel and Brock, 1970.)

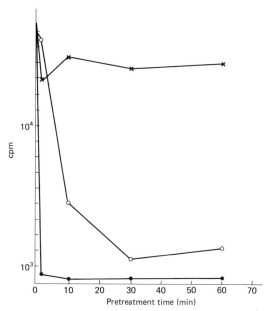

Figure 9.13. Thermostability of the photosynthetic system. A portion of the population from a habitat with a temperature of 45°C was collected, resuspended in spring water, distributed to vials, and pretreated at 70°C (closed circles), 60°C (open circles), and 45°C (X) for the stated times. The vials were then returned to 45°C and photosynthesis measured by the $^{14}CO_2$ technique. Values on the ordinate are cpm incorporated into the total population in each 5-ml vial (all vials were filled from the same uniform suspension, so that normalization was not necessary). (From Doemel and Brock, 1970.)

The lower temperature limit for *Cyanidium* was more difficult to define. One problem is that at temperatures below 40°C, an acidophilic *Chlorella* is present, and this organism looks identical to *Cyanidium* under the microscope. In culture, the two organisms can be readily distinguished, because the *Chlorella* lacks phycocyanin. Another distinction between the two is that the chlorophyll of the acidophilic *Chlorella* is readily extracted with acetone or methanol. whereas the chlorophyll of *Cyanidium* is not extracted by these solvents, but is with dimethylformamide (Volk and Bishop, 1968). Therefore, if a natural population is counted under the fluorescence microscope (both organisms showing red chlorophyll fluorescence) and then extracted with acetone, the *Cyanidium* cells retain their fluorescence (probably not only because they retain chlorophyll but because they have phycocyanin), whereas the *Chlorella* cells lose all fluorescence and are thus no longer visible. Another count after extraction will then give a measure of the *Cyanidium* population. To check that the latter cells are indeed *Cyanidium,* a further extraction with dimethylformamide can be made, after which no fluorescing cells should be seen.

The other low-temperature organisms of acid hot springs are readily distinguished from *Cyanidium* under the microscope. These include *Euglena, Chlamydomonas,* pennate diatoms, photosynthetic flagellates other than *Euglena,* and the filamentous alga *Zygogonium.*

At temperatures below 40°C, the other algae just mentioned appeared, although *Cyanidium* could still be cultured from locations where the temperature was as low as 30°C. In terrestrial habitats, where water availability (Smith and Brock, 1973) and other factors may have prevented the growth of some of the other eucaryotic algae mentioned above, *Cyanidium* was frequently seen at considerably lower temperatures. *Cyanidium* in culture will grow at temperatures as low as 20°C, albeit slowly.

In certain thermal gradients, the displacement of *Cyanidium* by other algae can be very sharp. One of the more dramatic examples is the large and very acidic hot spring Yubatake (pH 1.5) at Kusatsu, in Japan, where *Cyanidium* is replaced at a temperature around 40°C by the diatom *Pinnularia brauni* var. *amphicephala* (Schwabe, 1942). The diatom forms strikingly brownish mats, so that the border between it and *Cyanidium* is very obvious. (When I visited this spring in 1970, it appeared unchanged from Schwabe's description of 1942.) Obsidian Creek (Figure 9.7) presents an interesting example of the selective role of temperature for *Cyanidium*. As seen in Figure 9.14, as the temperature of this creek warms up to greater than 35°C, *Cyanidium* rapidly replaces other eucaryotic algae.

Absence of Temperature Strains in *Cyanidium*

As has been discussed in Chapter 8, the blue-green alga of alkaline hot springs, *Synechococcus,* exists as a series of temperature strains, each strain present at the appropriate temperature along the thermal gradient of the effluent channel (Peary and Castenholz, 1964; Brock, 1967). Also, temperature strains exist in *Sulfolobus* (Chapter 6). However, temperature strains of *Cyanidium* do not exist despite the fact that the organism lives over a temperature range of about 20°C (from 55°C to 35°C), and that relatively stable thermal gradients exist in some places which would seem to be selective for specific temperature strains.

To demonstrate the absence of temperature strains, Doemel carried out a large number of photosynthesis experiments using natural samples collected from habitats of different temperatures. Figure 9.12 shows the temperature optima for several populations taken from different habitat temperatures. More detailed data are presented in Table 9.4, and it can be seen that the temperature optimum was between 45°C and 50°C, irrespective of the source of the sample.

Since these experiments only measured the temperature optimum for

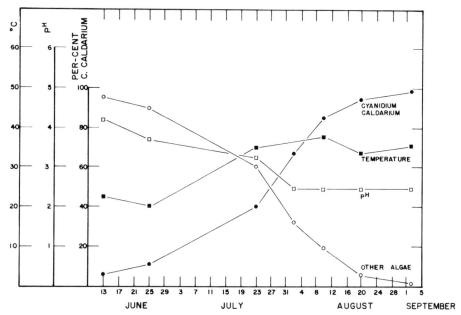

Figure 9.14. Changes in density of *Cyanidium* and other eucaryotic algae during the summer warm-up period in Obsidian Creek, showing the selective action of temperature. (From Doemel, 1970.)

photosynthesis, Doemel also did a number of cultural experiments to determine the temperature optimum for growth. Enrichment cultures were set up using samples collected from habitats of different temperatures, using incubation temperatures similar to the habitat. From these enrichment cultures, axenic cultures were obtained at the same incubation temperatures, and the temperature optima for growth then measured. As seen in Figure 9.15, all of the isolates had temperature optima at 45°C, despite the fact that they were obtained from habitats of widely differing temperatures, and had been kept in culture for many generations at habitat temperature.

The contrast between *Cyanidium* and *Synechococcus* is illustrated in Figure 9.16. It is not clear why temperature strains of *Cyanidium* do not occur. Since mutants of this organism are readily obtained, it would appear that genetic variability is not foreclosed. One possibility suggested by Doemel and Brock (1971b), is that the predominant habitat of *Cyanidium* is not the relatively stable aquatic environment but the rather unstable (as far as temperature goes) acid thermal soil. In these terrestrial habitats, the populations experience wide temperature fluctuations, so that the evolution of strains adapted to a specific temperature range would be of little advantage. Basically, the situation relates to the concept of stenothermal and

Table 9.4. Temperature Optimum for Photosynthesis of Natural Populations of *Cyanidium*[a]

Sample origin	Optimum temperature for photosynthesis			
	Nymph Creek	Cyanidium Creek	Roaring Mountain	Amphitheater Springs
56	49.5			
55				
54		45–48		
53	51			
52		45		
51				
50		50[4][b]		
49				
48	48.5	49–50[4][b]		
47				
46	49.5	47		
45	46	45–50		45
44				
43	48.5[2][b]			
42	48.5[2][b]			
41		46		
40	48.5			47.5[c]
39	52	45[3][b]		
38	48.5			
37		50.2		
36				
35				
34				
32			47[c]	
31				

[a]From Doemel (1970). All numbers in degrees Celsius.
[b]Brackets indicate the number of determinations.
[c]These samples were removed from soil.

eurythermal organisms. Stenothermal organisms have relatively narrow temperature optima, and are hence restricted to relatively narrow temperature ranges in nature. *Synechococcus* is a good example of a stenothermal organism. Eurythermal organisms, on the other hand, have wide temperature ranges for growth, and are thus not restricted to such narrow temperature ranges. *Cyanidium* is thus eurythermal, and hence adaptable to a wide range of temperatures, but at the same time it can only form large population densities at certain temperatures throughout its range. As will be discussed in the next section, *Cyanidium* shows its maximum population development in nature at temperatures of about 45°C, which is the optimum temperature for growth and photosynthesis.

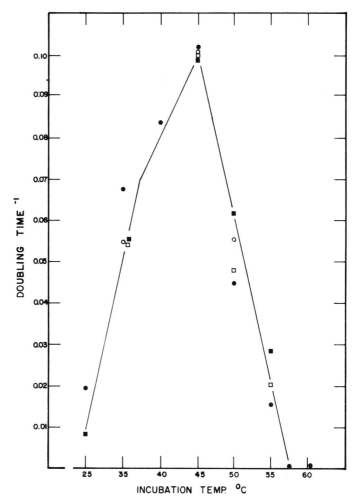

Figure 9.15. Optimum growth temperature for *Cyanidium* cultures isolated and maintained at various temperatures. Unialgal isolates which had derived from natural samples and which had been enriched and purified at the same temperatures were used: 35°C (closed circles), 40°C (open circles), 45°C (closed squares), and 50°C (open squares). Growth rates were determined at the temperatures listed on the abscissa, using autotrophic medium, 400–600-footcandles illumination, and 5% CO_2 in air. The doubling times were calculated from the exponential growth phase. (From Doemel, 1970.)

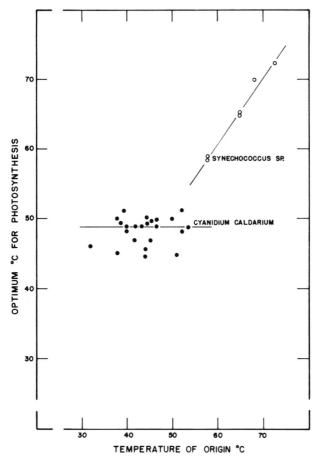

Figure 9.16. Comparison of temperature relationships of *Cyanidium* with those of the unicellular blue-green alga *Synechococcus,* which lives at the upper temperature limit in neutral and alkaline hot springs. Note that a number of temperature strains of *Synechococcus* are indicated, whereas temperature strains do not exist for *Cyanidium.* (From Doemel, 1970.)

Growth Rates of *Cyanidium* in Nature

Doemel carried out a number of studies to measure the growth rate of *Cyanidium* directly in its habitat. The procedure used was to quantify the rate of increase of population density with time on cleaned or "sterile" substrates placed in the habitat. This procedure does not give an actual growth rate, but an apparent growth rate, since it does not take into consideration cell losses that occur concomitantly with growth, as a result of washout, lysis, or predation (see Chapter 8). The fact that an increase in cell numbers occurs when a cleaned substrate is placed in the spring shows that growth is occurring at a rate faster than loss. Eventually a steady state must occur, in which loss rate balances growth rate, and the standing crop under steady-state conditions must be determined by the relative rates of these two processes.

Doemel developed a suction method that permitted sampling quantitatively a known area of stream bottom or artificial channel, and cell counts were made on the resultant suspensions. The cell density could thus be expressed as cell number per unit area.

Cyanidium is seen in nature on rock, gravel, pine needles, pine branches, and other solid substrates. Initially, portions of natural channels at Nymph Creek and Cyanidium Creek were cleared, and attempts were made to monitor the increase in cell number. Although the cell density did increase, accurate quantitation was difficult because of the irregular nature of the substratum. Thus, artificial channels were prepared. After a series of preliminary studies, it was found that wooden channels would serve as excellent *Cyanidium* substrata (Figure 9.6). As seen in Figure 9.17, when a

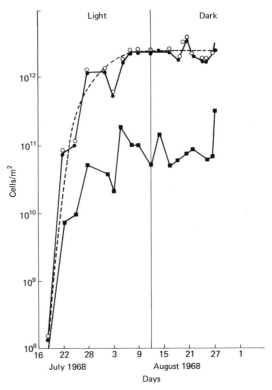

Figure 9.17. The rate of increase of *Cyanidium* on an artificial channel in Nymph Creek. The pine channel (4 × 6 feet × 3 inches) was placed on 16 July 1968. Known areas of the surface of the channel were sampled periodically to determine cell density. The total cells (open circles), total fluorescing cells (closed circles), and fluorescing mother cells (closed squares) were counted microscopically. On 12 August 1968, the channel was covered with an opaque shade, and sampling continued. (From Doemel, 1970.)

channel was placed in the stream, there was an initial exponential increase in cell number with little or no lag, and a doubling time of about 18 hours. A steady-state concentration of 2.5×10^{12} cells/m² was obtained. During the exponential increase and in the steady state, the reproductive or mother cells constituted about 10% of the total population. When the population had attained the maximum cell density, a portion of the channel was covered with a dark shade, and any changes in the population monitored. As seen in Figure 9.17 the population remained stable for at least 15 days after darkening, although (not shown) the color changed from a bright blue-green to a faded green. One year later the dark portion of this channel lacked any evidence of green color, and it appeared that the primary organisms were fungi [a filamentous fungus, possibly *Dactylaria gallopava* (see Chapter 3) and a yeast], with only a few moribund *Cyanidium* cells.

A similar channel was placed in Cyanidium Creek, and the results obtained are shown in Figure 9.18. On this channel, the rate of increase of population was about the same as at Nymph Creek, with a doubling time of 1 day. Prior to reaching the steady state, a portion of this channel was

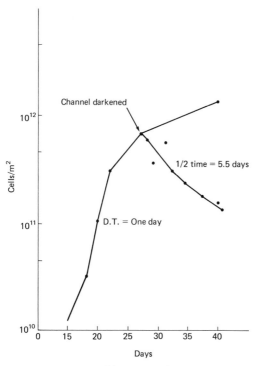

Figure 9.18. Rate of increase of *Cyanidium* on an artificial channel in Cyanidium Creek. Channel installed 10 July 1968. Only the total cell population was monitored. On August 12, one portion of the channel was covered with a dark shade and the other portion was left uncovered as a control. The doubling times for growth and washout are given on the graph. (From Doemel, 1970.)

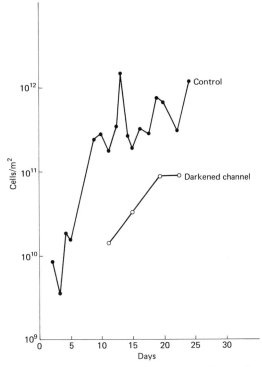

Figure 9.19. The rate of increase of *Cyanidium* on light and dark portions of a channel placed in Nymph Creek on 3 August 1968. (From Doemel, 1970.)

covered with a dark shade, and it can be seen in the figure that in this channel there was a marked decrease in population density in the darkened portion, with a half-time for loss of 5.5 days. Eventually, the population density decreased to a level about 100-fold lower than in the control population in the light.

In a third channel placed in Nymph Creek, one-half of the surface was darkened from the beginning, with the intent of measuring the settling rate on the channel (from cells washing into the channel from upstream). As seen in Figure 9.19, on the uncovered portion of the channel, the population density increased exponentially with a doubling time of 1–2 days, and a steady-state concentration of 1.1×10^{12} cells/m² was reached (similar to the cell densities on the other channels). On the darkened portion of the channel, the appearance of *Cyanidium* was considerably delayed, and the population density then increased exponentially at a doubling time of about 3 days, and reached a steady-state level which was 10% of the level in the uncovered portion of the channel. The organisms on the darkened portion of the channel lacked the characteristic blue-green color of those organisms seen on the light portion, but the cells fluoresced (indicating presence of chlorophyll) and the population would photosynthesize when placed in the light (measured with $^{14}CO_2$).

Although the rate of increase in cell density on these channels is obviously not a direct measure of growth rate, the results show that settling, washout, and lysis are relatively low, so that the rate of increase in cell density is a fairly reasonable estimate of growth rate. In fact, the growth rates obtained on these channels are similar to the growth rates measured for *Cyanidium* under (presumably optimal) conditions in culture. Thus, when colonizing a fresh substratum in nature, *Cyanidium* grows rapidly and unrestrictedly. The steady-state level reached is probably determined primarily by the availability of light, with self-shading eventually restricting growth. The effect of light intensity on growth will be discussed in a later section of this chapter.

Artificial channels were then used to determine the optimum temperature for growth of natural populations. Channels were placed at various

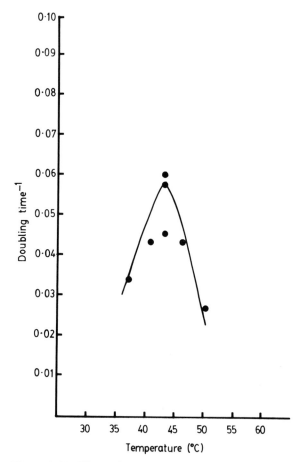

Figure 9.20. The optimum temperature for the development of natural populations of *Cyanidium,* as determined from growth rate measurements on artificial channels installed at different temperatures in Nymph Creek and Cyanidium Creek. (From Doemel and Brock, 1971b.)

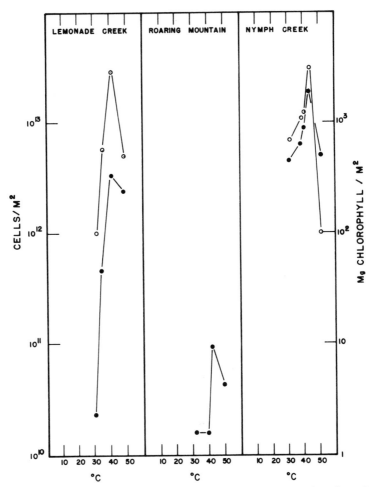

Figure 9.21. Population density of *Cyanidium* in natural effluents as a function of temperature. Samples of known area from natural substrate (Lemonade Creek) or artificial channels at Roaring Mountain (in place for 3 months) or Nymph Creek (in place for 1 year) were collected with a standard coring procedure and the average chlorophyll content (open circles) and cell density (closed circles) of triplicate samples were determined and plotted as a function of temperature. (From Doemel and Brock, 1971b.)

temperatures in Nymph Creek and Cyanidium Creek. As seen in Figure 9.20, there was a sharp temperature optimum for growth at about 42°C, which is about the optimum temperature found in the photosynthesis and culture experiments (see previous section).

Peak standing crop on the natural substrata also occurs at 45°C, the temperature optimum for growth. This is shown by the data in Figure 9.21, for three springs. This suggests a fairly close relationship between growth rate and standing crop.

Long-term variations in standing crop occur, but are almost exclusively due to changes in temperatures and flow rates of springs. Some of these long-term changes are presented by Doemel (1970). As noted, these acid springs are not nearly as constant in flow and temperature characteristics as neutral springs, which accounts for a lot of the long-term variability seen. When attempting to measure growth rates under natural conditions, one must have a modicum of luck in regard to the weather and the behavior of the springs. Fortunately, the differences seen in Doemel's work were sufficiently large so that minor variations in temperature and flow rate did not mask the overall picture.

Relationship to pH

The acidophilic character of *Cyanidium* has already been noted. An element of confusion was introduced when Allen (1959) claimed that *Cyanidium* did not require acidity for growth, but would grow at a neutral pH, although not as well as in acid media. However, Ascione et al. (1966) clearly showed that *Cyanidium* was an obligate acidophile. If a *Cyanidium* culture is inoculated into medium at pH 7, it does not grow immediately, but the pH of the medium rapidly decreases and growth only begins after the pH has decreased to less than 5.0. The usual medium used for the culture of *Cyanidium* is very poorly buffered at neutral pH, but if the pH is maintained at neutrality by periodic additions of alkali, no growth occurs.

The optimum pH for growth of autotrophic and heterotrophic cultures of *Cyanidium* is shown in Figure 9.22. A broad pH optimum is seen, but no growth occurs at pH 5 or above. The ability of *Cyanidium* to grow at pH near 0 is striking: Note that this is not acid tolerance, but the ability to actually grow readily at these pH values.

Interestingly, pH 7 was not toxic to *Cyanidium*. As seen in Figure 9.23, photosynthesis occurred about as well when a population of cells was adjusted to pH 7 as at pH 2. There was a sharp pH optimum for photosynthesis at pH 4, but the significance of this optimum is not known. The possibility that high pH values might have detrimental effects over a longer period of time was not investigated. Further work on the physiology of adaptation of *Cyanidium* to low pH would be desirable.

As would be expected from the laboratory results, in nature *Cyanidium* is restricted to habitats of low pH. The pH of habitats in which *Cyanidium* has been found is given in Figure 9.10 and 9.11.

Effect of Light Intensity

Cyanidium has been reported to be inhibited by light intensities that are about one-tenth of the intensity of full sunlight (Brown and Richardson, 1968; Halldal and French, 1958), yet in nature *Cyanidium* populations

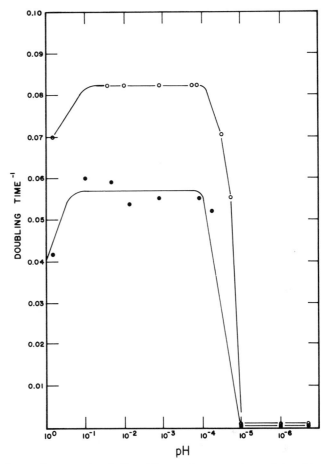

Figure 9.22. pH optimum for growth of *Cyanidium* cultures under autotrophic (closed circles) and heterotrophic (open circles) conditions (1% glucose in the dark). The pH of the medium was monitored at frequent intervals and adjusted with either 0.1 N NaOH or 0.1 N HCl to maintain the pH at a relatively constant value. The inverse of the doubling time is plotted as a function of pH. (From Doemel and Brock, 1971b.)

readily develop in full sunlight. It was therefore of interest to study the adaptation of *Cyanidium* in the field to various light intensities. To this end, an experimental channel was installed and neutral density filters were used which reduced light intensity by known amounts. The response to light of the populations that developed under these filters was then measured. The techniques used have been described in Chapters 7 and 8 and in Brock and Brock (1969).

As seen in Table 9.5, after 1 month there was a 2–3-fold increase in chlorophyll content per cell in the population that had developed under the densest filter (92% reduction in light intensity), suggesting some ability of *Cyanidium* to modify its photopigments in response to light intensity.

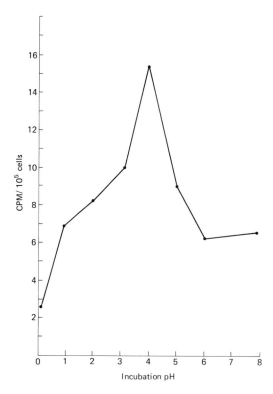

Figure 9.23. The pH optimum for photosynthesis of a natural population of *Cyanidium* collected from a terrestrial habitat at Roaring Mountain. The pH did not change significantly during the experiment. (From Doemel and Brock, 1971b.)

Unfortunately, a method was not available to measure phycocyanin in the adapted populations, but visual observations suggest that phycocyanin also changed, since the cells under one or two neutral density filters appeared to be a darker blue-green than the uncovered populations, even though chlorophyll per cell was less.

The effect of light intensity on photosynthesis of these populations is seen in Figure 9.24. The population that developed in full sunlight or at 47% of full sunlight showed no inhibition of photosynthesis at high-light inten-

Table 9.5. Chlorophyll Content of Natural Populations of *C. caldarium* as Related to Light Intensity[a]

Location	Condition	Percent light reduction	μg chlorophyll/cell
Cyanidium Creek[b]	No filters	0	1.76×10^{-7}
Cyanidium Creek	1 filter	53	1.02×10^{-7}
Cyanidium Creek	2 filters	83	1.03×10^{-7}
Cyanidium Creek	3 filters	92	3.03×10^{-7}
Nymph Creek	No filters	Naturally shaded	1.02×10^{-7}

[a]From Doemel (1970).
[b]In the Cyanidium Creek series, samples were taken 1 month after the filters were installed.

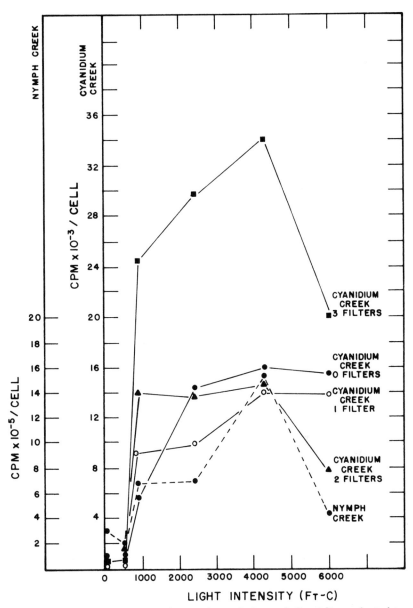

Figure 9.24. The photosynthesis of natural populations of *Cyanidium* adapted to different light intensities, as a function of light intensity. Naturally shaded channel in Nymph Creek and portions of a channel in Cyanidium Creek that were uncovered or covered by one (53% light reduction), two (83% light reduction), and three (92% reduction) neutral density filters. (From Doemel and Brock, 1971b.)

sity, whereas the populations developing at 17% and 8% of full sunlight showed considerable high-light inhibition. The population developing at 8% of full sunlight also showed an increase in photosynthesis per cell. However, since the chlorophyll content of this population is higher than the others, on a per chlorophyll basis the photosynthesis rate is the same as the others.

Almost certainly a similar adaptation to lowered light intensities occurs for the cells that are found in the deeper parts of the mats. This would be especially true for the mat at Nymph Creek, which is quite thick, and which additionally develops in a very shaded environment, where full sunlight rarely reaches the organisms (Doemel, 1970). In the lower portions of these thick mats, the *Cyanidium* cells appear yellowish or brownish, suggesting that they have lost their photopigments, probably because they are effectively dark. However, cells even from the depths of the mats often possess some photopigments, as shown by the fact that they fluoresce red when viewed under the fluorescence microscope. The question of whether cells from the depths of the mats are growing heterotrophically is discussed in the next section.

Heterotrophy of *Cyanidium:* Ecological Significance

Like many eucaryotic algae (and a few blue-greens) *Cyanidium* grows well heterotrophically in the dark. When growing heterotrophically in the dark, *Chlorella* continues to synthesize chlorophyll, but in *Cyanidium* chlorophyll and phycocyanin synthesis are virtually completely repressed. Rosen and Siegesmund (1961) showed that dark-grown *Cyanidium* possessed a rudimentary chloroplast, and in fact *Cyanidium* can be cultured indefinitely in the dark without losing its ability to synthesize photopigments upon being returned to the light. *Cyanidium* can grow on a large number of carbon sources, including mono- and disaccharides, mannitol, glycerol, ethanol, succinate, glutamate, and lactate (Allen, 1952). Doemel (1970) showed that the growth rate on glucose was about the same as the growth rate heterotrophically, and was independent of glucose concentration down to 0.1% glucose. The cell yield was markedly affected by glucose concentration, as would be expected. Doemel (1970) measured the incorporation of ^{14}C-glucose into *Cyanidium* cells grown either autotrophically, heterotrophically, or mixotrophically (light + glucose). As seen in Figure 9.25, heterotrophically grown cells incorporate ^{14}C-glucose much more readily than do autotrophically grown cells, and mixotrophically grown cells are intermediate. Interestingly, glucose uptake by all three types of cells was stimulated by light (not shown in the figure). However, although glucose is incorporated in autotrophically growing cells, there was no enhancement of autotrophic growth rate by glucose. There was also no inhibition of pigment

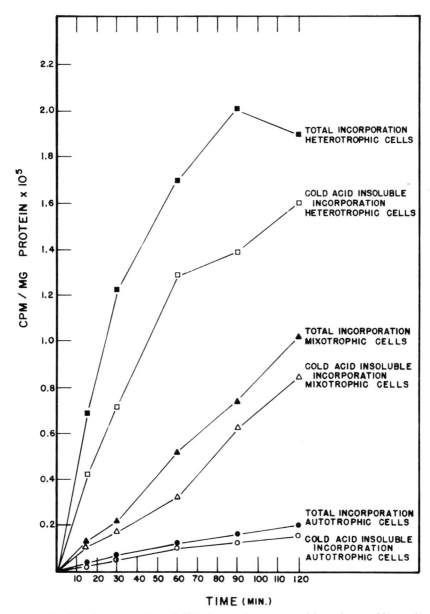

Figure 9.25. The incorporation of ^{14}C-glucose by autotrophic, mixotrophic, and heterotrophic cultures of *Cyanidium*. All cell suspensions were washed and suspended in mineral salts medium for the experiment. ^{14}C-glucose was added at 0.1 μCi/ml. (For details, see Doemel, 1970.)

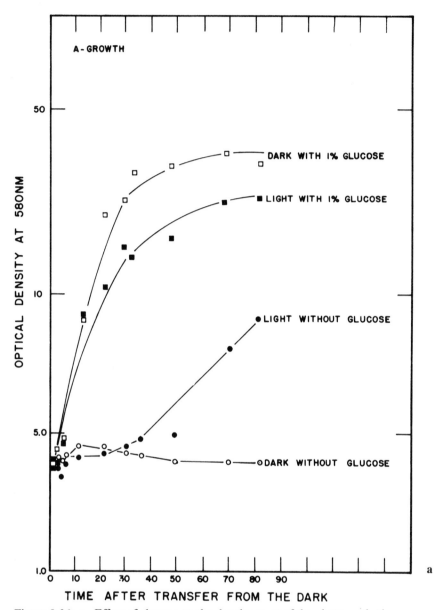

Figure 9.26a-c. Effect of glucose on the development of the photosynthetic process in *Cyanidium*. Cells were grown in the dark in 1% glucose-containing medium, washed, and resuspended in mineral salts medium. During subsequent growth, photosynthesis was measured by the $^{14}CO_2$ method, and chlorophyll content and cell density were monitored periodically. Changes in growth (part a), chlorophyll (part b), and photosynthetic ability (part c) are given as a function of time. (For details, see Doemel, 1970.)

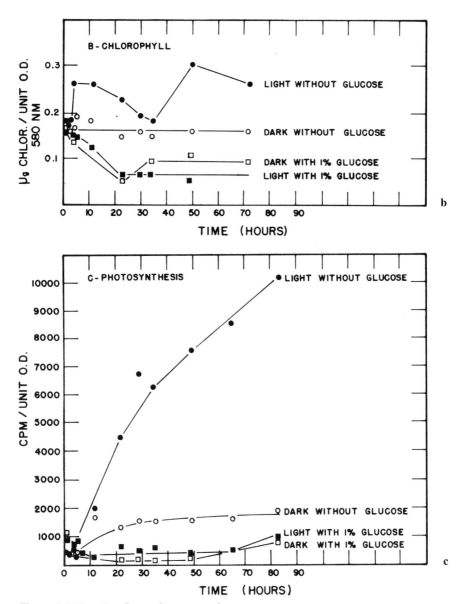

Figure 9.26 b and c. Legend see opposite page.

synthesis by glucose in autotrophically growing cells. Similar results were obtained with galactose or sucrose.

Although glucose has no effect on growth or pigment synthesis in autotrophically grown cells, it continues to affect growth and pigment synthesis when heterotrophically grown cells are transferred to the light. Figure 9.26 shows that when dark-grown cells were transferred to the light in the presence of glucose, heterotrophic growth continued and the synthesis of photopigments and the photosynthetic apparatus continued to be repressed. In contrast, if such dark-grown populations were transferred to the light without glucose, growth ceased immediately but the cells began to synthesize photopigments and to assimilate CO_2 at an increasing rate. Growth eventually resumed in this population at a slower rate after a 20–30-hr lag. Thus, although glucose has no effect on photopigment synthesis when added to autotrophically growing cells in the light, it does continue to repress photopigment synthesis when heterotrophically growing cells are transferred to the light. In the light, there is a gradual "escape" of such cells from the effects of glucose, so that eventually the photosynthetic apparatus is resynthesized and a return to the autotrophic state occurs. The mechanism whereby this "glucose memory" is effected is unknown. However, since glucose incorporation is very poor in autotrophically growing cells (Figure 9.25), it is likely that glucose repression involves uptake and conversion of glucose to some product, a process that the heterotrophically grown cells can continue to carry out after they are transferred to the light.

Doemel attempted to determine whether natural populations might grow heterotrophically under certain conditions. As noted in the previous section, light was essential for growth in the springs, but since the water of the springs is low in organic matter, it would be unlikely that sufficient organic matter would be present to permit simple heterotrophic growth. However, in the thick mats that develop in some habitats, and in the subsurface terrestrial environment, there was a possibility that heterotrophic growth could occur using organic materials synthesized by the overlying autotrophically growing cells. Doemel (1970) attempted to isolate heterotrophic *Cyanidium* from some of the thick mats at Nymph Creek, by enriching in glucose-containing medium in the dark, at several temperatures. In all cases, instead of *Cyanidium* he isolated heterotrophic bacteria and fungi. However, Gary Darland did succeed in isolating one *Cyanidium* culture heterotrophically. He obtained a yellowish colony containing cells resembling *Cyanidium* microscopically. When this colony was transferred to autotrophic medium and incubated in the light, a typical green culture of *Cyanidium* was obtained. No pigment-less variants were found in any isolation attempts. It would be desirable to determine whether the yellowish or brownish populations from the lower parts of the Nymph Creek mat are now able to assimilate glucose. Because of the wide variety of heterotrophic organisms present in these mats (Belly et al., 1973), simple ^{14}C-uptake experiments could not be done, but the use of autoradiography would permit a determination of whether natural heterotrophy occurred.

Nitrogen Nutrition of *Cyanidium*

In general, acid hot springs are high in ammonia (Chapter 2), and it is thus not surprising that *Cyanidium* uses ammonia quite well as a nitrogen source. Allen (1952) showed that *Cyanidium* is also able to use nitrate, casein hydrolysate, casein, and urea as nitrogen sources. Noguchi and co-workers (see Chapter 2) showed that many of the acid springs in Yellow-stone also had significant amounts of nitrate, so that it is not surprising that *Cyanidium* can use nitrate. Rigano et al. (1975) have studied the nitrate reductase of *Cyanidium* in some detail. These workers have shown that not all strains of *Cyanidium* are able to utilize nitrate, and they compared enzymes of the ammonia assimilation pathway in one nitrate-utilizing strain with a strain that could not utilize nitrate. The nitrate-assimilating strain exhibited glutamate dehydrogenase activity whereas the other strain lacked glutamate dehydrogenase but possessed high alanine dehydrogenase and L-alanine aminotransferase activities. This suggested that this strain incorpo-rated ammonia through reductive amination of pyruvate and then formed glutamate via alpha-ketoglutarate. Neither strain possessed glutamate syn-thase activity but both strains contained similar levels of glutamine synthe-tase. Further details on the utilization of glutamate and other amino acids by these two strains were reported by Rigano et al. (1976).

Although the strain differences reported by these workers are interest-ing, they are probably not surprising, since there is likely to be a variety of biochemical differences between various strains of *Cyanidium* once these differences are sought. Because of the ease of isolating *Cyanidium* from nature, it would be of considerable interest to pursue this whole area of biochemical heterogeneity.

Biogeography of *Cyanidium*

I discussed briefly in Chapter 6 the dispersal problems of organisms living in acid hot springs. Although thermal waters of neutral pH are fairly extensive, and many man-made neutral thermal waters exist, hot acid waters are much rarer, and man-made locations are almost nonexistent. The impressive thing about *Cyanidium* in this regard is that virtually every suitable habitat world-wide has been colonized by this organism. This is in contrast to the situation with *Synechococcus,* the blue-green alga that lives at the highest temperatures in neutral pH waters, which is missing from important thermal areas in Iceland and New Zealand (Chapter 8).

Our biogeographical work on *Cyanidium* arose out of a number of travels that I made. At the time, Doemel was in the midst of his *Cyanidium* work, and was culturing the organism routinely from natural samples. It was relatively easy for him to set up enrichment cultures for samples I mailed back, and the presence of *Cyanidium* could thus be readily confirmed.

Table 9.6 summarizes the locations where *Cyanidium* has been defi-nitely shown to exist. Also of some interest is one place where *Cyanidium* has been sought but not found, although conditions were apparently appro-priate: the Sulfur Banks, Hawaii Volcano National Park. The Sulfur Banks is an area of H_2S-rich fumaroles where sulfur oxidation leads to the

Table 9.6. Biogeography of *Cyanidium caldarium*

Geographical location	Number of cultures isolated
Presence of *Cyanidium* confirmed by culture	
United States	
Yellowstone National Park, Wyoming	34
Lassen National Park, California	13
The Geysers, Sonoma Co., California	5
Steamboat Springs, Nevada	3
Italy	
Solfatara and Agnano crater (Naples area)	6
Terme Valdieri (northern Italy)	2
Iceland	1
New Zealand (North Island, many thermal areas)	51
Japan (Gunma Prefecture, Kusatsu area)	14
Cyanidium confirmed only microscopically	
Indonesia (Geitler and Ruttner, 1936)	
El Savador, Central America (T. D. Brock, unpublished)	
Dominica (West, 1904; T. D. and K. M. Brock, unpublished, 1971	
Iceland (T. D. and K. M. Brock, unpublished, 1971)	

development of moist, acidic soils. It is the only hot acid area in the Hawaii thermal area, and a range of temperatures exists from almost ambient up to boiling. I visited this area in December 1969, and made visual observations and collected samples. Visibly green algal development was present, and I carefully determined the highest temperature at which any obvious green was present. The temperature was about 36–37°C, all hotter areas being devoid of green color. (The pH of these soils was around 2.9–3.1.) Microscopic examination of some of these soils that appeared to be algal-free failed to reveal any algae, and enrichment cultures set up at 45°C at pH 2 or 3.5 were unsuccessful. At the same time, the visibly green material at 36–37°C was examined microscopically and typical *Chlorella-Cyanidium* type cells were seen. Enrichment cultures of this material at 25°C and 35°C at pH 2 and 3.5 were successful, but the organism isolated was not *Cyanidium* but *Chlorella*. (Spectroscopic examination of unialgal cultures showed no presence of phycocyanin, but typical chlorophyll peaks of a green alga.)

From these results we concluded that *Cyanidium* was absent from the Sulfur Banks, even though a habitat of appropriate temperature and pH was present. Thus, there appears to be an uncolonized niche for *Cyanidium* at the Sulfur Banks. The question immediately arises as to why *Cyanidium* is absent. One possibility is that the habitat is not appropriate, even though it appears to be, due to presence of a toxic substance or absence of an essential nutrient. However, since the *Chlorella* grows well up to its upper temperature limit, this conclusion does not seem justified. Another explanation is that the habitat is too unstable for *Cyanidium,* the temperature occasionally rising to such high levels that the alga, if present, would be killed. The most interesting possibility is that the habitat is favorable for *Cyanidium* but that the organism is missing because it has not yet been inoculated. When it is considered that the Sulfur Banks (and Hawaii volcanoes as a whole) are relatively recent on the geological scale, and that Hawaii is very far from any suitable source of inoculum, with no intermediate stepping stones, this conclusion seems reasonable.[1] This possibility is further strengthened by the fact that the Hawaii flora and fauna are depauperate in many species, again because of isolation and difficulties of dispersal. In the case of *Cyanidium* at the Sulfur Banks, a simple test of the dispersal hypothesis would be to inoculate the empty niche with a pure culture and observe for growth over the next year or so. Although this experiment would be extremely easy, and harmful results are unlikely, there would probably be many who would object, being mindful of other detrimental results of unwarranted introductions of new species.

References

Adams, B. L., V. McMahon, and J. Seckbach. 1971. Fatty acids in the thermophilic alga, *Cyanidium caldarium. Biochem. Biophys. Res. Commun.* **42**, 359–365.

Allen, C. F., P. Good, and R. W. Holton. 1970. Lipid composition of *Cyanidium. Plant Physiol.* **46**, 748–751.

Allen, E. T. and A. L. Day. 1935. Hot springs of the Yellowstone National Park. Carnegie Inst. Wash. Publ. No. 466.

Allen, M. B. 1952. The cultivation of Myxophyceae. *Arch. Microbiol.* **17**, 34–53.

Allen, M. B. 1954. Studies on a blue-green *Chlorella.* Proc. 8th International Botanical Congress, Sect. 17, pp. 41–42.

Allen, M. B. 1959. Studies with *Cyanidium caldarium,* an anomalously pigmented chlorophyte. *Arch Mikrobiol.* **32**, 270–277.

Ascione, R., W. Southwick, and J. R. Fresco. 1966. Laboratory culturing of a thermophilic alga at high temperature. *Science* **153**, 752–755.

Belly, R. T., M. R. Tansey, and T. D. Brock. 1973. Algal excretion of ^{14}C-labeled compounds and microbial interactions in *Cyanidium caldarium* mats. *J. Phycol.* **9**, 123–127.

Brock, T. D. 1967. Microorganisms adapted to high temperatures. *Nature* **214**, 882–885.

Brock, T. D. and M. L. Brock. 1966. Temperature optima for algal development in Yellowstone and Iceland hot springs. *Nature* **209**, 733–734.

Brock, T. D. and M. L. Brock. 1969. Effect of light intensity on photosynthesis by thermal algae adapted to natural and reduced sunlight. *Limnol. Oceanogr.* **14**, 334–341.

---

[1]One isolated site where *Cyanidium* has been found is an acid hot spring at the summit of Mount Shasta, California (4316 m) (Wharton and Vinyard, 1977).

Brock, T. D. and M. L. Brock. 1971. Microbiological studies of thermal habitats of the central volcanic region, North Island, New Zealand. *N.Z. J. Mar. Freshwater Res.* **5**, 233–257.

Brown, T. E. and F. L. Richardson. 1968. The effect of growth environment on the physiology of algae: light intensity. *J. Phycol.* **4**, 38–54.

Collins, F. S., Holden, and Setchell, W. A. 1901. Phycotheca Borealis Americana Fasc. 18, No. 851. Cited by Tilden, 1910.

Copeland, J. J. 1936. Yellowstone thermal Myxophyceae. *Ann. N.Y. Acad. Sci.* **36**, 1–229.

Doemel, W. N. 1970. The physiological ecology of *Cyanidium caldarium*. Ph.D. thesis, Indiana University.

Doemel, W. N. and T. D. Brock. 1970. The upper temperature limit of *Cyanidium caldarium*. *Arch. Mikrobiol.* **72**, 326–332.

Doemel, W. N. and T. D. Brock. 1971a. pH of very acid soils. *Nature* **229**, 574.

Doemel, W. N. and T. D. Brock. 1971b. The physiological ecology of *Cyanidium caldarium*. *J. Gen. Microbiol.* **67**, 17–32.

Drouet, F. 1943. New species and transfers in Myxophyceae. *Am. Midland Nat.* **30**, 671–674.

Edwards, M. R. and E. Gantt. 1971. Phycobilisomes of the thermophilic blue-green alga *Synechococcus lividus*. *J. Cell Biol.* **50**, 896–900.

Edwards, M. R. and J. D. Mainwaring, Jr. 1973. Ultrastructural localization of phycocyanin in the acidophilic, thermophilic alga, *Cyanidium caldarium*. In *31st Ann. Proc. Electron Microscopy Soc. Am.*, C. J. Arceneaux (ed.). New Orleans, La.

Fredrick, J. F. 1976. *Cyanidium caldarium* as a bridge alga between Cyanophyceae and Rhodophyceae: evidence from immunodiffusion studies. *Plant Cell Physiol.* **17**, 317–322.

Geitler, L. 1958. Die Gattung *Cyanidium*. *Oesterr. Bot. Z.* **106**, 172–173.

Geitler, L. and F. Ruttner. 1936. Die Cyanophyceen der Deutschen Limnologischen Sunda-Expedition. *Arch Hydrobiol. Suppl.* **XIV**, 308–481.

Halldal, P. and C. S. French. 1958. Algal growth in crossed gradients of light intensity and temperature. *Plant Physiol.* **33**, 249–252.

Hirose, H. 1950. Studies on a thermal alga, *Cyanidium caldarium*. *Bot. Mag. Tokyo* **63**, 745–746.

Hirose, H. 1958. Rearrangement of the systematic position of a thermal alga, *Cyanidium caldarium*. *Bot. Mag. Tokyo* **71**, 347–352.

Ikan, R. and J. Seckbach. 1972. Lipids of the thermophilic alga *Cyanidium caldarium*. *Phytochemistry* **11**, 1077–1082.

Kao, O. H. W., M. R. Edwards, and D. S. Berns. 1975. Physical-chemical properties of C-phycocyanin isolated from an acido-thermophilic eukaryote, *Cyanidium caldarium*. *Biochem. J.* **147**, 63–70.

Kleinschmidt, M. G. and V. A. McMahon. 1970a. Effect of growth temperature on the lipid composition of *Cyanidium caldarium*. I. Class separation of lipids. *Plant Physiol.* **46**, 286–289.

Kleinschmidt, M. G. and V. A. McMahon. 1970b. Effect of growth temperature on the lipid composition of *Cyanidium caldarium*. II. Glycolipid and phospholipid components. *Plant Physiol.* **46**, 290–293.

Mercer, F. V., L. Bogorad, and R. Mullens. 1962. Studies with *Cyanidium caldar-*

ium. I. The fine structure and systematic position of the organism. *J. Cell Biol.* **13**, 393–403.

Möller, M. and H. Senger. 1972. Photosyntheseleistung synchroner Kulturen von *Cyanidium caldarium*. *Ber. Deutsch. Bot. Ges. Bd.* **85**, 391–400.

Nichols, K. E. and L. Bogorad. 1960. Studies on phycobilin formation with mutants of *Cyanidium caldarium*. *Nature* **188**, 870–872.

Nichols, K. E. and L. Bogorad. 1962. Action spectra studies on phycocyanin formation in a mutant of *Cyanidium caldarium*. *Bot. Gaz.* **124**, 85–93.

Peary, J. A. and R. C. Castenholz. 1964. Temperature strains of a thermophilic blue-green alga. *Nature (Lond.)* **202**, 720–721.

Rigano, C., G. Aliotta, and V. D. M. Rigano. 1975. Observations on enzymes of ammonia assimilation in two different strains of *Cyanidium caldarium*. *Arch. Microbiol.* **104**, 297–299.

Rigano, C., A. Fuggi, V. D. M. Rigano, and G. Aliotta. 1976. Studies on utilization of 2-ketoglutarate, glutamate and other amino acids by the unicellular alga *Cyanidium caldarium*. *Arch. Microbiol.* **107**, 133–138.

Rosen, W. G. and K. A. Siegesmund. 1961. Some observations on the fine structure of a thermophilic, acidophilic alga. *J. Biophys. Biochem. Cytol.* **9**, 910–914.

Schwabe, G. H. 1942. Thermalökologische Beiträge aus Kusatu. *Gesellschaft für Natur—und Völkerkunde Ostasiens. Mittelungen* **33C**, 25–56.

Seckbach, J. 1972. On the fine structure of the acidophilic hot-spring alga *Cyanidium caldarium*: a taxonomic approach. *Microbios* **5**, 133–142.

Seckbach, J., F. A. Baker, and P. M. Shugarman. 1970. Algae thrive under pure CO_2. *Nature* **227**, 744–745.

Seckbach, J. and R. Ikan. 1972. Sterols and chloroplast structure of *Cyanidium caldarium*. *Plant Physiol.* **49**, 457–459.

Smith, D. W. and T. D. Brock. 1973. Water status and the distribution of *Cyanidium caldarium* in soil. *J. Phycol.* **9**, 330–332.

Staehelin, L. A. 1968. Ultrastructural changes of the plasmalemma and the cell wall during the life cycle of *Cyanidium caldarium*. *Proc. R. Soc. Lond. B.* **171**, 249–259.

Tilden, J. E. 1898. Observations on some west American thermal algae. *Bot. Gaz.* **25**, 89–105.

Tilden, J. E. 1910. Minnesota algae. I. The Myxophyceae of North America and adjacent regions including Central America, Greenland, Bermuda, The West Indies, and Hawaii. Geological and Natural History Survey, Botanical Series. University of Minnesota, Minneapolis.

Troxler, R. F. 1972. Synthesis of bile pigments in plants. Formation of carbon monoxide and phycocyanobilin in wild-type and mutant strains of the alga, *Cyanidium caldarium*. *Biochem.* **11**, 4235–4242.

Troxler, R. F. and A. Brown. 1970. Biosynthesis of phycocyanin *in vivo*. *Biochim. Biophys. Acta* **215**, 503–511.

Troxler, R. F., A. Brown, R. Lester, and P. White. 1970. Bile pigment formation in plants. *Science* **167**, 192–193.

Troxler, R. F. and R. Lester. 1967. Biosynthesis of phycocyanobilin. *Biochemistry* **6**, 3840–3846.

Troxler, R. F. and R. Lester. 1968. Formation, chromophore composition, and labeling specificity of *Cyanidium caldarium* phycocyanin. *Plant Physiol.* **43**, 1737–1739.

Volk, S. L. and N. I. Bishop. 1968. Photosynthetic efficiency of a phycocyanin-less mutant of *Cyanidium*. *Photochem. Photobiol.* **8**, 213–221.

West, G. S. 1904. West Indian freshwater algae. *J. Bot.* **42**, 280–294.

Wharton, R. A. and W. C. Vinyard. 1977. Summit life, a preliminary report. *Shasta,* Newsletter of the Mount Shasta Resource Council, Mt. Shasta, Calif. **2**, 10–12.

Chapter 10
Life in Boiling Water

To me, the most fascinating discovery that we made during our Yellow-stone work was finding organisms that carry out their whole life histories in boiling water. I obtained the first inkling that living organisms might exist in boiling springs and geysers during our first summer of work, 1965. When we began our work in the summer of 1965, the intention was to study algal distribution along thermal gradients, but I was aware of Kempner's paper (Kempner, 1963) stating that 73°C was the upper temperature for life, and I planned to do some experiments relating to the upper temperature limits. I thought of bacteria, and the first experiments were an attempt to get bacterial growth at high temperature by enriching spring water with organic nutrients and incubating at various temperatures from 70°C up. As incubators and source of inoculum, we selected a series of pools with temperatures of 91°C, 84°C, 76°C, and 70°C in the White Creek area of the Lower Geyser Basin, calling them Pool A, B, C, and D. But at the very beginning of the experiment, I noticed that in the outflow channel of Pool A there was pink gelatinous stringy material (see Figures 3.2 and 3.3) at temperatures in the 80–90°C range. The pink material definitely appeared to be living, and at temperatures considerably above those that Kempner had defined as the upper temperature of life.[1] The incubation experiments subsequently led to the discovery of the genus *Thermus* (see Chapter 4) but in addition they led to the discovery of the pink bacteria. Within 3 days of discovery, we had

---

[1] I do not propose to deal in any more detail with Kempner's observations. His quick trip through Yellowstone did not permit him to make the kinds of detailed observations that we could carry out. Although his paper was erroneous, his approach to the question by the use of radioisotopes was pioneering, and his brief paper had considerable influence on my thinking and research planning over those first few years.

carried out chlorophyll, protein, and RNA assays on the material and showed that it was high in protein and RNA but devoid of chlorophyll. The material was obviously bacterial, and growing at such densities that one could harvest literally kilograms of material from the water. In subsequent years, I did harvest large amounts of the pink bacteria for various organic geochemists and led a number of other scientists to the site, but the only paper I know of that deals specifically with the organic chemistry of the pink bacteria is that of Bauman and Simmonds (1969) on the fatty acids and lipids (Table 10.1). Through the years, we did a variety of studies at Pool A, and it was featured in our movie with Encyclopaedia Britannica (Movie #3143, Ecology of a Hot Spring: Life at High Temperatures), but it was not until 1976 that I discovered that Pool A had an official name, Octopus Spring.

The presence of visible masses of bacteria in the outflow channels of hot springs is not exactly rare, but only a small percentage of the springs I have examined show such development. Other springs which look superficially to be similar show no *macroscopic* signs of bacteria, although virtually all show *microscopic* evidence of organisms (see below). In the Lower Geyser

Table 10.1. Percentage Composition of Fatty Acids in Bacterial Masses Collected at Two Hot Spring Locations[a]

Peak elution order	Identification	Percentage composition	
		Firehole Pool	Pool A
1	C_{14}	0.715	7.513
2	iso-C_{15}	4.102	0.520
3	C_{15}	0.413	1.664
4	iso-C_{16}	2.931	0.777
5	NI[b]	Trace	1.839
6	C_{16}	4.269	9.134
7	iso-C_{17}	21.144	9.587
8	ante-C_{17}	2.930	0.373
9	C_{17}	1.011	1.766
10	iso-C_{18}	2.011	1.943
11	C_{18}	9.756	17.77
12	iso-C_{19}	8.934	16.89
13	ante-C_{19}	0.231	0.385
14	C_{19}	2.863	0.943
15	iso-C_{20}	2.680	0.793
16	C_{20}	3.796	4.971
17	$C_{20:1}$	24.454	1.768
18	NI[b]	2.565	0.329
19	C_{21} cyclopropane	5.190	22.02

[a]Data of Bauman and Simmonds (1969).
[b]Not identified.

Basin, I have seen pink filamentous masses only in Octopus Spring but several other springs (e.g., Firehole Pool, Twin Butte Vista) show presence of yellowish, grayish, or whitish masses. Although I have not searched my records thoroughly, I know that I have seen pink masses in some other Yellowstone springs, probably in Heart Lake Geyser Basin. I have also seen filamentous masses in a spring in Iceland (Deildartunga), but not in New Zealand, although many New Zealand boiling springs show microscopic evidence of bacteria (see later). I assume that macroscopic development requires particular conditions of nutrient availability in the water, and perhaps in flow characteristics.

I have never seen filamentous masses in the boiling pools that are the sources of water for the springs, but at Octopus Spring we obtained the development of small amounts of macroscopically visible pink material on strings that had been immersed in the source.

We first began to insert microscope slides in boiling springs in an attempt to show that these bacteria were really growing. I began at Octopus Spring by putting slides in the effluent channel, but growth was so rapid that at temperatures of 85–86°C the slides were heavily covered in 2 days. So I moved up the thermal gradient and placed slides in the source. The rate of development there was not as fast as in the outflow channel, but it was still rapid. Surprisingly, I did not put slides in boiling pools until August 1967, 2 years after we had discovered the pink bacteria. (We had been occupied during the previous time on our blue-green algal studies.) In the summer of 1967, I had a lead article in press in *Science* (Brock, 1967) and wanted to illustrate the high-temperature bacteria. When the bacterial development was rapid on the slides immersed in the source of Octopus Spring (temperature 91°C), it was natural to immerse slides in other pools. By the end of the summer of 1967, we knew that virtually every boiling hot spring of neutral to alkaline pH in the Lower Geyser Basin had bacteria in the source, at temperatures that never dropped below boiling. Some of these springs were superheated, with temperatures of 94–95°C (Figure 10.1). Photomicrographs of a few of the kinds of organisms seen in some of the boiling springs are given in Figure 10.2.

Bacterial Growth Rates above 90°C

Although these studies were fairly convincing, and provided some evidence that bacteria could grow in boiling water, this concept was so new that I knew we would have to do much more to convince others. At this time, Tom Bott joined my laboratory as a postdoctorate, and expressed an interest in providing more definitive proof that the bacteria were growing. Although the arrangement of bacterial cells in microcolonies (Figures 10.2 and 10.3) on microscope slides suggested that growth actually took place in the spring, it was important to provide independent evidence of growth and

Figure 10.1a,b. Superheated springs that showed good bacterial growth. (a) Step-brother, in the White Creek Valley (see map of the Lower Geyser Basin in Chapter 2). This spring had a temperature over boiling and showed periodic eruptions. The log was used to hold a slide rack. (b) Geyserino, near Great Fountain Geyser (see map in Chapter 2). This spring was really a small geyser, but never showed eruptions higher than that pictured. The round, bead-shaped objects around the edge (foreground) are geyser eggs, an indication of periodic splashing with silica-rich water.

to determine growth rates under these extreme conditions. To demonstrate that organisms were growing on the slide surface, some slides were treated with germicidal ultraviolet radiation at intervals of between one and two generations. Any organisms that had attached to the slide during the interval since the last irradiation would thus be killed and could not produce progeny. Thus, the rate of increase of cell number on the irradiated slides would give an estimate of passive attachment, and when this number was deducted from the rate of increase of cell number on the unirradiated slides, the growth rate could be calculated. Slides were placed in Plastisol-coated racks which were immersed in the springs at least 1 meter below the water surface. At intervals the racks were raised to the surface, and slides were transferred from the rack to petri dish lids submerged about 1 cm in the water. Irradiation was done using a Mineralight Ultraviolet source with the filter removed, and the lamp was powered by an inverter connected to a storage battery. Tom Bott got quite good at carrying a heavy storage

Figure 10.1b. Legend see opposite page.

battery through the field, even if he had to do it in the middle of the night. At intervals, irradiated and unirradiated slides were removed in duplicate, air dried, and the number of cells per microscope field or per microcolony was determined, using a water-immersion phase contrast objective. Usually 100 fields or microcolonies were counted on each slide, and the values for the duplicates were averaged.

Table 10.2 presents data obtained for unicellular bacteria of a number of springs in the Lower Geyser Basin (see also Figure 10.3). The data provide conclusive evidence that growth occurred, for in every instance the counts on the irradiated slides were much lower than on the controls. Counts on the bottoms of irradiated slides were similar to those of unirradiated controls, a result to be expected because germicidal wavelengths of ultraviolet radiation should not pass through the glass.

Growth rates of bacteria in Octopus Spring are presented in Figure 10.4. Because rod-shaped bacteria grow in the form of microcolonies in this spring, data are presented as number of cells per microcolony. Generation times calculated from the exponential phase of growth of these two experiments are 2.5 and 3.0 hours. In other springs, where growth does not occur in discrete microcolonies, results are expressed as numbers of cells per microscope field. Calculated generation times for all the springs studied are summarized in Table 10.3. It can be seen that all of the generation times are

Table 10.2. Bacterial Counts on Irradiated (UV) and Unirradiated Microscope Slides. Irradiated Slides Were Treated Eight Times at Approximately Equal Intervals during the Immersion Period[a]

Treatment	Immersion time (hr)	Average number of cells per field	
		Top of slide	Bottom of slide
Geyserino			
UV	74.5	3.6	889
UV	74.5	1.4	431
Control	74.5	948	1385
Porcupine			
UV	138.75	1.4	149.0
Control	138.75	158.1	118.1
Stepbrother			
UV	72.0	8.2	136
UV	72.0	1.6	7.4
Control	72.0	42.4	34.0
Boulder Experiment 1			
UV	74.5	0.56	144
UV	74.5	0.88	120
Control	74.5	250	125
Boulder Experiment 2			
UV	72.0	9.4	889
UV	72.0	1.6	961
Control	72.0	1361	626
Boulder Effluent[b] Experiment 1			
UV	12.0	2.62	32.2
Control	12.0	30.7	24.7
Boulder Effluent Experiment 2			
UV	12.0	4.0	29.6
UV	12.0	0	26.5
UV	12.0	0	31.3
Control	12.0	32.0	23.2
Octopus Spring[b] Experiment 1			
UV	53.0	2.0	32.5
Control	53.0	10.5	15.8
Octopus Spring Experiment 2			
UV	33.0	2.5	39.0
UV	33.0	2.2	39.2
Control	33.0	37.9	44.8

[a]Data of Bott and Brock (1969).
[b]Counts for Octopus Spring and Boulder Effluent are number of cells per microcolony rather than number of cells per field.

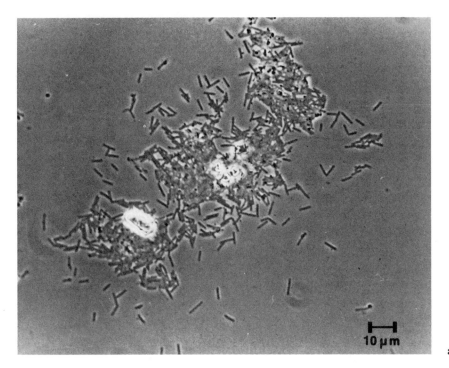

a

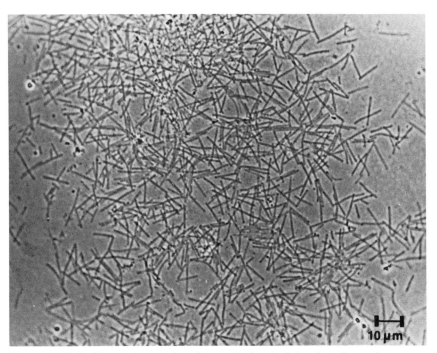

b

Figure 10.2a,b. Photomicrographs of bacteria that had developed on microscope slides immersed in boiling springs. (a) Stepbrother (see Figure 10.1a). (b) Geyserino (see Figure 10.1b).

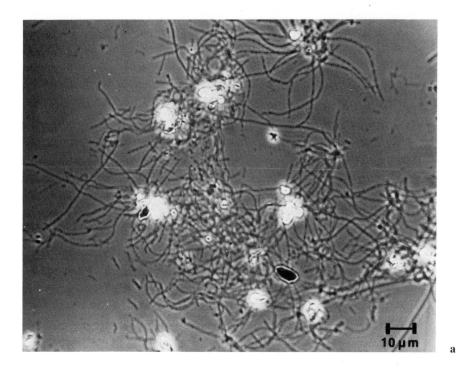

a

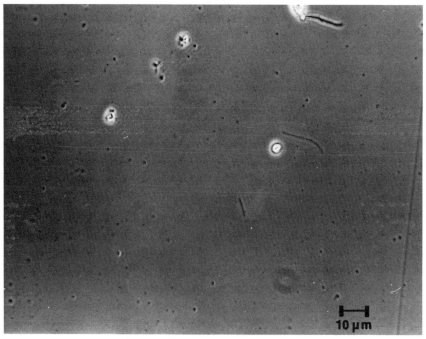

b

Figure 10.3a,b. Photomicrograph by phase contrast of bacteria developed on a slide immersed in Boulder Spring for 3 days, with periodic UV irradiation of one side. (a) Unirradiated side of slide; (b) irradiated side of slide.

fairly rapid, of the order of a few hours. Thus, growth is not only occurring in boiling water, but is occurring at surprisingly rapid rates.

Some of the springs had filamentous bacteria as well as rods. We used another technique to quantify growth rates of such filaments, and the data obtained are also given in Table 10.3. It is interesting that the growth rates of both rods and filaments in the same spring are about the same. Irradiation studies were also carried out with the filamentous forms. Not only were the numbers of filaments on the irradiated slides considerably lower than on the unirradiated slides, the lengths of the filaments on the irradiated slides were always short, resembling those of unirradiated slides that had been immersed in water for only short periods of time. On the unirradiated slides, the lengths of filaments became progressively longer with increasing

Table 10.3. Summary of Bacterial Generation Times for Various Springs[a]

Experiment	Temperature range[b] (°C)	pH	Generation time (hr)	
			Rods	Filaments
Geyserino				
1	93.0–94.5	8.5	4.0	
2	92.5–94.5		5.0	
Porcupine				
1	92.8–94.8	8.6	6.0	
2	94.2–95.2		6.0	
Stepbrother				
1	91.8–92.5	8.65	5.5	
2	91.9–93.0		3.5	
Boulder				
1	90.0–91.8	8.9	7.5	7.0
2	90.5–91.5		4.75	
3	90.3–91.8		3.0	
Boulder Effluent				
1	78.0–80.5	8.9	2.2[c]	1.6
2	79.0–80.5		2.2[c]	
3	79.8–81.8		2.1[c]	
Octopus				
1	90.5–92.0	8.1	3.0[c]	4.0
2	89.0–91.0		2.0[c]	1.0

[a]Data for rods of Bott and Brock (1969). Data for filaments, unpublished.

[b]Temperature ranges listed encompass measurements made during the experiments, which usually extended over 2 to 3 days. Measurements of pH were made with a glass electrode at the temperature of the spring.

[c]These values were calculated on the basis of rate of increase in number of cells per microcolony. All others are from counts of number of cells per microscope field or lengths of filaments per microscope field.

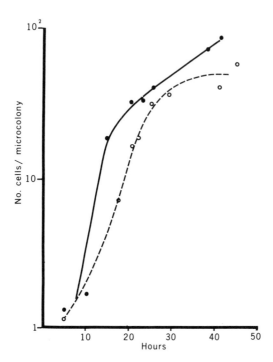

Figure 10.4. Growth curves of bacteria in Octopus Spring, temperature 91°C. Results of two independent experiments. (From Bott and Brock, 1969.)

time of incubation, suggesting that the short filaments, derived from the water, were growing after becoming attached to the slides. The growth curves for these filamentous organisms were exponential, suggesting that growth was occurring throughout the lengths of the filaments, as in other filamentous bacteria, rather than only at the tips, as in the filamentous fungi.

Upper Temperature for Life

Because of Yellowstone's altitude, water boils at about 92°C, and even superheated springs rarely exceed 95°C. It was of interest to learn whether bacteria lived in boiling springs at even higher temperatures. Both New Zealand and Iceland have lots of springs at considerably lower altitiudes than Yellowstone, with temperatures of 99–100°C (even 101°C for a rare superheated spring). On one of my four trips to Iceland (1970) I surveyed a number of boiling springs with temperatures of 97–98°C and pH values on the alkaline side (most nonacidic springs in Iceland have pH values over 9). Every spring that I sampled had rod-shaped or filamentous bacteria, similar to those seen in Yellowstone.

In New Zealand we were able to carry out an even more extensive survey, and could find a number of springs with temperatures even higher than those in Iceland. The data for the New Zealand springs of neutral to alkaline pH are given in Table 10.4. As can be seen, every spring examined had bacteria, and the temperatures of some of these springs are 99–101°C. The series of springs at Rota-a-Tamaheke and Orakei Korako are noteworthy partly because of the high temperatures and partly because we immersed microscope slides in these springs and retrieved them after a week or so, to obtain some better idea that growth was indeed taking place at the temperatures involved. My most striking impression of the Roto-a-

Table 10.4. Distribution of Bacteria in Some Very Hot Springs in Central Volcanic Region, North Island, New Zealand, Related to Temperature and pH [a]

Location		Temperature (°C)	pH	Bacteria
Rotorua (Whakarewarewa)				
Roto-a-Tamaheke Spring	353	99–101.5	8.25	Present
	354	99–100	7.45	Present
	351	98.5–100	8.58	Present
	344	95.5	3.05	Absent
	363	93–97	4.3	Absent
	359	95–97	3.05	Absent
	364	100	7.65	Present
Ngaratuatara		98.5	8.25	Present
Waiotapu (Tourist Reserve)				
Loving Sister		98	3.25	Absent
Cadmium yellow pool		72	2.25	Absent
Greyish pool		81.5	2.45	Absent
Near Frying Pan Flat		95.5	4.05	Absent
Yellow turbid pool		82.5	2.65	Absent
Weather Pools		74–76	2.1	Absent
Near Weather Pools		99.5	2.05	Absent
Champagne Pool		64–77.5	5.55	Present
Orakei Korako				
Fred and Maggie Spring		98–99.5	6.95	Present
Manganese Pool		97.2–99.6	7.4	Present
The Cauldron		88.8–90.8	6.5	Present
Dreadnought Geyser		98.6–99	7.0	Present
Near Dreadnought Geyser		97.5	6.65	Present
Artist's Pallette Spring	774	84	6.35	Present
	773	96	6.65	Present
	775	97	8.2	Present
	770	97.8–100.8	7.85	Present
Taupo Spa				
Near Witch's Caldron		88.5	3.2	Doubtful
Waikite Springs				
Largest spring		92.9	6.95	Present
Wairakei (Thermal Valley)				
Near Champagne Cauldron		94.2	4.55	Absent
Grey mud pool		93.5	2.11	Absent

(*continued on p. 314*)

Table 10.4. Distribution of Bacteria in Some Very Hot Springs in Central Volcanic Region, North Island, New Zealand, Related to Temperature and pH[a] (*continued*)

Location	Temperature (°C)	pH	Bacteria
Te Kohanga	95	3.9	Absent
Red Pool	87.5	2.35	Absent
Large grey pool	91	2.3	Absent
Wairakei (Geothermal Field)			
Natural mud pool	85.5	2.6	Absent
Natural sulphur pool	90.5	2.55	Absent
Tikitere			
Devil's Bath	44	2.45	Present
Hell's Gate	53	2.25	Present
Baby Adam	87	6.15	Present
The Inferno	67	3.05	Present
Sulphur Bath	76.8	6.05	Present
Feature E	81.5	6.15	Present
Feature H	86.5	3.75	Doubtful
Feature I	83	2.0	Present
Devil's Cauldron	72	6.0	Doubtful
Near Devil's Cauldron	61	2.0	Present
Mud spouter	98	2.75	Absent
Grey pool	87	2.2	Absent
Red brown pool	91.5	3.4	Absent
Greyish green pool	64.5	6.1	Present
Watery grey pool	86	6.45	Present
Adjacent spring	79	5.9	Present
Reddish black pool	72	6.25	Present
Ketetahi Springs			
Large bubbling spring	87	6.8	Doubtful
Small black bubbler	74.7	7.4	Present
Grey turbid pool	91.8	3.9	Absent
Grey pool	91.2	6.65	Absent
Black bubbling pool	91.2	5.95	Doubtful
Black bubbling pool	89	3.75	Absent
Lake Rotokawa			
Spring 141-4	93	2.9	Absent
Spring 141-6	98.5–99.5	2.95	Absent
Spring 141-7	86–88.5	2.90	Absent
Spring 142-1	88.8	2.45	Absent
Spring 142-2	52	2.48	Present
Spring 142-3	74	2.6	Absent
Spring 142-4	95	2.6	Absent
Waimangu			
Ngapuia-o-te-papa	98	8.9	Present

[a]From Brock and Brock (1971a).

Tamaheke area was of the large number of alkaline springs depositing sulfur around their rims, since S^0-depositing springs of neutral pH are rare in Yellowstone. Microscopic examination of rocks and twigs from the springs, or of microscope slides immersed for 1 week, showed bacteria present in every case. The bacteria were thin, filamentous types (about 0.5 μm in diameter, length 10 to greater than 100 μm) and mostly occurred in rosettes with individual cells attached to the centers of tiny mineral particles. Some filaments had spheres protruding at the distal end. The slides showed quite heavy coatings of bacteria after 1 week, similar to those found at somewhat lower temperatures in Yellowstone.

At Orakei Korako many of the springs had dark mineral deposits, possibly sulfide minerals, and the bacteria were attached primarily to these particles. In the same springs, light-colored minerals, perhaps silica, did not show heavy bacterial attachment. In one spring at the Artists Pallete at Orakei Korako, which had a temperature fluctuating between 97.8°C and 100.8°C (pH 7.85), microscope slides had heavy bacterial accumulations after 9 days. The bacteria in this spring were long straight rods, and again were attached primarily to mineral particles.

I have made observations of fumaroles in Italy and Iceland with temperatures far over 100°C, and have never found any evidence of living organisms. However, there is no liquid water in these fumaroles, the steam issuing in a superheated condition. It would appear that liquid water is necessary for development of living organisms.

Thus, we can conclude that the upper temperature of life has not yet been defined. The presence of bacteria in superheated springs at temperatures of 99–100°C in New Zealand, and the evidence that they are growing rapidly, suggests that certain kinds of bacteria may perhaps be able to live in any boiling habitat that contains liquid water.

Limits of Microbial Existence: Temperature and pH

I have been careful in the above discussion to specify that the springs under study were of neutral or alkaline pH. In acid springs, the situation may be somewhat different, as there is some reason to believe that bacteria do not thrive in boiling springs of low pH. The situation is not as clear-cut as indicated in the paper by Brock and Darland (1970), because since that paper was published, the organism *Sulfolobus* has been found, and there is good reason to believe that many of the acid springs studied by Brock and Darland and thought to be devoid of bacteria probably had *Sulfolobus*. At the time that the Brock and Darland work was done, the existence of *Sulfolobus* was unknown, and as I have discussed in Chapter 6, the unusual morphology of *Sulfolobus* fooled me.

Despite this disclaimer, the evidence still seems good that bacteria do not live in *boiling* acid springs, at least in a manner that can be detected by the techniques available. In Yellowstone, where we have done the most work, there is no question that even *Sulfolobus* does not live in boiling acid springs, temperature of 92°C, although as noted above, boiling springs in Yellowstone of neutral pH are rich in bacteria. I discussed in Chapter 6 the fairly strong evidence that the upper temperature limit for *Sulfolobus,* both in nature and in culture, is about 90°C. Admittedly, this is rather close to the 92°C that water boils at, and it is conceivable that the absence of *Sulfolobus* from boiling springs is not due to temperature, but to some other factor. I also noted in Chapter 6 the observations of *Sulfolobus*-like organisms from an acid spring in Iceland which had a temperature of 97°C, but successful cultures could not be obtained.

The only piece of direct evidence I have to suggest that temperature is the limiting factor for *Sulfolobus* at high temperature concerns observations made on Evening Primrose Spring, in the Sylvan Springs area. In 1970 and before, this very acid, sulfur-rich spring had a temperature of 90°C and was devoid of life. Kathie Brock and I not only did a very detailed microscope examination in the summer of 1970, but we also did isotope experiments to see if any incorporation of organic compounds could occur. We found no microscopic evidence of organisms and no uptake of labeled amino acids or sugars. Interestingly, sometime between the summer of 1970 and the summer of 1971 a change took place in Evening Primrose, and its temperature dropped from 90°C to less than 80°C (see Figure 6.7b). When Jerry and Anne Mosser studied Evening Primrose in the summer of 1971, after its temperature had dropped, it was teeming with *Sulfolobus,* and continued to yield viable *Sulfolobus* over the next 4 years. Since *Sulfolobus* was absent when the temperature was 90°C, and colonized the spring when the temperature dropped, it seems reasonable to conclude that there is a real temperature barrier that this organism cannot overcome.

Darland and I suggested that high acidity may add an additional environmental stress that makes microbial growth at very high temperatures impossible. Hot acid is potentially much more lethal than heat alone, since hot acid will tend to cause hydrolysis of various ester bonds. I discussed in Chapters 5 and 6 the interesting biochemical characteristics of the plasma membrane of *Thermoplasma* and *Sulfolobus,* two organisms that live under hot acid conditions. These bacteria contain ether-linked hydrophobic groups in their phospholipids instead of ester linkages.

The Bacteria of Boulder Spring

Since organisms living in boiling water are of great biological interest, it was of considerable interest to obtain some detailed knowledge of their structure and function. The logical approach would be to isolate pure cultures and study their characteristics, but cultivation of these high-temperature bacteria has not been possible (see later in this chapter). We thus elected to study these organisms directly in their natural environment. Although a number of springs could have been used as study areas, we concentrated on two, Boulder Spring and Octopus Spring, because these springs were favorable for detailed and undisturbed long-term study, and because they differed in some important ways. The work on Boulder Spring occupied us the longest, and the data are the most complete. Bott began work at Boulder Spring when he did the first growth rate studies (see above), and it was a natural evolution from this work to attempt to study the organisms in more detail. At about this same time, Bott had also developed a cover slip technique which permitted measurement of uptake of radioactive compounds by the bacteria, and this technique ultimately proved amenable to use with the Boulder Spring Bacteria.

Boulder Spring (Figure 10.5) is a large superheated spring (93–93.5°C) in the Lower Geyser Basin of Yellowstone National Park. Chemical analyses are given in Table 10.5. Boulder Spring has two superheated sources which are connected; both sources contribute water to the single effluent channel, also shown in Figure 10.5. The rapid flow rate in the effluent channel shows that water in the source pools exchanges relatively rapidly. Most of our

Figure 10.5. General view of Boulder Spring. The upper source is erupting and the lower source is to the left behind a large boulder.

Table 10.5. Chemical Analysis of Boulder Spring[a,b]

Component	U.S.G.S., 1968[c]	Allen and Day (1935)	Present work
Al	0.45	2	
Fe	0.02	1	
Mn	0.01		
Cu	≤0.01		
Ca	1.0	2	
Mg	0.016	ND[d]	
Sr	<0.02		
Na	322	327	
K	8.9	11	
Li	2.6		
NH_4^+	<0.1		
HCO_3^-	153	89	
CO_3^{2-}	58	102	
SO_4^{2-}	28	21	
$S_2O_3^{2-}$		9	
Cl^-	299	298	
F^-	31	22	
NO_2^-	0.03		
NO_3^-	<0.1		
PO_4^{3-}	1.6		
B_2O_3		15	
SiO_2	211	173	
Sulfide			3.1–3.2
pH			8.9

[a]From Brock et al., (1971).
[b]Values expressed as micrograms per milliliter.
[c]U.S. Geological Survey data provided by Robert Fournier, Menlo Park, California.
[d]None detected.

studies were done at the site shown in Figure 10.6 just downstream from the lower source pool, at a temperature of 90–91.5°C.

Methods

Although the bacteria occur in large numbers on the walls of the spring and in the bottom sediment, it was convenient for isotope studies to have organisms that had developed on a uniform substrate. The cover slip technique described by Bott and Brock (1970) was therefore used. The cover slips (20 × 25 mm, No. 2) were held vertically in small vinyl rings which were tied to plastic-coated test tube racks, the latter being immersed in the spring at the site shown in Figure 10.6. The type of holder used offered very little restriction to the flow of water across the cover slips. The cover slips were immersed in the pool usually for 10 days; during this time, bacteria colonized the cover slips and formed a relatively dense covering. About 100 μg of protein per cover slip was present at this time, as assayed by the method of Bott and Brock (1970). Isotope incorporation on these cover slips was measured by placing them in shell vials (22 $\times$ 70 mm) containing 8 ml of fluid (spring water or synthetic salt solution with additives as required). The vials were closed with cork stoppers which had been wrapped in autoclavable plastic film. Provided that closure was firm, fluid loss due to evaporation during 1 hr of incubation was minimal, even if temperatures above boiling were used. Usually two cover slips were placed in a vial, and they were kept apart by means of a small piece of glass tubing in the bottom of the vial.

After all cover slips for an experiment had been placed in vials, and after 5 to 10 minutes for temperature equilibration, the radioactive isotope was injected usually in 0.8 ml. After incubation, 0.9 ml of Formalin (40% formaldehyde solution) was added to stop uptake and the vials were returned to the laboratory for processing. For the controls, the Formalin was added before the isotope at zero time. The cover slips were removed from the vials, washed in three changes of tap water, and dried. If only [14]C

Figure 10.6. Colonization site at Boulder Spring. The lower source pool is erupting in the background. Racks for cover slips were suspended from the log shown.

were to be counted, each cover slip was placed on a planchet holder in a Nuclear-Chicago gas glow counter: both sides of the cover slip were counted, and the two values were added. With ^{3}H-labeled compounds, single cover slips were placed in vials, covered with liquid scintillation fluid, and counted with a Beckman liquid scintillation counter. For some of the later experiments in which simultaneous ^{14}C and ^{3}H labeling was used, both isotopes were counted at appropriate settings in a liquid scintillation counter.

Temperatures for incubation were obtained in the pool or runoff channel of Boulder Spring or in other nearby springs. To obtain temperatures above the boiling point at the altitude of Yellowstone, medicinal grade magnesium sulfate (Epsom Salts) was added to a boiling water bath. Most experiments were done directly in the field but a few were done in our West Yellowstone laboratory. Temperatures were measured either with conventional mercury thermometers or with a Yellow Springs Instrument Co. thermistor and bridge, care being taken to measure the temperatures precisely at the sites where incubations were carried out. The temperature of the liquid in the vials was usually about 1°C below that of the surrounding water.

Electron microscopy was carried out on bacteria that had colonized either glass microscope slides or Mylar strips (5-mil thickness, Dupont Chemical Co.) immersed in the spring. Methods of fixation, embedding, and sectioning were those of Brock and Edwards (1970). After 6 to 20 days of immersion, the slides or Mylar strips were removed from the spring and immediately immersed in the fixative. The glass slides were then washed in buffer, and the cells were scraped from the surface with a sharp razor blade and embedded in agar which was then further processed. The Mylar strips were fixed, washed, cut into small squares that were further processed, and then sectioned directly, the organisms thus being seen essentially *in situ* attached to the Mylar surface.

Morphology

Two shapes of bacteria are seen on cover slips or glass slides incubated in Boulder Spring: rods and filaments. Figure 10.7 shows photomicrographs of two microcolonies of rods and two filaments, one of which is quite long. On slides immersed for long periods of time, dense coatings of bacteria developed and spherical cells were also seen. Similar spheres were also seen at the ends of filaments by either light microscopy or electron microscopy. Gram stains performed on both dried and Formalin-fixed slides revealed that the organisms were gram negative.

Figure 10.8 shows, at low magnification, electron micrographs of both filamentous and rod-shaped cells. Note that in filaments no cross walls are seen. An unusual feature of both rods and filaments is their narrow diame-

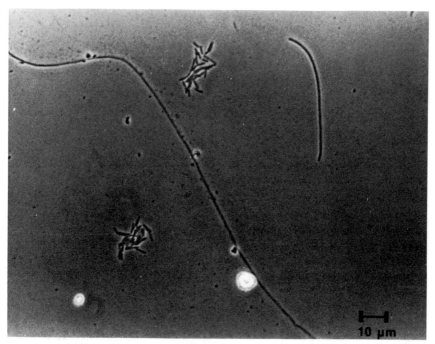

Figure 10.7. Morphology of Boulder Spring Bacteria. Microcolony of rods and a long filament.

ter, 0.15 to 0.2 μm. The swellings at the ends of the filaments (Figures 10.8 and 10.9) presumably develop into spheres. The same structural features seen in organisms fixed and sectioned *in situ* on Mylar strips were found in organisms fixed and scraped from glass slides before they were embedded.

An enlarged view of a single rod-shaped cell is seen in Figure 10.12. The plasma membrane is sharply defined (see also Figures 10.10 to 10.12). An unusual structural feature of the Boulder Spring organisms is the cell wall, which is different from that of either gram-negative or gram-positive bacteria. In many sections, the wall is composed of two diffuse layers of low electron density separated by a thin band of somewhat more electron-dense material (Figures 10.11 and 10.12). In some sections (Figures 10.8 to 10.10), the outer layer and the thin band are missing, and the inner layer has a striated appearance. This appearance is the result of a number of regular channels or tubes which are present in the inner wall layer. Figure 10.14 shows such channels in profile in a spherical cell, and Figure 10.13 shows them in transverse grazing section. In the latter the regular array of channels (arrows) is obvious. Many of the sections contain structures resembling ribosomes (Figures 10.10 and 10.12). In Figure 10.10, ribosomes are shown especially well.

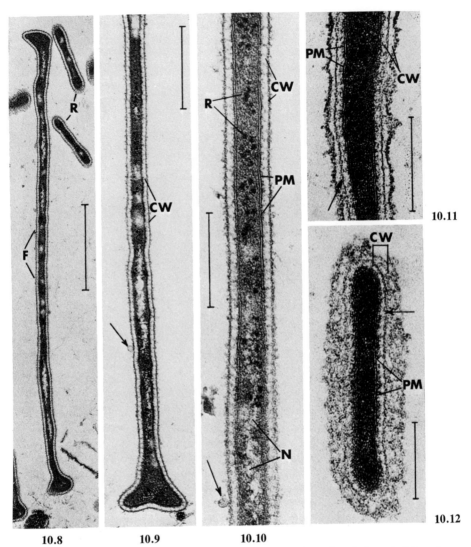

Figure 10.8. Electron micrograph showing overall view of rod-shaped (R) and filamentous (F) bacteria from Boulder Spring. Bar represents 1 μm. (From Brock et al., 1971.)

Figure 10.9. Filamentous bacterium with broad tip and striated cell wall (CW). Bar represents 0.5 μm. (From Brock et al., 1971.)

Figure 10.10. Enlarged view of a single filamentous bacterium. PM, plasma membrane; CW, cell wall; R, ribosomes; N, nucleoplasm. Bar represents 0.3 μm. (From Brock et al., 1971.)

Figure 10.11. Portion of a filament. CW, cell wall with inner and outer light regions separated by a dense layer (arrows); PM, plasma membrane. Bar represents 0.25 μm. (From Brock et al., 1971.)

Figure 10.12. Enlarged view of a rod-shaped bacterium. PM, plasma membrane; CW, cell wall with inner and outer regions and a dense middle layer (arrow). Bar represents 0.15 μm. (From Brock et al., 1971.)

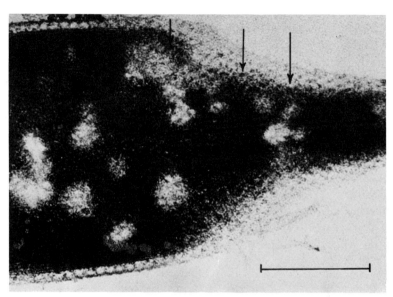

Figure 10.13. Portion of an enlarged swollen tip of a filamentous bacterium. Grazing transverse section through channels in wall (arrows). Bar represents 0.25 μm. (From Brock et al., 1971.)

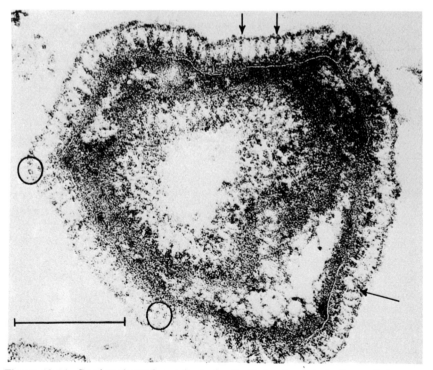

Figure 10.14. Section through a spherical cell. Note channels in the wall (arrows and circles). Bar represents 0.25 μm. (From Brock et al., 1971.)

Conditions for Isotope Uptake

The ability to take up isotopes was inhibited virtually completely by 4%
formaldehyde, and a formaldehyde control incubated under conditions
identical with those of the experimental vials was always run. Table 10.6
shows that mercuric bichloride, streptomycin, hydrochloric acid, and
sodium azide inhibit incorporation of ^{14}C-acetate. Since uptake of acetate
was linear with time over at least 2 hr, incubations were usually terminated
at 1 hr. An experiment was performed in which cover slips colonized for
various periods from 1 to 10 days were used in an isotope experiment under
identical incubation conditions. The amount of isotope taken up increased
progressively with colonization time. Microscopic observations showed
that the amount of organisms on the cover slips also increased in parallel.
Thus, the amount of isotope uptake (under fixed conditions of incubation
time and isotope concentration) is proportional to the amount of bacterial
biomass on the cover slips.

The amount of radioactivity obtained on the cover slips varied with the
compound used, its specific activity, and the concentration employed.
After preliminary experiments, the amount of labeled compound added was
varied to obtain reasonable but not excessive radioactivity of the cover
slips. In most experiments, radioactivity per cover slip (summation of the
two sides) ranged between 1000 and 10,000 counts/min.

Considerable variability in isotope uptake was noted for replicate cover
slips incubated under presumably identical conditions, and considerable
variability in uptake was also often noted between one side of a cover slip
and another, whether one or two cover slips were incubated in a vial. We
attribute this variability to differences in degree of colonization of different
surfaces. In a number of temperature optima experiments involving use of
two labeled compounds in a single vial, it was noted that the extent of
incorporation of both isotopes at various temperatures was virtually the
same on a single cover slip, suggesting that the variability between cover

Table 10.6. Effect of Inhibitors on Uptake of ^{14}C-Acetate[a]

Inhibitor	Counts per minute per cover slip
None	10,128
Formaldehyde (4%)	228
Mercuric bichloride (10 μg/ml)	263
Hydrochloric acid (0.1 N)	604
Streptomycin sulfate (1,000 μg/ml)	903
Sodium azide (1,000 μg/ml)	4,885

[a]Incubations for 1 hr in unaerated Boulder Spring water without sulfide
supplement at 87°C, with 1 μCi of ^{14}C-acetate per ml. Data are averages of
two cover slips per vial. (From Brock et al., 1971.)

slips was due to differential colonization. Although this variability did not affect those experiments where large differences were obtained (such as sulfide stimulation and inhibitor studies), it did affect conclusions on temperature optima. Therefore, the temperature optima experiments were repeated many times, with a variety of labeled compounds. The results from all experiments with a single compound were then averaged to obtain a more meaningful temperature optimum.

Sulfide Stimulation of Isotope Uptake

In early experiments, little or no incorporation on cover slips occurred when spring water taken from the surface of the pool was used, whereas with water collected from a deep location there was significant incorporation. This suggested that the organisms might be dependent on an oxygen-labile substance in the spring water, possibly sulfide. Table 10.7 summarizes an experiment with ^{14}C-acetate in which deep water was either used directly or aerated for 30 minutes with a small aquarium pump. Significant isotope incorporation occurred in unaerated water, whereas little occurred in aerated water. However, if sulfide were added to the aerated water, the ability of the organisms to incorporate radioactivity was restored. The optimal sulfide concentration was about 13 μg of sulfide/ml.

Sulfide stimulates incorporation of a variety of radioactive isotopes (Table 10.8). Although sulfide stimulates incorporation of all compounds studied, incorporation of some is stimulated more than others. The most dramatic stimulation was seen with leucine, acetate, and thymidine.

Isotope incorporation does not require the use of Boulder Spring water, provided that sulfide is present. Table 10.9 shows that isotope incorpora-

Table 10.7. Sulfide Stimulation of ^{14}C-Acetate Uptake[a]

Condition	Counts per minute per cover slip
Unaerated spring water	1424
Plus formaldehyde (4%)	69
Aerated spring water	282
Plus sulfide (1.3 μg/ml)	302
Plus sulfide (3.3 μg/ml)	958
Plus sulfide (6.5 μg/ml)	2070
Plus sulfide (13 μg/ml)	3280
Plus sulfide (33 μg/ml)	1210
Plus sulfide (65 μg/ml)	205
Plus sulfide (130 μg/ml)	43

[a]Incubations for 1 hr at 87°C with 0.3 μCi of ^{14}C-acetate per ml. Sulfide was added from freshly prepared stock solutions of Na$_2$S·9H$_2$O in aerated Boulder Spring water; concentrations given are for sulfide sulfur. (From Brock et al., 1971.)

tion occurs in water from a nonsulfur spring (Octopus Spring) or in a synthetic salts solution. Some incorporation (not shown) even occurs in demineralized water, provided sulfide is present, suggesting that at least in short-term experiments only sulfide is necessary to promote uptake.

Boulder Spring water contains about 3 μg of sulfide (expressed as S^{2-}) per ml, and both sources contain about the same concentration. As shown in Table 10.10, brief aeration results in a marked decrease in sulfide concentration, as expected. It might be noted that the sulfide concentration of Boulder Spring is considerably higher than that of many Yellowstone boiling alkaline springs. We have done sulfide assays on more than 100 Yellowstone hot springs; those in the Fairy Creek-Sentinel Meadow area, where Boulder Spring is located, are considerably higher in sulfide than most springs in the Lower Geyser Basin. Studies by us using similar techniques in Octopus Spring, a spring low in sulfide (Table 10.10), reveal that sulfide does not stimulate isotope incorporation by bacteria in this spring and in fact inhibits it (Brock and Brock, 1971b). Thus, it cannot be concluded that all bacteria of high-temperature springs are benefited by sulfide.

Sulfide stimulation of incorporation of labeled compounds occurs with a variety of sulfide compounds that are relatively soluble. Sulfide compounds stimulatory for ^{14}C-acetate uptake include sodium, aluminum, calcium, and antimony sulfides (Table 10.11). Of these, the first three readily hydrolyze in water and produce sulfide (S^{2-}) and hydrosulfide (HS^-) ions, and antimony sulfide forms a water-soluble thioanion. Highly insoluble compounds such as copper, lead, and zinc sulfides do not promote isotope incorporation in these short-term experiments. A number of other reduced sulfur compounds were also tested. Thiosulfate and elemental sulfur did not promote uptake, but sulfite and metabisulfite did (Table 10.10). All of the stimulations shown in Table 10.11 were confirmed on repeated experiments.

It thus seems reasonable to conclude that these bacteria are able to use sulfide and certain other reduced sulfur compounds as energy sources. Direct experiments to demonstrate sulfide oxidation by the bacteria have not been attempted, since sulfide itself rapidly oxidizes to sulfur spontaneously. Indirect evidence suggesting sulfide oxidation by at least some of the bacteria in Boulder Spring can be obtained from microscopic observations. Some of the bacterial filaments in natural sediments and attached to cover slips have sulfur globules present on their outer surfaces. (Sulfur globules of this type could of course also be formed spontaneously, the bacterial filaments merely serving as a nonspecific site of deposition.)

Sulfide itself becomes deposited on the cover slips. Cover slips which had been immersed for a number of days were brownish; when acidified, the brown color disappeared and an odor of H_2S could be detected. Sulfide assays showed that cover slips immersed 10 days had about 330 μg of sulfide per cover slip. The deposit is probably not iron sulfide because total iron assays show only 18 μg of iron per cover slip. Microscopic study

Table 10.8. Sulfide Stimulation of Uptake of Various Labeled Compounds[a]

Labeled compound μCi/ml	Counts per minute per cover slip		
	No sulfide	Sulfide	Sulfide + 4% formaldehyde
[14]C-glucose (0.1)	823	1,366	776
[14]C-glutamic acid (0.1)	326	3,626	289
[14]C-leucine (1.0)	1,084	16,083	
[14]C-leucine (0.075)		1,028	166
[14]C-acetate (0.1)	564	6,847	71
[14]C-bicarbonate (1.0)	337	1,341	301
[14]C-lactate (0.12)	956	3,234	15
[3]H-thymidine (1)	5,196	24,712	671
[3]H-phenylalanine (0.06)	3,012	3,868	214

[a]Incubations for 1 hr at 87°C in aerated Boulder Spring water. Sulfide was added at 13 μg/ml from fresh stock solution of $Na_2S \cdot 9H_2O$. (From Brock et al., 1971.)

Table 10.9. Stimulation by Sulfide of Incorporation of ^{14}C-Acetate by Using Water of Various Sources[a]

Source of water	Counts per minute per cover slip	
	No sulfide	Sulfide
Boulder Spring water, aerated	143	5591
Octopus Spring water	221	5658
Synthetic salts medium	345	4227

[a]Incubation for 2 hr at 87°C with 0.1 μCi ^{14}C-acetate per ml. Sulfide was added from fresh stock solution of $Na_2S \cdot 9H_2O$ at final concentration of 13 μg of sulfide/ml. The salts medium contained nitrilotriacetic acid, 1 g; $FeCl_3$, 2.8 mg; $CaSO_4 \cdot 2H_2O$, 0.6 g; $MgSO_4 \cdot 7H_2O$, 1 g; NaCl, 0.08 g; KNO_3, 1 g; $NaNO_3$, 6.9 g; Na_2HPO_4, 1.1 g; water, 1000 ml; pH 8.2. (From Brock et al., 1971.)

Table 10.10. Sulfide Concentration of Spring Waters[a]

Water	Sulfide concentration (μg/ml)[b]
Boulder upper source	3.21
Boulder lower source	3.18
Boulder cover slip colonization site	3.25
Aerated 30 minutes	0.55
Octopus Spring source	0.096

[a]From Brock et al., (1971).
[b]Values are expressed in terms of S^{2-} ion.

Table 10.11. Effect of Various Sulfides and Other Reduced Sulfur Compounds on Incorporation of ^{14}C-Acetate[a]

Compound	Counts per minute per cover slip
None	1002
Inactive sulfide compounds	
Zinc sulfide (ZnS)	660
Nickelous sulfide (Ni$_3$S$_4$)	470
Molybdenum sulfide (MoS$_2$)	833
Mercuric sulfide (HgS)	736
Lead sulfide (PbS)	437
Cuprous sulfide (Cu$_2$S)	689
Cobaltous sulfide (CoS)	827
Carbon disulfide (CS$_2$)	44
Cadmium sulfide (CdS)	689
Bismuth sulfide (Bi$_2$S$_3$)	437
Arsenic pentasulfide (AS$_2$S$_5$)	798
Active sulfide compounds	
Sodium sulfide (Na$_2$S)	3262
Ferrous sulfide (FeS)	1292
Calcium sulfide (CaS)	1844
Aluminum sulfide (Al$_2$S$_3$)	2115
Antimony sulfide (Sb$_2$S$_3$)	2394
Other reduced sulfide compounds	
Sulfur (S^0)	964
Sodium sulfite (Na$_2$SO$_3$)	2680
Sodium metabisulfite (Na$_2$S$_2$O$_5$)	1929
Sodium thiosulfate (Na$_2$S$_2$O$_3$)	650

[a]Incubations for 1 hr at 87°C with 0.1 μCi of ^{14}C-acetate per ml in aerated Boulder Spring water. All compounds were added to give 12 to 13 μg of sulfur/ml. (From Brock et al., 1971.)

showed that the sulfide deposit was present as tiny amorphous spheres of a size just at the limit of resolution; no metal sulfide crystals were visible.

pH Optimum

Boulder Spring water was adjusted to various pH values, and the amount of sulfide-stimulated uptake of ^{14}C-acetate was determined. The pH optimum was 9.2, a value near that of the natural spring water (8.9).

Temperature Optimum

All of the above experiments involved incubation at 85–89°C at a site about a meter downstream from the cover slip colonization site. The temperature optimum for the sulfide-stimulated uptake of a variety of labeled com-

pounds was next determined. A number of separate experiments were done
with each isotope. To average the results of all these experiments, the
percentage of maximal incorporation was calculated for each temperature
for each experiment, and then the percentages within a series of temperature intervals were averaged for all of the experiments with that isotope.
The data are plotted in Figure 10.15.

It can be seen that essentially the same graph was obtained for each
isotope, a relatively broad temperature optimum in the range of 80–90°C.
Note that in all cases, significant isotope uptake took place at the temperature of the colonization site, 90.5–91.5°C. It should also be emphasized that
Formalin controls, which were always included, never showed uptake at
the optimal temperature greater than 5–10% of the maximum, thus, even a

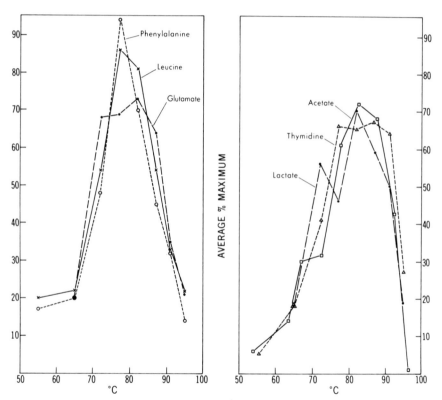

Figure 10.15. Temperature optima for the incorporation of radioactive compounds
by bacteria from Boulder Spring. Data are averages within temperature intervals of
percentage of maximal incorporation obtained in different experiments. The number
of separate experiments done with each compound was as follows: ^{14}C-acetate, 11;
^{14}C-leucine, 7; ^{3}H-phenylalanine, 6; ^{3}H-thymidine, 6; ^{3}H-glutamic, 5; ^{14}C-lactate, 4.
All experiments were done in aerated Boulder Spring water to which 13 μg/ml of
sulfide was added. For actual radioactivity data typical of those obtained in these
experiments, see Table 10.8. (From Brock et al., 1971.)

20% incorporation can be considered significant. However, at temperatures greater than 95°C, only low isotope uptake occurred, usually of the same level as the Formalin controls.

Significance

Despite the fact that cultures of these bacteria have as yet not been obtained, we now know quite a bit about their properties. Although the *in situ* growth rates showed that the organisms were able to multiply at reasonable rates at temperatures over 90°C, they did not show that the organisms were optimally adapted to such temperatures. The work with radioactively labeled compounds makes possible a study of the environmental optima for these organisms.

A variety of labeled compounds are taken up by the Boulder Spring bacteria, but significant uptake occurs only when sulfide is present. Studies using inhibitors have shown that uptake is indeed due to vital processes of the organisms. These organisms resemble bacteria such as *Beggiatoa,* which also incorporate organic compounds well when sulfide is present. Such bacteria have been termed mixotrophs. In addition to sulfide, several other reduced sulfur compounds will promote uptake of labeled compounds, although elemental sulfur and thiosulfate do not. Interestingly, Boulder Spring water contains a small amount of thiosulfate (Table 10.5), and it is conceivable that in long-term studies evidence for thiosulfate stimulation might be obtained.

Probably the most significant aspect of the present work is the determination of temperature optima for the uptake of labeled compounds. Because of the inherent variability in this study, a large number of experiments had to be done. The temperature optima for all the compounds used were in about the same range, 80–90°C. The optimal temperatures are somewhat lower than the temperature of the colonization site, 90.5–91.5°C, although in all cases, significant uptake did occur at these temperatures, but not at 95–97°C. It would be of considerable interest to know the temperature optima of the bacteria living in boiling sulfide springs at Rotorua, New Zealand, where, because of the lower altitude, temperatures of 99–100°C are obtained. Since the bacteria of Boulder Spring are unable to incorporate labeled compounds at such temperatures, one might predict that the New Zealand bacteria have higher temperature optima.

Since there are two morphological types of bacteria in Boulder Spring, it is not certain whether our data represent the physiological properties of one or both types. We attempted to ascertain this by using microscopic autoradiography, but this study was thwarted because the sulfide deposit on the cover slips reacted with the photographic emulsion. Although the sulfide deposit could be removed by exposing the cover slips to HCl fumes, this treatment also caused the organisms to fall off. However, since little uptake of labeled compounds occurred in the absence of sulfide, it seems likely that both kinds of bacteria require reduced compounds.

It should be emphasized that not all bacteria inhabiting Yellowstone boiling springs required reduced sulfur compounds. Studies in Octopus Spring, a spring low in sulfide, have shown that good uptake occurs in the absence of sulfide and that sulfide does not stimulate uptake but actually inhibits it (see below). Thus we must consider the possibility that a diversity of bacterial species inhabit these extreme environments.

The cell envelope structure of these bacteria is different from all other bacteria, either mesophilic or thermophilic. No morphologically distinct peptidoglycan layer was seen outside the plasma membrane. Instead, a rather thick diffuse layer was seen, within which a subunit structure was often distinctly visible, and connections frequently occurred between this outer layer and the plasma membrane. The thick outer layer usually consisted of two parts, the outer part of which was sometimes missing. Bacteria of similar structure have also been found in Octopus Spring, a spring low in sulfide.

The Bacteria of Octopus Spring

As I noted in the beginning of this chapter, our first idea of the existence of bacteria in boiling water came from observations on a spring which we called Pool A, which has a more-or-less official name of Octopus Spring. This spring has provided a study site for a wide variety of our work, both on algae and bacteria, and is mentioned in some detail in other chapters (Chapters 7, 8 and 11; see also location on Figure 2.2). When we began to study the high-temperature bacteria, it was natural to begin with the large masses of pink material living in the outflow channel at temperatures over 80°C. We carried out a large number of radioisotope experiments with this material, attempting to obtain uptake of some labeled compound. Almost always the results were either completely negative or equivocal. I finally concluded that the pink masses were actually rather moribund, a reasonable conclusion when it was found that they represent growth over a period of months. Apparently these organisms hold on tenaciously in the boiling water, so that they do not wash out readily, and since they lack predators or grazers, there is little turnover. The large masses that eventually build up thus probably represent nutrient-starved material which is in poor shape to show any metabolic processes.

After Tom Bott had developed the cover slip technique used at Boulder Spring (see previous section), it was natural to attempt to use this technique with the Octopus Spring bacteria. Instead of working in the outflow channel, we used the source, which has an almost constant temperature of 91°C, just a degree below boiling.

Bacteria from Octopus Spring incorporated radioactive amino acids, thymidine and lactate, but incorporation of labeled sugars was negligible. Detailed studies with radioactive leucine (Table 10.12) showed that isotope uptake was inhibited not only by formaldehyde but also by streptomycin, novobiocin, sodium azide, Na_2S, and mercuric bichloride. Uptake of radioactive leucine is also inhibited by nonradioactive leucine and by yeast extract. It is clear, therefore, that isotope uptake is due to a vital process and not to passive uptake or adsorption. Attempts to stimulate uptake of leucine, acetate, or bicarbonate by bubbling with methane, nitrogen, CO_2, or air during the incubation period were unsuccessful.

The optimum pH for uptake of leucine was between 7 and 7.5 (Figure 10.16) irrespective of whether spring water or synthetic salts adjusted to various pH values were used. In determination of the temperature optimum at Octopus Spring, both spring water at its pH and synthetic salts solution at pH 7 were used as incubation media. Although the amount of isotope uptake was much greater in the latter, the temperature optima for both media were the same. A number of experiments were done to determine the temperature optimum for the incorporation of different labeled compounds. Figure 10.17 shows results of a single experiment in which five compounds were run in parallel. Only 5–10-minutes equilibration at the various temperatures was allowed, so that there was no possibility of adaptation to the new temperatures. We assume therefore that the optima represent those that the bacteria had in the spring itself. The optima for all compounds are virtually the same, 90°C. In other experiments, optima ranging from 86°C to 90°C were found. During a 10 year period we measured the source temperature of Octopus Spring many times and have found that it ranged from

Table 10.12. Effect of Inhibitors on Uptake of ^{14}C-Leucine by Bacteria from Octopus Spring[a]

Conditions	Counts per minute per cover slip
Control	4622
+ HCHO (4%)	364
+ Streptomycin (1000 μg/ml)	429
+ $HgCl_2$ (100 μg/ml)	317
+ NaN_3 (1000 μg/ml)	153
+ Novobiocin (100 μg/ml)	100
+ Na_2S (150 μg/ml)	998
+ Na_2S (15 μg/ml)	2234
+ Yeast extract (0.1%)	35
+ Nonradioactive leucine (1000 μg/ml)	87

[a]Incubation at 86°C for 1 hr in synthetic D medium at pH 7.0 with 0.06 μCi/ml ^{14}C-leucine. (From Brock and Brock, 1971b.)

Figure 10.16. pH optimum, for the bacteria of Octopus Spring for the uptake of ^{14}C-leucine. Incubation was at 86°C for 1 hour with 0.06 μCi/ml. The pH values were obtained by adjusting the spring water with 1 M HCl or 1 M NaOH. (From Brock and Brock, 1971b.)

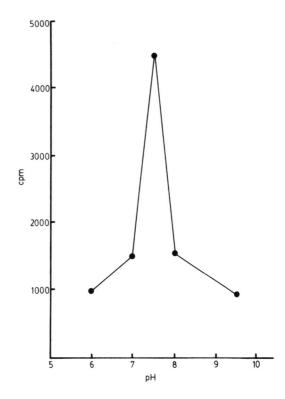

90.5°C to 91.5°C. Thus the temperature optimum of these bacteria is virtually identical to their environmental temperature.

In Retrospect

The biggest gap in our understanding of bacteria of boiling water is in the culture of these interesting creatures. Clearly such organisms should be of considerable biochemical and evolutionary interest. Despite numerous attempts, I was never able to obtain consistent growth at temperatures of 90°C or over. Occasionally, limited growth occurred, but transfers never were successful. Since similar things could be said about a variety of bacteria living at more moderate temperatures (e.g., the spirochete *Treponeoma pallidum*, rickettsia, chlamydia), I don't necessarily feel that our inability to culture these organisms raises any doubt about their existence. Almost certainly, we did not try the proper medium or conditions. At Boulder Spring we spent quite a bit of time trying to obtain cultures in flasks to which we introduced constantly at a very slow rate H_2S gas, designed to maintain a sulfide concentration close to that of the spring. These culture

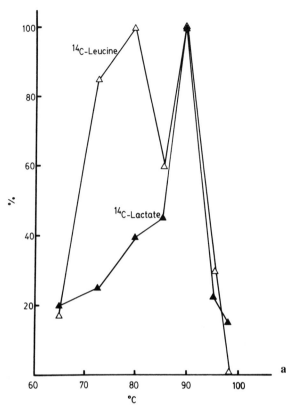

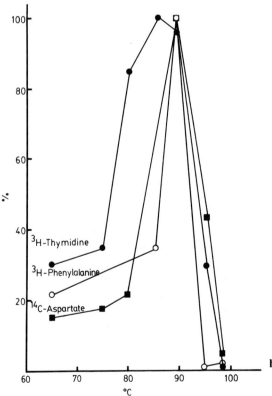

Figure 10.17. Temperature optima for the bacteria of Octopus Spring for the uptake of various labeled compounds. (a) leucine, lactate; (b) thymidine, phenylalanine, aspartate. All labeled compounds were present at 0.06 μCi/ml. The results are plotted as percentage of maximum incorporation for each isotope. The 100% values in cpm/coverslip were as follows: ^{3}H-thymidine, 492; ^{3}H-phenylalanine, 2224; ^{14}C-aspartic, 840; ^{14}C-lactate, 92; ^{14}C-leucine, 11,-545. Specific activities of the various radioactive compounds are as follows: thymidine, 0.25 mCi/0.00368 mg; phenylalanine, 1 mCi/2 mg; leucine, 0.05 mCi/0.025 mg; aspartic, 3.65 mCi/mmole; lactate, 0.1 mCi/2 mg. Incubation was in synthetic medium at pH 7. Temperatures over boiling (92.5°C) were obtained using boiling water baths with saturated MgSO$_4$. (From Brock and Brock, 1971b.)

attempts were done in reflux flasks at boiling temperature and all that resulted was a massive precipitate of elemental sulfur, arising from the gradual oxidation of the H_2S. Perhaps these organisms are obligate aerobes, and insufficient oxygen was present in our culture media at these high temperatures. Oxygen solubility at boiling is listed in the handbooks as zero, but certainly some oxygen is carried down into the springs from the air, as a result of turbulence. On the other hand, perhaps the organisms are obligate anaerobes, and we had either the wrong electron acceptor or had sufficient oxygen contamination to prevent their growth.

These studies were done in two springs (Boulder, Octopus) that had slightly alkaline pH values. At such pH values, *Thermus* grows at an upper temperature of 79–80°C, but I have been unsuccessful in culturing any bacterium at neutral or alkaline pH at temperatures above this. A large number of enrichment cultures using various media and inocula from various springs, at temperatures of 80–90°C, have failed. It was thus a surprise to see how easy it was to cultivate *Sulfolobus,* the bacterium of hot, acid waters, at temperatures of 80–85°C (although not at 90°C). After we had cultivated *Sulfolobus,* our attention turned to it for detailed ecological and physiological studies (see Chapter 6), and further work on the bacteria from Boulder Spring and Octopus Spring was dropped.

I think ultimately that the organisms of boiling water will be culturable, and I would hope that someone would go to Yellowstone or New Zealand and initiate a detailed study. New Zealand might be a more favorable location for such work than Yellowstone, partly because the springs are at higher temperatures, and partly because there are good laboratory facilities in the Rotorua area that could probably be used. I can't guarantee that the organisms cultured from boiling water would be of any practical significance, but they will amost certainly turn out to be entirely new types, with interesting biochemical and physiological properties.

References

Allen, E. T. and A. L. Day. 1935. Hot springs of the Yellowstone National Park. Carnegie Institution of Washington Publication No. 466, Washington, D.C., 525 pp.

Bauman, A. J. and P. G. Simmonds. 1969. Fatty acids and polar lipids of extremely thermophilic filamentous bacterial masses from two Yellowstone hot springs. *J. Bacteriol.* **98**, 528–531.

Bott, T. L. and T. D. Brock. 1969. Bacterial growth rates above 90°C in Yellowstone hot springs. *Science* **164**, 1411–1412.

Bott, T. L. and T. D. Brock. 1970. Growth and metabolism of periphytic bacteria: methodology. Limnol. Oceanogr. **15**, 333–342.

Brock, T. D. 1967. Life at high temperatures. *Science* **158**, 1012–1019.

Brock, T. D. and M. L. Brock. 1971a. Microbiological studies of thermal habitats of the central volcanic region, North Island, New Zealand. *N.Z. J. Mar. Freshwater Res.* **5**, 233–257.

Brock, T. D. and M. L. Brock. 1971b. Temperature optimum of non-sulphur bacteria from a spring at 90°C. *Nature* **233**, 494–495.

Brock, T. D., M. L. Brock, T. L. Bott, and M. R. Edwards. 1971. Microbial life at 90 C: the sulfur bacteria of Boulder Spring. *J. Bacteriol.* **107**, 303–314.

Brock, T. D. and G. K. Darland. 1970. Limits of microbial existence: temperature and pH. *Science* **169**, 1316–1318.

Brock, T. D. and M. R. Edwards. 1970. Fine structure of *Thermus aquaticus*, an extreme thermophile. *J. Bacteriol.* **104**, 509–517.

Kempner, E. S. 1963. Upper temperature for life. *Science* **142**, 1318–1319.

Chapter 11
Stromatolites: Yellowstone Analogues

Stromatolites are important types of geological structures, found primarily in the Precambrian Era, but present to lesser extent throughout the rest of earth history up until the present. Stromatolites are important geologically because they provide a useful fossil record of the Precambrian, and because many mineralized deposits are found within or associated with them. A number of definitions of "stromatolite" have been used, but a recent definition (Walter, 1976) attributable originally to Stanley Awramik and Lynn Margulis (*Stromatolite Newsletter,* February 1974, p. 5) follows: "Stromatolites are megascopic organosedimentary structures produced by sediment trapping, binding, and/or precipitation as a result of growth and metabolic activity of organisms, primarily blue-green algae."

The paper by Barghoorn and Tyler (1965) can well be said to have issued in the modern era of paleomicrobiology. This seminal paper presented evidence from thin sections that filamentous microorganisms, perhaps blue-green algae, were well preserved in 2-billion-year-old rocks of the Gunflint formation. Following this paper, there has been an explosion of papers describing microfossils from a wide variety of Precambrian sedimentary rocks, some as old as 3 million years. I do not propose to review all of this work here, most of which is not relevant to the Yellowstone project. However, of considerable importance for my own thinking along these lines was the statement in the original Barghoorn and Tyler paper: "The black chert is characterized by discontinuous anastomosing pillars oriented roughly perpendicular to the gross structure of the algal dome. . . . The general appearance is that of a nest of thimbles, strikingly similar in morphology to structures associated with certain modern algal growths (Figure 2, Parts 1 and 2)." The figures referred to were thin sections of siliceous sinter from currently forming geyserite deposits in Yellowstone

National Park, showing concentric banding of algal remains. About 1967 Barghoorn came to Indiana University for a seminar and I spoke to him further about this analogy, which he generally confirmed. I have the distinct impression that he said that the Gunflint chert may have been a hot spring deposit, although subsequently this idea has not found any specific mention in the literature. I mentioned the Barghoorn and Tyler paper specifically in my paper in *Nature* (Brock, 1967a) which discussed the upper temperature limit for blue-green algal development, and used it as evidence that hot spring algae may have been around since the Precambrian. This reference focused attention on the Yellowstone hot spring algae as of some interest to paleomicrobiologists.

I did nothing specific about Yellowstone analogues of stromatolites, and did not even mention stromatolites in my paper on vertical zonation in hot spring algal mats (Brock, 1969). In 1970 I had a visit from a Minnesota geologist, David Darby, who was interested in the Gunflint and in stromatolites, but when I took him around Yellowstone, he expressed little interest in the algal mats, and concentrated on geyser deposits. Such deposits are laminated, and as was later shown (Walter, 1972) present an analogue for the nonbiogenic formation of fine-grained Precambrian formations, such as the Gunflint or Biwabik banded iron formations (associated with the cherts, but distinctly different from the fossiliferous deposits). An Australian geologist, Malcolm Walter, read my 1969 paper, was aware of Weed's early work (Weed, 1889), and became interested in looking at the Yellowstone material. He was coming to the U.S. on a fellowship through Yale University and wrote me about visiting Yellowstone. I was delighted, as I had been interested in relating our work more closely to the stromatolite work. Malcolm first visited Yellowstone during the summer of 1971. Although he used our facilities, he was from the beginning an independent investigator, but I was happy that a collaboration began between him and my graduate student John Bauld, a fellow "Aussy." Malcolm came back for the summer of 1972, and also returned several times in off-seasons on his own. In 1972 we were also delighted to have a visit of 1 week from the eminent geologist and paleobiologist Preston Cloud, who is very knowledgeable about Precambrian paleobiology, and who expressed great interest in the things that Malcolm was doing.

By Awramik and Margulis's definition, many of the Yellowstone algal mats are stromatolitic. They are megascopic, organic, cause sediment trapping and binding, cause precipitation of minerals, and are a result of growth and metabolic activity of organisms, primarily blue-green algae. Although there may be a wide variety of stromatolitic structures in Yellowstone, our work concentrated on two types: (1) conical-shaped structures, which are related to Precambrian Conophytons, and which are produced almost exclusively by a species of *Phormidium*; and (2) the flat, laminated mats produced by an association of the photosynthetic bacterium *Chloroflexus* and the blue-green alga *Synechococcus*. The Conophytons are of interest because they are extremely common in the Precambrian, although they are absent from the subsequent fossil record. It had thus been thought that they had died out, but Malcolm found them living and growing in Yellowstone. The *Chloroflexus-Synechococcus* mats are of interest because the dominant organism responsible for their structure proved to be a photosynthetic bacterium rather than a blue-green alga. If it could be shown that a photosynthetic bacterium alone could produce stromatolites, then it would cast doubt on the common interpretation that the presence in the fossil record of stromatolites indicated the presence of blue-green algae. Since blue-green algae produce O_2 during photosynthesis, and photosynthetic bacteria do not, and since blue-green algal-produced photosynthesis was the dominant source of O_2 in the atmosphere from the Precambrian, it became important to study the *Chloroflexus* mats in some detail. During Malcolm's two summers, he concentrated his work on the Conophytons, and after he returned to Australia, we completed some of the Conophyton work and did detailed studies on the *Chloroflexus* mats. It was about this time that Bill Doemel, my former graduate student, now a faculty member at Wabash College, contacted me about spending another summer in Yellowstone. I suggested the stromatolite problem, he was enthusiastic, and together with undergraduate students he brought from Wabash College, we initiated and carried to completion during the summers 1973 through 1975 a number of studies on Yellowstone stromatolites. In the rest of this chapter, I will review in some detail both Walter and Doemel's work.

Siliceous Algal and Bacterial Stromatolites in Hot Springs and Geyser Effluents of Yellowstone National Park[1]

Growing algal and bacterial stromatolites composed of nearly amorphous silica occur in the alkaline effluents of hot springs and geysers in Yellowstone National Park, Wyoming. Several points of interest have emerged from our studies: (1) The stromatolites are primarily siliceous. (2) Photosynthetic flexibacteria are very abundant in the stromatolites. (3) There are growing Conophyton stromatolites, a distinctive group previously thought to have become extinct near the Precambrian-Cambrian boundary. The immediate relevance of these results is twofold: (1) They indicate that bacteria may build stromatolites and, thus, that some fossil stromatolites, especially those in the Archean, can be bacterial rather than algal. This has important implications in atmospheric evolution, since bacterial photosynthesis does not release oxygen. (2) They prove that stromatolites can be

[1]Some of this section is taken directly from Walter et al. (1972, 1976).

primarily siliceous, and thus support interpretations of the primary origin of the silica in Precambrian iron formations (some of which include siliceous stromatolites built by a microbiota comparable with that in hot spring effluents).

The silica is precipitated nonbiogenically, due to the cooling and evaporation of the water. It encrusts the bacterial and algal filaments, building tubules and eventually a solid rock permeated by microbial filaments. All of the samples analyzed by x-ray diffractometry are nearly amorphous opaline silica.

As noted in Chapter 8, microbial mats occur in Yellowstone at temperatures up to about 70°C. Animal grazers are restricted to temperatures less than 50°C (Brock, 1970; Brock et al., 1969) and have no effect on most of the structures described. Mats at temperatures above about 30°C consist of fine filamentous cyanophytes ("blue-green algae") and bacteria, whereas those in cooler parts of the effluents are formed by coarse filamentous cyanophytes (*Calothrix* sp. and *Mastigocladus* sp.). The *Calothrix* mats are crudely laminated and are generally stratiform with wavy surfaces and infrequent small domes or, rarely, clusters of contiguous domes, but no columnar forms are known. In contrast, the higher temperature mats are usually finely laminated, and in a number of springs they form small columns with a relief of up to 3 or 4 cm and a width of 1 or 2 mm up to 2 or 3 cm. Their maximum relief is governed by the depth of water in which they grow. Columns have not been found at temperatures above 59°C, a temperature near that at which the mats reach their maximum thickness (Brock, 1967b).

There are two types of columns, flat-topped and conical. Both are vertical and apparently do not branch. They form beds from 1 cm to more than 10 cm thick. Lenticular gas cavities occur between the laminae of many columns. The flat-topped columns (Figure 11.1) are known from the effluents of two hot springs: "Column Spouter" (Fairy Creek meadows) and near "Weeds Pool" (Fountain-Paint Pots area); they were subsequently recognized in a number of other springs which are mostly unnamed.

The conical columns (Figure 11.2) are built from acutely conical laminae with their apexes directed upward. They have been found in the effluents of several springs and geysers at temperatures up to 50°C, and in several pools with temperatures in the range of 30–50°C. They occur in a wide variety of hydrological regimes ranging from subaqueous and almost stagnant through subaqueous with rapidly flowing water to alternately subaerial and subaqueous with rapidly flowing water; the pH of the water ranges from 7.5 to 9.0. Their shape bears no obvious relationship to directions of water flow; column transverse sections are subcircular even in areas of rapid unidirectional water flow. Conical stromatolites are known from the following hot springs and geysers in Yellowstone National Park: "Conophyton Pool," Column Spouter, and two unnamed springs (Fairy Creek meadows);

Figure 11.1. (a) Flat-topped columnar stromatolites in an outflow channel of Column Spouter, Fairy Creek meadows. The columns are about 1 cm wide. (b) Cross-section through a single column. Note the laminations and the area of decomposition (hollow area).

Figure 11.2. (a) Small conical stromatolites from Tromp Spring. The structures are about 1 cm high. (b) Similar structures in a spring (Walter's Column Spouter) in the Fairy Meadows area. Note pine needle for scale.

a

b

Queens Laundry Spring (Sentinel Meadows); "Weed's Pool" (on the flank of the Fountain-Paint Pots group); Grand Prismatic, Tromp Spring, and an unnamed spring (Midway Geyser Basin); White Crater Spring (Shoshone Geyser Basin); and Main Spring (Jupiter Terrace, Mammoth area). In addition, they have been seen in a number of other small, unnamed springs in Yellowstone. I have also collected similar structures from a spring at Hveravellir in central Iceland.

Conditions of Formation of Stromatolitic Columns

The stromatolites described here occur in thermal waters with temperatures between 32°C and 59°C and with pH values of 7–9. It was demonstrated by silicon carbide (carborundum) marking that the stromatolites are actively growing in the environments in which they were observed. Filamentous cyanophytes of the genus *Phormidium,* which are an essential component of the biotas of both the flat-topped and conical columnar stromatolites, are apparently absent from waters hotter than 59°C [the filamentous photosynthetic bacterium *Chloroflexus aurantiacus* forms stratiform and nodular stromatolites at higher temperatures (see later)]. Below about 32°C, coarsely filamentous cyanophytes such as *Calothrix coriacea* dominate the stromatolitic mats, which are stratiform or nodular. Animals are present at temperatures up to 48–51°C (Brock, 1970; Castenholz, 1969; Wickstrom and Castenholz, 1973). These include rotifers, ostracods, spiders, and flies. Although some of these organisms graze the microbial mats (Brock et al., 1969), they apparently are not effective in shaping the stromatolites we have studied, as morphologically comparable stromatolites also form at higher temperatures than those at which the animals can survive. Encrustations of silica may protect some of the stromatolites from grazing animals.

The flat-topped columns were studied during 1971–1972 in three outflow channels: two at Column Spouter (Fairy Creek meadows) and one immediately above Weed's Pool (Fountain-Paint Pots area). Stromatolites from all these localities are siliceous. Column Spouter erupts approximately hourly. During eruptions the columns are submerged by several centimeters of water with a temperature of up to 47°C. Eruptions last about 45 minutes. For about 15 minutes between eruptions there is very little water flow and during this period the stromatolites, which occur near the margins of outflow channels, are exposed to the atmosphere, and the temperature of any water flowing around the columns drops markedly, commonly to 20–25°C. Similar conditions prevail in the channel near Weed's Pool.

In 1971–1972 the conical stromatolites were abundant in 10 springs and geysers. In nine, silica is precipitating from the thermal waters; in the tenth (Main Spring, Jupiter Terrace, Mammoth area), calcium carbonate is precipitating in the form of aragonite. The various waters range in temperature from 32°C to 50°C and in pH from 7 to 9. As most previous studies of stromatolites have attempted to relate their morphology to environmental influences, especially to directional water currents, particular attention was paid to these factors in Yellowstone. The waters in which conical stromatolites were growing ranged from intermittently flowing in outflow channels, through gently and irregularly flowing sheets of water in shallow pools, to nearly still in pools up to 30 cm deep. Measured flow rates varied considerably. In the outflow channels of Column Spouter the cones were submerged in several centimeters of water flowing at 5–20 cm/second at peak flow, but were exposed to the atmosphere between eruptions. In Conophyton Pool the cone's were continually submerged in 2–10 cm of water in which current directions and velocities are variable, depending on wind direction and intensity as well as on the rate of outflow from the spring (which discharges continually). In this pool there are large differences in flow rates between the water surface and the bottom: on one occasion flow at the surface was 3–6 cm/second, but 1 cm beneath the surface no flow was detected. Flow rates of 1–10 cm/second were measured at the tips of 2–3-mm-high cones on a windy day. Flow around the cones in White Crater Spring was measured only once and was about 1 cm/second. On a windy day, flow in Weed's Pool was 5–6 cm/second at the surface and 0.5–1 cm/second around the cones 5–10 cm below the surface; this pool (Figure 11.3) is 30 cm deep, and is the deepest pool known to contain conical stromatolites (Figure 11.4).

Microbiology

The microbiology of the flat-topped columnar stromatolites is very poorly known because before they could be studied in detail slight changes in outflow patterns destroyed them all. They were constructed by a cyanophyte which in one sample was identified as *Phormidium truncatum* var. *thermale,* accompanied by the photosynthetic filamentous bacterium *Chloroflexus aurantiacus*. The unicellular cyanophyte *Synechococcus lividus* was also present. Other filamentous cyanophytes were sporadic.

The conical stromatolites were studied in 10 hot springs and geysers. In each case the organisms were essentially the same: the dominant organism was the filamentous cyanophyte *Phormidium tenue* var. *granuliferum*. Next most abundant was the bacterium *Chloroflexus aurantiacus*. The unicellular cyanophyte *Synechococcus* was usually present, with *S. lividus* much more abundant than *S. minervae*. Other filamentous cyanophytes were sporadic; these included forms resembling *Pseudanabaena* and *Isocystis*. Organisms resembling myxobacteria were present in some samples and *Spirillum*-like bacteria and spirochetes occurred rarely. The organisms of the calcareous cones from the Mammoth area were somewhat different from that of the siliceous forms. The *Phormidium* species present was

Figure 11.3. Photograph of Weed's Pool in the Fountain Paint Pots area of the Lower Geyser Basin. This pool is the deepest pool containing conical stromatolites, and some of the structures in this pool are over 10 cm high (see Figure 11.4).

a

b

Figure 11.4a,b. Conophytons in Weed's Pool (see Figure 11.3). About 10 cm high. Note the distinct ridges. (a) Several Conophytons; (b) Single large structure. Photographed *in situ*.

slightly coarser (trichomes 1.0–1.7 μm wide) than *P. tenue* var. *granuliferum* (trichomes 0.8–1.2 μm wide), although in other ways it was indistinguishable.

P. tenue var. *granuliferum* was usually much more abundant in the cones than in the mats between the cones. Conversely *Chloroflexus* was less abundant in the cones than in the mats. *P. tenue* var. *granuliferum* was most abundant in the tiny tufts and in the smallest, most recently developed cones. The ridges and spines frequently present on the cones had a microbiota that was indistinguishable from that of the cones generally.

To further the analysis of the morphogenesis of the conical stromatolites, the relative photosynthetic activities of natural populations of *Phormidium* and *Chloroflexus* in the stromatolites from Conophyton Pool were determined. To do this, light-stimulated $^{14}CO_2$ fixation (photosynthesis) was measured under the prevailing environmental conditions. The results are presented in Table 11.1. These data show that *Phormidium* was responsible for about 90% of the total $^{14}CO_2$ fixation in the tufts; the remainder (DCMU-insensitive photosynthesis) was attributable to *Chloroflexus*. Incubation under a Wratten filter which excluded light of wavelengths suitable for algal photosynthesis resulted in an inhibition of algal photosynthesis in the tufts similar to that observed in the presence of DCMU. A statistical analysis of 40 samples that included both tufts and the surrounding mat showed that the mean chlorophyll-a content of the tufts, when standardized per unit protein, was about 1.5 times that of the mat from which they arise. This is consistent with the microscopic observation of more dead cells in the mat than in the tufts.

Table 11.1. Photosynthetic $^{14}CO_2$ Fixation in Conical Tufts from Conophyton Pool[a]

Component	cpm/μg Chl-a	Percent total fixation	Relative abundance (%)
1. *Phormidium* + *Chloroflexus*	9419	100	100
2. *Chloroflexus* (+ DCMU)	681	7	9
2a. *Chloroflexus* (88A filter)[b]	(1295)	(13)	9
3. *Phormidium*[c]	8738	93	91

[a]It is assumed that *Phormidium* and *Chloroflexus* account for all of the fixation of $^{14}CO_2$. The amount of fixation is expressed as counts per minute (cpm) standardized to the chlorophyll-a (Chl-a) content of the samples. The relative abundances of *Phormidium* and *Chloroflexus* in the samples are expressed as percentages, assuming that these two groups of microorganisms are the only components of the tufts. Lengths of filaments were estimated on homogenates by the method of Bott and Brock (1970). Prevention of algal photosynthesis was by addition of DCMU (row 2) or incubation under an infrared filter (row 2a). Each value for $^{14}CO_2$ fixation is the average of four incubations. (From Walter et al., 1976.)
[b]No dark controls. The percentage of $^{14}CO_2$ fixation given in parentheses was corrected only for absorption (formaldehyde).
[c]Values for $^{14}CO_2$ fixation by *Phormidium* were obtained by subtracting component 2 from component 1.

It is apparent that *Phormidium* is primarily responsible for the construction of the cones. *Chloroflexus* may play some role by utilizing the excretory products of *Phormidium* as a carbon source (Bauld and Brock, 1974), thereby affecting the physical properties of the mats and cones (e.g., by reducing the cohesiveness of the aggregates of filaments). However, we have demonstrated in the laboratory that *Chloroflexus*-free cultures of *P. tenue* var. *granuliferum* are capable of independently forming conical structures (Figure 11.5); these grow rapidly under ideal laboratory conditions. Conversely, natural *Chloroflexus* mats are frequently nodular but are never conical (see later).

Figure 11.5. Conical structures formed by a pure culture of *Phormidium tenue* var. *granuliferum*. Cultures were grown under fluorescent lights in a 37°C incubator room where the heat of the lights raised the temperature of the culture vessels to about 40°C. Erlenmeyer flasks containing a thick layer of agar medium were inoculated with a suspension of filaments. Sufficient inoculum was used so that a confluent growth of the alga occurred after about 4 days of incubation. At this time, liquid medium was carefully added to fill the flasks, and the organisms were allowed to grow up into the liquid. Within 24–48 hours after addition of the liquid medium, small nodes of algal growth appeared on the surface of the solid medium at the bottoms of the flasks. To stimulate the periodicity of sunlight a light-dark cycle of 12 hours was used, the light being controlled by an automatic timer. This resulted in the formation of two laminae per 24 hours. Figures show cones as they appeared 12 days after addition of liquid growth medium to the preformed algal mat.

Controls of Stromatolite Morphogenesis and Lamination Production

Biological Influence

Many species of *Phormidium* are present in Yellowstone waters (Cope-land, 1936) but apparently only *P. tenue* var. *granuliferum* constructs conical stromatolites of the kind described here. The flat-topped columnar stromatolites seem to be built by a different species, *P. truncatum* var. *thermale*. [This correlation of cyanophyte morphotypes with stromatolite morphotypes has a validity independent of that of the cyanophyte taxonomy; Copeland's (1936) taxonomy and nomenclature are used for convenience, with the understanding that the named cyanophyte morphotypes may not represent true genetic taxa]. These facts clearly suggest biological control of stromatolite morphogenesis, an interpretation that is supported by the growth of the conical stromatolites in a variety of hydrological regimes (including a laboratory flask).

Effects of Light

Light may be expected to play a key role in stromatolite accretion since these structures are constructed by photosynthetic microorganisms. Previous studies of marine stromatolites (Monty, 1967; Gebelein, 1969) have shown that the orientation of cyanophyte filaments within stromatolites can be controlled by light and that lamination can be produced by the diurnal variation in intensity of light. In these examples, laminae with vertical filaments form during the day whereas those with horizontal filaments form at night. Two interrelated aspects of our study involved investigating the role of light in morphogenesis and the temporal significance, if any, of the lamination. It was difficult to predict the effects of light because both bacteria and cyanophytes (which may respond differently) are abundant in the stromatolites studied and furthermore the filamentous bacterium present *(Chloroflexus)* is only facultatively phototrophic (Pierson and Castenholz, 1974).

To investigate stromatolite growth in the dark, a long-term experiment was conducted from August 1971 to April 1972 in which a large light-proof black plastic box was used to shade about a square meter of cones in Conophyton Pool. The box did not prevent water movement; however, deprivation of light did prevent algal and bacterial photosynthesis and photoheterotrophism. The darkened stromatolites and those outside the box were marked with silicon carbide powder. After 7 months the uncovered columns had accreted 0.5–2.0 mm and the new deposit was finely laminated. The columns under the box had a very uneven and almost completely unlaminated coating up to 0.5 mm thick above the silicon carbide. These results demonstrate that the conical stromatolites will not accrete in the dark. In another experiment, run for 3 months, an infrared filter was used to prevent photosynthesis by the cyanophytes but not by the bacteria. This also stopped the accretion of the stromatolites. It is apparent from these two studies that light of the wavelengths used for algal photosynthesis is necessary for accretion.

The intensity of incident sunlight is subject to two major variations, diurnal and seasonal, which may be reflected in patterns of stromatolite accretion. Like the diurnal laminae of some marine stromatolites (Monty, 1967; Gebelein, 1969) the laminae of the flat-topped columns in Yellowstone consist of alternations of vertical and prostrate filaments. The laminae composed of vertical filaments are pale and 30–150 μm thick, whereas those composed of prostrate filaments are dark and 20–60 μm thick. A 4-day experiment at Column Spouter demonstrated that these laminae also form diurnally, with the filaments standing vertically during the day and lying prostrate at night. Other experiments in the same hot spring showed no accretion over the same period. A 12-month experiment in Column Spouter during which silicon carbide layers were emplaced on three occasions also indicated that the lamination is diurnal. However, laminae do not form every day. During summer about 70% of the days are represented by pairs of laminae (pale and dark), but very few laminae formed during winter. The following results were derived from a single specimen in which the three marked layers were visible. Between 2 August 1971 and 20 April 1972 only 27 laminae formed, mostly near the beginning and end of this period (i.e., not during winter). Winter is represented by a 1-mm-thick, dark, coarsely and diffusely laminated layer. Between 20 April and 20 June 1972 (61 days), 45–49 laminae formed. Between 20 June and 21 July (31 days), 20 laminae formed. In other examples there was either no growth during winter or else *Calothrix coriacea* overgrew the columns (because of a temperature drop resulting from changing water-flow patterns). Monty (1967) has observed that lamination is not strictly diurnal in Bahamian stromatolites either.

No clearly diurnal lamination was found in the cones. One example from Column Spouter shows possibly diurnal lamination, but was overgrown by *Calothrix coriacea* before the accretion experiment was completed. Three scales of lamination are usually discernible in the cones. Laminae about 1 μm thick are ubiquitous; these consist of single layers of silica-encrusted filaments and their thickness is determined by that of the filaments (0.7–1.5 μm). Up to six such laminae can form in 1 day. The other two scales of lamination may be distinct or diffuse and are not always present. Laminae 5–15 μm thick occur frequently. They are difficult to count but those from marked stromatolites in Conophyton Pool may be crudely diurnal. Frequently there are more laminae than days elapsed. The thickest laminae can be termed macrolaminae. In one experiment 7–8 macrolaminae 35–60 μm thick formed in 8½ months. Thus, formation of laminae in the cones is not clearly controlled by diurnal variations in light intensity.

Suggested Morphogenesis of the Conical Stromatolites

The following morphogenetic sequence (Figure 11.6) is a provisional interpretation. Everywhere in this section the word "filaments" refers to *P. tenue* var. *granuliferum*.

(1) A flat mat forms by the random gliding of filaments over their substrate.

(2) On agar, the *Phormidium* from conophytons often forms pronounced whorls, which probably initiate centers of conophyton formation. Randomly gliding filaments then become entangled, forming knots 100 μm or so wide and high, which project above the surface of the mats. Tangling due to gliding is a well-known phenomenon (Castenholz, 1969). A possible physiological reason for knot formation is given in the next section.

(3) Filaments gliding over the mats may encounter the knots and be deflected upward over them, toward the light source (the sun). This reorientation may initiate positive phototopotaxis and additional upward gliding may result. It is known that in at least some species of *Phormidium* there is no phototopotactic response when filaments are perpendicular to the light rays (as in a flat mat under water), but phototopotaxis does occur when filaments are in other orientations (Haupt, 1965). Self-shading within the knots of filaments may also initiate positive photophobotaxis. Thus tufts of erect filaments form above the knots.

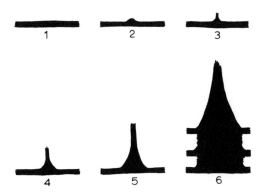

Figure 11.6. Proposed morphogenetic sequence of the Yellowstone conical stromatolites: (1) A mat of horizontal filaments; (2) Development of a knot of filaments due to tangling during gliding; (3) A tuft of erect filaments forms over the knot (due to phototaxis); (4),(5) Tuft enlarges; (6) Occasional lateral filament movement produces bridging laminae (some of which are suspended above the substrate). Silicification occurs continually during the growth process. (From Walter et al., 1976.)

(4) As a result of steps 1–3, the substrate is now dotted with vertical protuberances. Thus, increasing numbers of randomly gliding filaments are diverted up the sides of the tufts, which therefore become enlarged. At this stage the tufts are about 1–2 mm high. Tuft formation as described in stages 1–4 occurred within a 3-month period (from August to November 1972) in a small area of Conophyton Pool.

(5) There will be a strong tendency for the most actively gliding filaments to be concentrated in the tufts, because they are the most likely to encounter them. The most phototopotactic filaments would probably glide the farthest, to the tips of the tufts. Those that exhibit weaker phototopotaxis would be left at the base or on the flanks of the tufts. Thus the tufts become pointed, and maximum accretion is concentrated at the tips. In cone formation, the process of phototopotaxis is augmented by cohesion among the mucus-coated filaments (the filaments cohere like the hairs of a wet brush).

(6) The cones accrete upward at rates determined by the supply of nutrients and solar energy. Variations in these factors, particularly in light intensity, produce intermittent accretion, and therefore lamination.

(7) If the directions of gliding of the filaments in a flat mat with protruding knots are not entirely random, then more filaments may impinge on each knot from some directions than from others. Thus lateral ridges will form on the knots. Once a slight ridge has formed it will enlarge because it will deflect more and more filaments upward. This explanation is consistent with the occurrence of ridges converging on small tufts as well as on larger cones.

(8) Local entanglement of filaments can occur on cone flanks producing knots that may develop into spines.

(9) At cone tips, filaments tend to stand vertically, forming tufts or thickenings within the laminae. Filaments accumulate fastest at cone tips. The tips, therefore, are less silicified than the rest of the cone and so mechanically weaker, rendering them more susceptible to structural deformation.

The three essential processes in the morphogenetic sequence suggested here are gliding, phototaxis, and cohesion, all of which demonstrably occur in filaments of *P. tenue* var. *granuliferum*.

Studies on the Reasons for Node Formation by *Phormidium*

The striking conically shaped structures formed by *Phormidium tenue* var. *granuliferum* were of interest initially because of their relationship to Precambrian Conophytons, but the question soon arose as to why this organism produced such characteristic structures. The phenomenon of Conophyton formation is basically a morphogenetic one, and although the model described in the previous section provides a convincing picture of how cones grow, it does not suggest why *this Phormidium* comes together to form nodular structures in the first place. My initial (erroneous) hypothesis was that nodes might be an adaptation to a low-nutrient environment, the cooperative interaction of a large number of filaments leading to more efficient uptake of nutrients. Don Weller, an undergraduate student from Wabash College, had spent the summer of 1973 working with Doemel as a technician and returned in 1974. We felt that he was able to carry out a research project on his own, and suggested a study of the mechanism of node formation in the *Phormidium* Conophytons. He compared photosynthesis in intact nodes with that in nodes dispersed by gentle homogenization and found that homogenization strongly reduced photosynthesis, but this lost activity could be restored by addition of agents lowering the oxidation-reduction potential. Although bacterial photosynthesis can be stimulated by lowered oxidation-reduction potential, this had not been previously reported for blue-green algae. We hypothesized that formation of compact conical structures by the alga restricts oxygen penetration and creates the microaerophilic conditions apparently preferred by this algae, thus permitting effective growth in an oxidizing environment.

Most of the algal material for these experiments was collected at temperatures between 42°C and 47°C from the eastern effluent of Tromp Spring, a small pool in the Midway Geyser Basin of Yellowstone National Park, but a few experiments were done with material collected from several other nearby springs. A large cork borer was used to remove cylindrical cores of material from areas covered with nodes approximately 1–4 mm in height and 0.5–1.5 mm in diameter. The cores were transported in water-filled vials to the dissection site. Micro-scissors were used to remove each node at its base. The dissected material was stored in spring water until homogenization or distribution to 5-ml screw cap vials. Homogenates were prepared by gently dispersing the sample in spring water with a plastic homogenizing tube and Teflon homogenizer. The amounts of sample and water were calculated to give the same amount of material per vial as when intact nodes were used.

For experiments involving stimulation by sulfide or other reducing agents, all manipulations of material were done anaerobically. Immediately after dissection, the material was placed in spring water and continuously bubbled with nitrogen gas to remove and exclude oxygen. The nodes were then transferred to 5-ml serum bottles and gassing was continued for 10 minutes with nitrogen. Homogenization was accomplished by transferring dissected material to a plastic tube, bubbling with nitrogen for 10 minutes, and homogenizing under flowing nitrogen. Subsamples of anaerobic homogenates were introduced by injection through the stoppers of vials previously gassed for 10 minutes with nitrogen. Any reagents added were prepared in nitrogen-gassed water and injected through the serum stopper. Aerobic controls were prepared from the anaerobic materials by exposing open vials to atmospheric oxygen for 10–15 minutes during which the vials were frequently shaken to insure through aeration.

For field studies, following the dissection and preparation of vials (which took between 2 and 5 hr for each experiment), photosynthesis was measured with $^{14}CO_2$ by our standard methods. Some vials received DCMU [3-(3,4-dichlorophenyl)-1,1-dimethylurea], an inhibitor of algal but not bacterial photosynthesis, to give a final concentration of 1×10^{-6} M in the incubation vial.

Photosynthesis in Intact and Dispersed Nodes

In initial studies to determine why this alga forms nodes during growth, experiments were performed comparing photosynthesis in intact and dispersed material. As seen in Figure 11.7, photosynthesis (normalized to chlorophyll content) was completely inhibited in dispersed suspensions as compared to intact nodes. Although the dispersion process was gentle, and microscopic examination did not reveal detectable cell breakage, experiments were set up to see if dispersion might induce cell leakage. Only an insignificant amount of incorporated radioactivity was released upon dis-

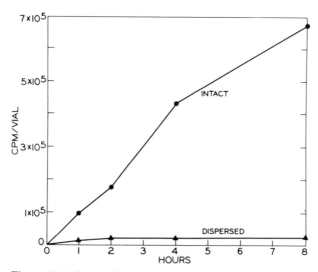

Figure 11.7. Rate of light-dependent uptake of $^{14}CO_2$ by intact and dispersed nodes. Each vial contained about 1.8 μg/ml chlorophyll-a and 6 μCi/ml NaH$^{14}CO_3$. Counted with a liquid scintillation counter. All incubations in air. (From Weller et al., 1975.)

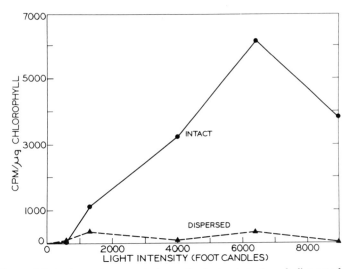

Figure 11.8. Effect of light intensity on photosynthesis by intact and dispersed nodes. Each vial contained about 1.5 μg/ml chlorophyll-a and 0.6 μCi/ml NaH^{14}CO$_3$. Incubation time, 1 hr. Counted with gas-flow counter. All incubations in air. (From Weller et al., 1975.)

persion of intact nodes. Also, supernatants of centrifuged dispersed material did not show detectable chlorophyll, which might have been released during cell breakage. We concluded that the dispersion process itself was not damaging the organism.

Another consequence of dispersion was that algal filaments from the interior of the structure, grown at a low-light intensity due to self-shading, would be exposed to bright light when evenly dispersed in the suspension, so that light inhibition of photosynthesis might occur. An experiment was therefore performed in which the rate of photosynthesis was measured at various light intensities. As seen in Figure 11.8, intact material showed significantly higher photosynthesis at all light intensities than did dispersed material, suggesting that the main inhibitory effect of dispersion was not exposure to bright light.

Stimulation of Photosynthesis by Sulfide

Because of the compact nature of the *Phormidium* nodes, it seemed possible that oxygen diffusion might be inhibited, allowing partial anaerobic conditions to develop within the structures. Consequently, a series of experiments was performed to measure photosynthesis under conditions of lowered redox potential. Because the photosynthetic bacterium *Chloroflexus* was also present in these structures, and because this organism shows a significant stimulation of photosynthesis by sulfide, it was essential to measure differentially bacterial and algal photosynthesis. This was done by adding DMCU to some of the vials in each experiment to inhibit algal photosynthesis without inhibiting bacterial photosynthesis. Control experiments had shown that, at a concentration of 1×10^{-6} M, DCMU inhibited greater than 90% of algal photosynthesis, without detectably affecting bacterial photosynthesis. It was thus possible to calculate algal photosynthesis by subtracting the photosynthetic rate measured in the presence of DCMU from that measured in its absence. As seen in Table 11.2, algal photosynthesis was stimulated significantly by anaerobic conditions (N_2 atmosphere) and even more strikingly by sulfide. Although the intact nodes also showed stimulation by N_2 and sulfide, stimulation was considerably greater in the dispersed material. Addition of N_2 and sulfide restored photosynthetic ability of the dispersed material to levels higher than those of intact nodes in air, confirming that the dispersion process was not damaging the algal filaments. Thus, the principal effect of dispersion is probably to aerate the material and a probable secondary effect is to dilute the endogenous sulfide levels in the intact nodes, since sulfide is present in the natural nodes (see below).

To further investigate the sulfide stimulation of photosynthesis in this blue-green alga, an experiment with a series of sulfide concentrations was performed. As seen in Figure 11.9, sulfide stimulation is concentration dependent, with an optimum at about 2.3 μg/ml sulfide (72 μM). A higher concentration was inhibitory, but significant stimulation occurred even at quite low concentrations (around 0.1 μg/ml). Although not shown in the figure, bacterial photosynthesis was also stimulated by sulfide, with an

Table 11.2. Effect of Anaerobic Conditions and Sulfide on Algal Photosynthesis[a]

Intact nodes	cpm/μg chlorophyll-a
Air	220
N_2	473
N_2 + S^{2-}	2449

Dispersed nodes	cpm/μg chlorophyll-a
Air	0[b]
N_2	70
N_2 + S^{2-}	585

[a]Bacterial photosynthesis (DCMU resistant uptake) has been deducted. Sulfide concentration, 0.67 μg/ml (20 μM).
[b]Uptake of radioactivity in light was the same as uptake in the dark. (From Weller et al., 1975.)

Figure 11.9. Stimulation of algal photosynthesis by sulfide. All vials contained dispersed material. Each vial contained about 1.6 μg/ml chlorophyll-a and 0.4 μCi/ml NaH^{14}CO$_3$. Incubation time, 1.5 hr. All preparations and incubations were done in a nitrogen atmosphere. Sulfide concentrations are those actually measured by chemical assay at the time sulfide additions were made. Bacterial photosynthesis (DCMU resistant uptake) has been deducted. In the same experiment, uptake of ^{14}C by the alga in air instead of N$_2$ was lower (158 cpm/μg chlorophyll-a. (From Weller et al., 1975.)

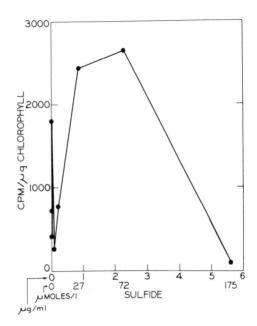

optimum around 1 μg/ml, but inhibition did not occur at the high-sulfide concentration.

Effect of Lowered Oxidation-Reduction Potential on Photosynthesis

In order to determine if the stimulation of algal photosynthesis by sulfide was due to the ability of this agent to lower the redox potential, several other substances which lower redox potential were also tested. As seen in Figure 11.10, stimulation was observed with all of these agents, although the concentration that gave best stimulation varied among the different agents. Both sulfide and sulfite were toxic at the highest concentration, whereas the two organic reducing agents, thioglycolate and cysteine, were nontoxic but less effective, since significant stimulation required considerably higher concentrations. Two other reducing agents, ascorbic acid and thiosulfate, did not show significant stimulation at any of the concentrations tested. Bacterial photosynthesis was not stimulated significantly by any of the agents tested except sulfide (data not shown).

Redox potential measurements with an Orion redox electrode of the agents used above showed that all of the agents had potentials that were either negative or near zero at concentrations that stimulated algal photosynthesis. Because of the range of compounds that stimulated photosynthesis, it seems likely that the effect is due to the lowering of oxidation-

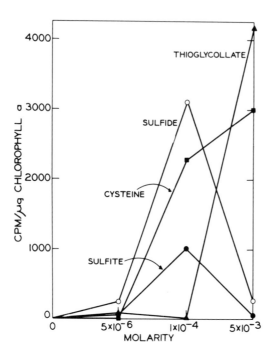

Figure 11.10. Stimulation of algal photosynthesis by reducing agents. All vials contained dispersed material. Each vial contained about 1 μg/ml chlorophyll-a and 0.4 μCi/ml $NaH^{14}CO_3$. Incubation time, 2 hr. All preparations and incubations were done in a nitrogen atmosphere. Bacterial photosynthesis has been deducted. In the same experiment, uptake of ^{14}C by the alga in air instead of N_2 was 18 cpm/μg chlorophyll-a. (From Weller et al., 1975.)

reduction potential rather than to a specific effect of these compounds on algal photosynthesis.

Sulfide in the Natural Mats

In considering sources of reducing agents in the natural environment, attention was directed to the nodes themselves. Although the water of the springs studied is low in sulfide, it seemed possible that sulfide formed by sulfate-reducing bacteria might be present in the nodes. Consequently, sulfide assays were performed on nodes of different sizes to obtain some idea of the range of sulfide concentrations present. Concentrations of sulfide were found in these nodes that were similar to those stimulating photosynthesis in dispersed material.

Stimulation of Photosynthesis in Axenic Cultures

Experiments were performed to determine if axenic cultures of the *Phormidium* would also show stimulation of photosynthesis by lowered oxidation-reduction potential. Preliminary experiments with 3 of the 15 cultures isolated showed stimulation by anaerobic conditions and by sulfide similar to those seen in the natural material. More detailed studies were performed with a single culture, strain 52-12. It was found that aeration greatly inhibited photosynthesis, and both sulfide and cysteine strongly stimulated photosynthesis above the levels seen with N_2 alone. The concentrations of sulfide and cysteine which stimulated photosynthesis with the pure culture were similar to those stimulating photosynthesis in the natural material.

Significance

Cultures of this *Phormidium* grow rapidly and profusely under anaerobic conditions, although anaerobic conditions are not required for growth. However, a constant observation during the isolation and culture of these strains has revealed the tendency of this organism to grow as compact nodes or clumps on agar or in liquid culture. On agar, single hormogonia may glide away from a clump, but soon form tight whorls or knots which eventually develop into small nodes. In liquid culture, these nodes can grow to quite a large size. It seems reasonable that in these nodes oxygen diffusion is restricted and reducing conditions may develop even though the alga would produce O_2 during photosynthesis. These observations may well explain the reason why this *Phormidium* forms, apparently specifically, Conophyton-like structures. It may initially form knots and clusters because of the benefit it obtains from restricting O_2 penetration, but then as the size of the nodes increases, upward migration occurs by the mechanism described earlier, resulting in the formation of conical structures. The formation of Precambrian Conophytons could have been carried out by blue-green algae with similar characteristics. Indeed, it is conceivable that the first blue-green algae to evolve might have been unusually O_2 sensitive, so that a mechanism such as node formation might have been essential for survival. It is well established that the first enzyme in the photosynthetic carbon cycle for CO_2 fixation, ribulosediphosphate carboxylase, is affected by O_2, and O_2 inhibits CO_2 fixation. This O_2 inhibition seems to be an inherent property of the RuDP carboxylase, so that an ability to avoid O_2 toxicity could have been of evolutionary significance.

Bacterial Stromatolites

While the above work dealt with a type of stromatolite that showed considerable vertical relief, providing a modern analogue for the important Precambrian Conophyton, the rest of this chapter deals with a laminated mat with little or no vertical relief, and with no obvious Precambrian analogue (Doemel and Brock, 1977). Its primary interest rests in the fact that the dominant structural component is a photosynthetic bacterium, *Chloroflexus aurantiacus,* rather than a blue-green alga. It has been conventionally assumed that Precambrian stromatolites were formed by filamentous blue-green algae, yet the existence of the *Chloroflexus* mats shows that they can also be formed by filamentous bacteria. The presence in the fossil record of stromatolites has conventionally been equated with the appearance in the atmosphere of O_2, since it is well established that the initial rise in O_2 in the Precambrian was due to blue-green algal photosynthesis. *Chloroflexus* mats are thus of interest partly because the organism does not produce O_2 and, additionally, because it resembles morphologically a thin blue-green alga, and if seen in the fossil record could well be classified by the unwary as such an organism.

The distinction between the *Chloroflexus* mats under study in this section and the Conophyton-like stromatolitic structures described above should be emphasized. The Conophyton-like structures are formed by an association of a filamentous blue-green alga *Phormidium tenue* and *Chloroflexus,* but in most cases the alga is dominant and is capable of forming the structures in the complete absence of the bacteria. The *Chloroflexus*-dominated mats are generally flat rather than conical, although nodular structures are sometimes formed. Detrital silica derived from the siliceous sinter of the geyser basins is frequently incorporated into the *Chloroflexus* mats, being carried onto the mats with runoff from occasional thunderstorms. Between storms, the mats are relatively undisturbed, and *Chloroflexus* migrates rapidly around and on top of the detrital material that falls on a mat. *Chloroflexus* is thus a sediment-trapping organism as are the stromatolite-forming blue-green algae.

Morphology of *Synechococcus-Chloroflexus* Mats

A typical *Synechococcus-Chloroflexus* microbial mat is in a "tidepool" of Octopus Spring (Figure 11.11) and in the stream flowing from this pool.

Figure 11.12 illustrates the distinctive laminations of these mats. To better understand the distribution of microorganisms within the mat, microscopy was done on vertical sections (prepared with a razor blade) of fresh cores embedded in 2% agar. Because of the opacity of such sections, visualization was done with vertical fluorescence illumination. The *Synechococcus* could be observed directly with this microscopic technique because of the autofluorescence of chlorophyll-a. The sections were then stained with a 0.1% solution of the fluorescent dye acridine orange; with this stain, bacteria fluoresce either red or green. Also samples were removed from the various laminations and were examined directly with phase and phase-fluorescence illumination.

In the upper layer, 0.2–1 mm in thickness, *Synechococcus* and *Chloroflexus* predominate. Often there was a second layer, 0.1–0.6 mm thick, immediately beneath this upper layer, which was similar except that it was a dark blue-green and the autofluorescence of the *Synechococcus* was much more intense. Beneath this region, *Synechococcus* was rarely observed, even empty cells, but there was an abundance of *Chloroflexus* in this next layer, 0.8–1.4 mm thick. Below about 3–5 mm, most of the filamentous bacteria appeared moribund and unicellular rods were concentrated in bands having sulfide-forming activity, as revealed by the ferrous

Figure 11.11. The Octopus Spring mat. The source is in the background and the mat is in the foreground, in the ''backwater'' behind the sinter dike.

Figure 11.12. A macrophoto-graph of a complete core, showing laminations. This is one from the Octopus Spring mat. Note the surface nodules (N), the upper dark layer (D), and the laminae below this layer. Often there are bands of sinter (Si) in the core (coarse white zones). (From Doemel and Brock, 1977.)

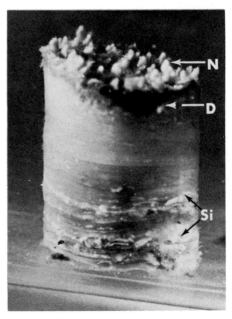

ammonium sulfate technique (Doemel and Brock, 1976), suggesting that at least some are sulfate-reducing bacteria.

The Role of *Chloroflexus* in the Structure of the Mat

Since microscopy of isolated nodes, colonies, and streamers revealed that all of these structures were composed predominantly of bacterial filaments with some *Synechococcus,* the composition of these structures was further investigated with ^{14}C-bicarbonate incorporation, chlorophyll analysis, and autoradiography. These observations with colonies, nodules, and stream-ers demonstrate that a large proportion of the population in these structures is photosynthetic bacteria. Since bacterial filaments are abundant while other bacteria are few and since *Chloroflexus* has been isolated from these structures, this is indirect evidence supporting the structural role of *Chloroflexus.*

If the mat is held together by *Chloroflexus,* then removal of the *Synecho-coccus* should not significantly alter the structure of the mat. Preliminary experiments demonstrated that when a microbial mat was darkened, within 24 hours the mat turned orange in color and this orange layer was enriched in *Chloroflexus.* Also, when the photosynthetic efficiency of microbial mats as a function of light intensity was determined with and without DCMU, at low-light intensities *Chloroflexus* was apparently more effective than *Syne-chococcus* at utilizing the light (Figure 11.13). This observation implied that

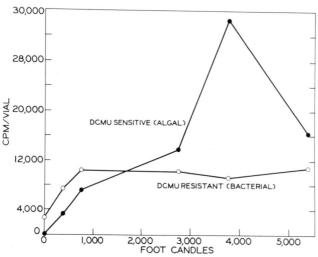

Figure 11.13. Photosynthetic rate as a function of light intensity. Experiment done on homogenates prepared from one colony collected at Toadstool Geyser. Average protein concentration per vial, 314 μg.

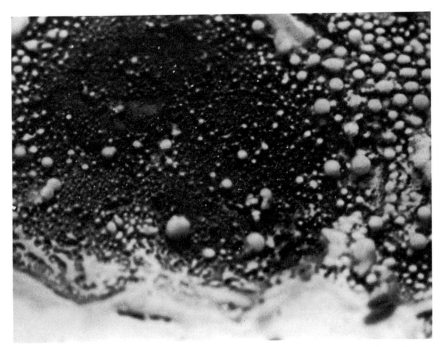

Figure 11.14. Pinkish-orange nodules of *Chloroflexus* which developed in 24 hours on top of a silicon carbide circle laid down in Toadstool Geyser. A neutral density filter over the mat had reduced the light intensity to 90% of full sunlight. Diameter of circle on photograph, about 85 mm. (From Doemel and Brock, 1977.)

at low-light intensities, *Synechococcus* would not be able to maintain a population and *Chloroflexus* would predominate. To test this, an area of the mat at 56°C in the stream flowing from Toadstool Geyser was covered by two layers of a dense nylon cloth held on a frame which reduced the incident light by 98%. At the same time, a circle of 120 mesh silicon carbide was placed on the mat as a marker and substratum. Within 4 hours small nodules had formed on the surface of the mat and on the surface of the silicon carbide and within 24 hours a number of pinkish-orange nodules were present (Figure 11.14). Microscopy showed that *Synechococcus* was absent from the nodules and internodular regions, yet the nodules were similar to normal nodules. Uptake of ^{14}C-bicarbonate by homogenates of these nodules incubated with and without DCMU had equivalent radioactivity, suggesting that the algae were not contributing significantly to the primary production of these nodes. Furthermore, autoradiograms prepared of this material indicated that all of the incorporation of ^{14}C-bicarbonate was by bacterial filaments. Together these observations add direct evidence to support the notion that *Chloroflexus* is the structural component of the mat and that it furthermore accounts for the growth of the mat.

Light Penetration into Mats

The depth of light penetration into the mat was estimated directly by use of an artificial light source and a photometer. The top green layer and an underlying orange layer of similar thickness were used. These sections were placed on glass slides and the reduction of light energy by each of the sections was determined. The results of a typical core are shown in Table 11.3A. The green layer reduced the radiation by 94% and the orange layer reduced the radiation by about 91%. To check for attenuation by photosynthetic bacteria, in some experiments the white light was passed through an infrared (IR) filter before passing through the mat. Although the reduction of infrared radiation was greater in some cores, the difference was small so that the reduction of both tungsten and IR radiation was approximately the same. Thus, in these compact mats, light attenuation is not only by absorption by pigments, but also by scatter. Indeed, because there was no difference in attenuation by tungsten and IR radiation, it seems likely that scattering causes more attenuation than pigment absorption.

On a bright cloudless day, light intensities may reach 8400 footcandles at noon. Since the thin layer of water overlying the mats, 1–5 cm at the Octopus Spring mat, does not significantly reduce the light intensity, the

Table 11.3. Reduction of Radiant Energy by the Green and Orange Layers of a Microbial Mat[a]

A. Light source	Light alone	With green layer	Percent reduction	Light alone	With orange layer	Percent reduction
White	1.0	0.057	94	1.0	0.097	91
Infrared	0.76	0.025	97	0.89	0.080	91

B. Location	Calculated light intensity (footcandles) at the bottom of	
	Green Layer	Orange Layer
Mat (50°C)	344	45
Mat (49°C)	258	23
Mat (54°C)	258	46
Outflow (59–65°C)	462	42
Outflow (50–54°C)	258	18
Mat (58°C)	258	36

[a]Samples from Octopus Spring (50°C). In part A, the values are gcal/cm². The values for Part B were calculated by assuming a light intensity of 8400 footcandles (maximum intensity at midday). The light intensity passing through the green layer (0.2–0.5 mm thick) was determined by subtracting the radiant energy attenuated by the green layer from the total energy. The light intensity at the bottom was determined by multiplying the light passing through the green layer by the percentage reduction by the orange layer and subtracting this from the light passing through the green layer (total distance = 0.4–1 mm). (From Doemel and Brock, 1977.)

light energy passing through the green layer and reaching the orange layer can be calculated (Table 11.3B). About 250–350 footcandles pass through the green layer, which is probably sufficient to support autotrophic growth, but only 18–45 footcandles pass through both layers, about 1 mm thick. Thus, below the green and the top orange layers, the mat is essentially dark, implying that the light penetrates only about 1–2 mm into the mat.

Effect of Light Intensity on Algal and Bacterial Photosynthesis

Although light does reach the orange layer of the mat, *Synechococcus* is apparently restricted to the green layer and even cell ghosts of *Synechococcus* are not present in the orange layer. If the green layer of a core is homogenized, and the rate of photosynthesis as a function of light intensity is determined, the photosynthetic rate of *Synechococcus* is significantly reduced below that of *Chloroflexus* at intensities less than 1000 footcandles (Figure 11.13). At maximum light the rate of photosynthesis by *Synechococcus* is reduced, with the optimum being at about 4000 footcandles. Probably only a thin upper layer of cells is adapted to high light since there is a distinct darker blue-green band immediately below the surface and also since *Synechococcus* cells below the thin upper layer autofluoresce more strongly when excited with blue light. On the other hand, photosynthesis by *Chloroflexus* is light saturated at all intensities above about 100 footcandles. *Synechococcus* may be restricted to the upper layer simply because it cannot efficiently utilize low-light intensities.

Chloroflexus is present in both the green layer and in the orange layer of the mat (Bauld and Brock, 1973). However, the peak concentration of bacteriochlorophyll-c is apparently highest in the upper region of the mat where *Synechococcus* is also present (Bauld and Brock, 1973). The *Chloroflexus* in these upper regions also appears to incorporate ^{14}C-bicarbonate more efficiently (Table 11.4) than the *Chloroflexus* in the lower regions. On a protein basis, 57% of the ^{14}C-bicarbonate incorporated by *Chloroflexus* is in the nodes and in the upper region of the mat. Furthermore, if it is conservatively assumed that 50% of the protein in these regions is derived from *Synechococcus*, the contribution of *Chloroflexus* in these regions increases to 73% of ^{14}C-bicarbonate incorporation by *Chloroflexus* in all layers. These observations, together with the observations on light attenuation in the mats, suggest that the primary photic zone is restricted to the upper 0.5–1 mm of mat. Since *Chloroflexus* apparently is the structural component of the mat and accounts for the growth of the mat, the growth region probably coincides with the region where *Chloroflexus* is most active.(See Chapters 7 and 8 for discussion of light adaptation by *Chloroflexus* and *Synechococcus*.)

Table 11.4. Photosynthetic Efficiency of *Synechococcus* and *Chloroflexus* in the Various Layers of a Microbial Mat[a]

	Uptake of ^{14}C-bicarbonate			
	Algal photosynthesis		Bacterial photosynthesis	
Layer	cpm/unit Chl-a	cpm/μg protein	cpm/unit Bchl-c	cpm/μg protein
Nodules on mat surface	53,500	50.3	32,100	6.55
Blue-green upper layer	92,300	60.1	34,100	14.2
Orange layer below blue-green layer	No incorp.	No incorp.	6,780	11.4
Pale orange layer	No incorp.	No incorp.	1,580	4.4

[a]Nodules, 0.2–0.5-mm projections from the mat surface, were harvested from the Octopus Spring mat by aspiration with a Pasteur pipette fitted with a rubber bulb. The upper 1.5-mm portion of a microbial core, lacking nodules, was sectioned into three 0.5-mm layers: a blue-green layer, an orange layer, and a pale orange layer. Photosynthesis was measured on homogenates of each layer. (From Doemel and Brock, 1977.)

Diurnal Growth and Response of *Chloroflexus* to Reduced O_2 Concentrations

In some experiments, covers that excluded all light were placed over the mats. The covers were placed at 0930 hours; within 5 hours (1415) the mats were noticeably pinker than the mats in full light, indicating that the *Chloroflexus* had responded to darkness by moving up through the algal layer. This experiment was repeated several times with similar results. Because of the rapid response, it seemed likely that similar migration of *Chloroflexus* would occur every night. To confirm this, observations were made throughout the night by using a flashlight and taking color photographs with electronic flash. Analysis of the photographs showed that the mat did become pink at night and became green again during the next morning. Confirmatory evidence was obtained by placing thin sheets of lens paper on mats either in the evening or early in the morning. *Chloroflexus* migrated up through the mesh of the paper and microscopy of the paper could be used to determine if differential migration occurred. Lens papers removed in the morning after overnight placement generally had predominantly *Chloroflexus,* whereas in papers placed in the morning and removed at the end of the day, *Synechococcus* was predominant.

It seems possible that the upward migration of *Chloroflexus* in response to low light or darkness, and the lack of migration during daylight, could account for the origin of the laminated mats. It has been shown (Pierson and Castenholz, 1971) that *Chloroflexus* is capable of moving at rates of 30–150 μm/hour, and hence an accretion rate of 0.1 mm/day (100 μm/day), as measured by carborundum marking, is possible. It is of interest that work with blue-green algal stromatolites (Gebelein, 1969) has shown that upward migration of algal filaments occurs in the day, followed by horizontal growth at night. Work on blue-green algal stromatolites in Yellowstone hot springs (Walter et al., 1976) has shown a similar migration pattern. Thus, the migration of the bacteria is exactly opposite to that of the blue-green algae, yet the result is the same, the formation of a laminated mat.

The upward migration of *Chloroflexus* at night could be a response to low light or to reduced O_2, since under the low-light conditions the algae are not photosynthesizing and hence will not produce O_2. These alternative hypotheses were tested by incubating cores collected from the Octopus Spring mat in vials containing either spring water or culture medium to which in some instances a suspension of DCMU had been added to a final concentration of 10^{-5} M. Vials containing cores were also incubated without DCMU in the light and dark. Since DCMU inhibits algal production of O_2, this experiment should test the response of *Chloroflexus* to reduced O_2. The cores were incubated from 1 to several days at 56°C in a stream from Octopus Spring. Within 1 day, the surfaces of the cores incubated in dark vials and in light or dark with DCMU were orange-red, due to the surface accumulation of *Chloroflexus,* while the surface of cores incubated in light vials remained a light green. The oxygen concentration in the DCMU vials incubated in the light was about 0.7 mg/l compared to 1.5 mg/l in the controls without DCMU. Algal incorporation of ^{14}C-bicarbonate in the DCMU vials did not occur. It is apparent from these observations that the absence of algal O_2 production rather than the light stimulates the migration of *Chloroflexus*.

The response of *Chloroflexus* to O_2 was further tested by placing cores in vials containing culture medium and then incubating them at 56°C in the dark. One set was incubated with air being bubbled continuously; the other set was not bubbled. After 24 hours, the surface of both sets of cores was red, suggesting that oxygen did not repress movement. However, pure oxygen was not used in these experiments. The production of oxygen was tested indirectly by measuring the oxygen in the water over the algal mat at 6:00 A.M. and at 10:00 A.M.. At 6:00 A.M., 0.14–0.17 mg O_2/l was detected and at 10:00 A.M. 0.30 mg/l was detected. Clearly oxygen levels are low, lower than in most aquatic habitats at normal temperatures. The samples were cooled to ambient temperature before adding the reagents so it is doubtful that oxygen was lost from the system; if there is an error, oxygen levels should be elevated because of introductions of air. Although O_2 was low in the water, in a dense algal mat oxygen levels in the microenvironment at the surface may be quite high.

Mechanism of Mat Growth

Observations of the Octopus Spring mat over many years have shown that the thickness of the mat has remained constant. However, if the mat is in steady state, the growth rate will equal the death rate and no change would be obvious. To study the growth of the microbial mat, the growth region has to be isolated in some fashion from the remainder of the mat. The growing region of the mat was isolated by placing substrata on the mat that were presumably neutral. In this study, plastic and glass plates, wooded channels, glass microscope slides, and silicon carbide chips were used. Each of these substrata eventually supported dense populations of microorganisms but the initial colonization of the horizontal plates and channels was on the order of weeks compared to the rapid colonization of vertical slides and silicon carbide chips which occurred within hours. The plates and channels differed from the other two substrata in that they could not be colonized from the underlying mat. Colonization had to proceed either by migration from the surrounding mat or by the sedimentation of organisms suspended in the water. Always, on horizontal substrata, colonization proceeded from the borders of the substrata inward until they were covered.

Glass slides placed vertically into the mat were rapidly colonized and although not suitable for the study of growth they provided an estimate of the migration rate. On slides present in the spring for 5 days, vertically oriented filaments were present to about 13.5 mm above the surface of the mat. Assuming that cells present at that level migrated from the mat, the migration rate, 0.11 mm/hour, is similar to the rate of gliding by *Chloroflexus* on agar (Pierson and Castenholz, 1974).

The most suitable substratum for the study of mat growth was silicon carbide. When placed onto the mat, silicon carbide is rapidly colonized by *Chloroflexus* from beneath. Within 4–5 hours orange material can be seen on the surface. The only observable effect of the silicon carbide is that the mat surface sometimes appears to be darker in color than the surrounding mat and can be readily distinguished from the surrounding mat for several weeks. However, the mats above the silicon carbide are level with the surrounding mat, suggesting that growth is not stimulated. As a result of continued growth, the silicon carbide becomes buried deep within the mat (Figure 11.15) and serves as an excellent marker of mat growth and decomposition (see below).

In many stromatolitic systems, the active or passive deposition of inorganic material, particularly silicates or calcium carbonate, contributes significantly to the increase in mat thickness (Golubic, 1973). Although siliceous sinter is always a component of the microbial mat at Octopus Spring, it appears to be minor (see Figure 11.12). The absence of a significant inorganic contribution would be suggested if the height were a function of the organic matter in the mat. Figure 11.16 illustrates that in the

Figure 11.15. Vertical section of a core from Toadstool Geyser, showing a silicon carbide (C) layer. The silicon carbide had been placed on the surface of the mat 29 days before and is now at a depth of 4 mm into the mat. The silicon carbide layer is about 0.2 mm thick. (From Doemel and Brock, 1977.)

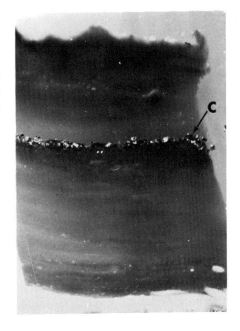

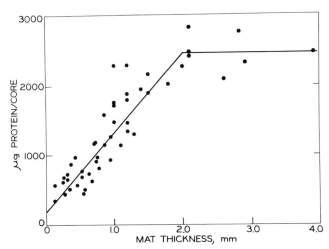

Figure 11.16. The relationship between protein concentration and thickness of mat. Measurements were made of the thickness and protein concentration of the microbial mat that accumulated on the surface of silicon carbide layers in various mats in the White Creek area after 31 days. Diameter of cores, 8.4 mm. (From Doemel and Brock, 1977.)

upper 2 mm there is a direct correlation between mat height and protein concentration; in mats thicker than 2 mm, there no longer is a correlation. Not only does this observation support the essential organic nature of these mats but it also points again to the limits of the growth and photic zone, since height would not necessarily correlate with protein in the lower (decomposing) system. It was shown earlier that essentially no light penetrated below 2 mm in the mat.

Growth Rates

The rate of increase of the mat on the surface of silicon carbide should provide an estimate of the actual growth rate if washout of cells, decomposition, and settlement were minimal. Silicon carbide was placed at a number of different stations in a small backwash area of Octopus Spring. Cores were collected at 8-, 14-, 21-, and 27-day intervals and the protein was determined. Stations having the same average temperatures were averaged together. Figure 11.17 summarizes the data from these experiments. The apparent doubling time ranges from 5 to 13 days with an apparent optimum at 54°C. The relatively rapid increase of the 54°C and 56°C stations followed by a slow period of increase suggests that these may be approaching a steady-state level where growth is balanced by decomposition.

Longer term studies extending over a year support this hypothesis (Table 11.5). Protein concentration appears to be maintained at a relatively stable level, although height above silicon carbide continues to increase. A similar conclusion can also be drawn from Figure 11.16 which shows that above 2 mm, although height continues to increase, the protein concentration remains relatively stable. These observations suggest that at steady state the growth rate must be balanced by an equivalent rate of washout, decomposition, or predator consumption. In this mat, there is little washout

Table 11.5. Long-Term Growth of the Octopus Spring Mat[a]

Station	Date	Days	Height above silicon carbide (mm)	μg protein/ core (above silicon carbide)	Height increase μm/day	Protein increase μg/day
45–7	7/5/74–8/5/74	31	1.0	1540	32	50
·	8/5/74–5/26/75	295	5.3	1790	18	6.1
5	7/5/74–8/5/74	31	1.4	1220	45	39
	8/5/74–5/26/75	295	5.3	1750	18	5.9

[a]A silicon carbide layer was placed at each station on 5 July 1974. On 5 August 1974 the height and protein of the microbial mat above this layer of silicon carbide were measured on a core 8.4 mm in diameter. A second layer of silicon carbide was placed at each station on 5 August 1974 and the height and protein were determined on 6 May 1975. (From Doemel and Brock, 1977.)

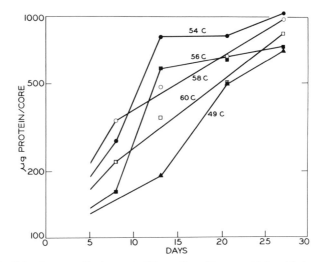

Figure 11.17. Growth of the Octopus Spring mat. Time after silicon carbide added. Data for several stations at each temperature have been pooled. Apparent doubling times as calculated from the graphs: 60°C, 10–11 days; 58°C, 13 days; 56°C, 6 days; 54°C, 5 days; 49°C, 10 days. (From Doemel and Brock, 1977.)

and eucaryotic organisms are absent, so in this system, growth can only be balanced by bacterial decomposition.

Figure 11.17 also suggests that there may be an optimal temperature of about 54°C. Short-term studies at more temperatures in the Octopus Spring mat support this conclusion. The data points are obtained by measuring the height over silicon carbide after 31 days; then, a second layer of silicon carbide was placed and the height over this layer was measured after 31 days. Assuming that the rate of increase is constant during this period, the average μm/day was calculated. Despite considerable variability in growth rates, the optimum was between 52°C and 56°C.

In other *Chloroflexus-Synechococcus* microbial mats in springs and streams in the White Creek Valley and in the vicinity of the Firehole Lake Loop Road, although the apparent average growth is sometimes less than at Octopus Spring, there is a similar temperature optimum (Table 11.6). The mats at Ravine Spring, Buffalo Pool, and Sulfide Spring all have apparent growth rates that are apparently considerably less than the growth rates at either Octopus Spring or Twin Butte Vista.

In the Ravine Springs group the efficiency of photosynthesis is also reduced compared to Octopus Spring. The uptake per unit chlorophyll-a or bacteriochlorophyll-c at Octopus Spring is about 20 times greater than at Spring 74-6 at the same temperature. Similar lower photosynthetic efficiencies were observed in the mats in other of the Ravine Springs group. Since the chemistry of these springs has not been studied, it is not clear what factors are reducing the rate of growth.

Table 11.6. The Growth Rate of Microbial Mats in Alkaline Thermal Springs, as Determined by Silicon Carbide Marking

Spring	pH of pool	Temperature (°C) at station	Range of growth rate (μm/day)	Average growth rate (μm/day)
Buffalo Pool	9.3	59	14–19	17
Spring 74-3 Main Pool	8.0	64	9.2–32	22
Main Pool		66	3.9–9.0	6.4
Spring 74-4 Main Pool	9.2–9.5	62	3.4–18	11
Ravine Spring Main Pool	8.0	63	15–28	23
Effluent		67	11–48	30
Effluent		62	17–26	23
Spring 74-6 Main Pool	7.7	65	18–23	21
Effluent		70	3.6	3.6
Effluent		62	13–14	14
Effluent		57	23	23
Twin Butte Vista Pool	8.0	60	15–29	22
Effluent		61	9.5–12	11
Effluent		57	33–55	44
Effluent		56	88–110	94
Serendipity Spring	9.4	66	5.6	5.6
Effluent		58	13	13
Effluent		58.9	14	14
Effluent		60	27	27
Sulfide Spring				
Pool	6.8	46	(only 1 determination)	14
Pool		56		12
Pool		47		14
Effluent		48		14

(From Doemel and Brock, 1977.)

Decomposition of Microbial Mats

The *in situ* rates of decomposition of microbial mats were determined with a variation of the procedure used to measure the growth of the mat. At the completion of a study of mat growth, another layer of silicon carbide was placed onto the mat immediately over the earlier layer. At intervals further cores were collected and measurements of height and protein were made on both the mat that had grown on the surface of the top layer of silicon carbide and on the mat between the first and second layer of silicon carbide. Decreases in protein and height of the material between the two silicon carbide layers provides a measure of decomposition rate. In some cases, additional layers of silicon carbide were made. Thus, at one station, growth of the upper microbial mat and decomposition of several lower layers could be measured. Figure 11.18 is a diagram of a core from one of these stations. Malcolm Walter placed the first two layers of silicon carbide

Figure 11.18. Long-term silicon carbide marking of a mat. This is a diagram of a core collected from Station 45-7, Octopus Spring, on 5 August 1974. Silicon carbide layers (shaded portions) were placed on the mat at various dates as indicated. Bands of microbial mat (unshaded) are lettered. The upper layer (E) accumulated between July and August 1974. The bands of silicon carbide were about 0.7 mm in thickness. The thickness of all layers is to scale. (From Doemel and Brock, 1977.)

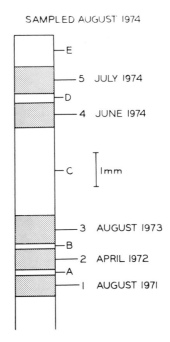

SAMPLED AUGUST 1974

(1 and 2) in August 1971 and in April 1972, respectively. Additional layers of silicon carbide were added in August 1973 and in June and July 1974. The bands of microbial mat between layers 1, 2, and 3 have decreased in size considerably, assuming that these bands were at least as thick as band C. The silicon carbide technique thus enables the long-term study of decomposition of a microbial mat *in situ*.

Decomposition Rates

The rate of decomposition of mat over a period of 1 year is illustrated in Figure 11.19. Layer A decreased to about 30% of its original thickness in about a year. This layer had accumulated during 9 months and was quite thick. Two thinner layers of mat, B and C, no longer can be distinguished within 9 months. These observations suggest that within the mat, the deposition of organic materials is minimal and also that although the rate of decomposition for at least a portion of the material is quite slow, eventually decomposition is complete.

The observations of various decomposing bands seemed to indicate that within a relatively short time, there was a significant decrease in both height and in protein. As seen in the inset to Figure 11.19, initially there is a rapid decrease of material, with an apparent half-time of about 1 month. Then, there is a slower period of decomposition, with an apparent half-time of 12 months. This same biphasic process of decomposition was observed on all of the mats examined—a rapid initial rate of decomposition followed by a slower rate.

The decomposition of the Octopus Spring microbial mat appears to be a function of temperature (Figure 11.20). The decomposition rates of bands of mat which had accumulated between June and July 1974 were measured 31 days after the placement of a covering layer of silicon carbide. Fourteen different sites on the Octopus Spring mat with temperatures ranging between 42°C and 70°C were studied. All of the mats were covered with water and insects were absent. Although there is considerable scatter of points, the optimum temperature for decomposition appears to be between 52°C and 56°C (Figure 11.20).

In all of these studies of cores, the height and protein content of the mat over the underlying siliceous sinter appear to remain constant, at least during a year. This suggests that the growth rate approximates the decomposition rate. Table 11.7 summarizes observations of decomposition in a

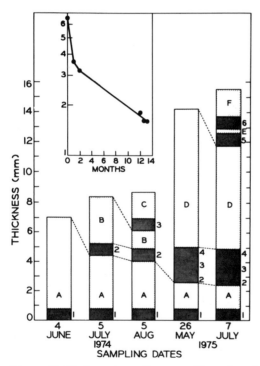

Figure 11.19. Long-term decomposition of a microbial mat. Silicon carbide layer No. 1 was placed onto the Octopus Spring mat in August 1973. Layer No. 2 was placed on 4 June 1974; layer No. 3 on 5 July 1974; layer No. 4 on 5 August 1974; layer No. 5 on 26 May 1975; and layer No. 6 on 10 June 1975. The layers of mat are indicated by letters. The dates on the graph are the times when cores were taken for analysis. The rate of decrease in thickness of layer A is shown in the inset. (From Doemel and Brock, 1977.)

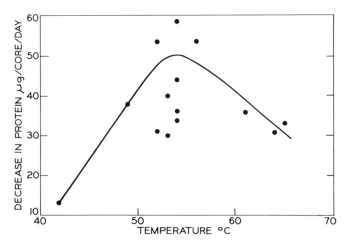

Figure 11.20. Temperature optimum for decomposition rate in the Octopus Spring mat. The decomposition of a mat layer which accumulated between 5 June 1974 and 5 July 1974 was measured 31 days after another layer of silicon carbide was added. Assuming the decomposition rate to be constant, the decrease per day was calculated. (From Doemel and Brock, 1977.)

number of microbial mats. Considering the inherent variability of the mat, it does appear that, in general, growth rates approximate decomposition rates.

Significance of Decomposition Study

The technique of silicon carbide marking has permitted long-term measurements of *in situ* decomposition rates. The thickness and protein content between successive silicon carbide layers decrease with time, and the rate of decrease is a measure of the decomposition rate. Ultimately, two silicon carbide layers merge into one. We have also used a pine pollen horizon, added to the mats around the end of June each year, as a further measure of decomposition.

The decomposition rates measured in this study always followed the same pattern: initially there was a rapid rate of decomposition for the first 2 to 4 weeks, followed by a considerably slower rate which lasted through the subsequent year. Presumably, the initial rapid rate reflects the breakdown of the most readily decomposable materials, and the slower subsequent rate is due to the breakdown of more stable substances. As far as we can tell, there is no organic material remaining after about a year of decomposition, so that the process goes essentially to completion. This does not necessarily mean that all of the organic carbon of the mats has been mineralized, since some diffusion of partially degraded materials from the mat into the water may occur. However, because of the compact nature of these mats,

Table 11.7. Growth and Decomposition Rates for a Large Number of Stations, Measured over a 31-Day Sampling Period[a]

Spring	Station	Temperature (°C)	µg protein/core					
			Growth		Decomposition		Net growth or decomposition day⁻¹
			Increase 31 days	Increase day⁻¹	Decrease 31 days	Decrease day⁻¹	
74-3	1-3	64	+547	+18	−356	−11	+7
Ravine Spring	1-5	63	+222	+7	−237	−8	−1
	2-7	62	+388	+13	−415	−13	0
	2-8	51	+354	+11	−170	−6	+5
74-6	2-1	65	+241	+8	−520	−17	−9
	2-10	62	+203	+7	−288	−9	−2
	2-11	57	+919	+30	−419	−16	+14
Octopus Spring	3-4	42	+701	+23	−399	−13	+10
	5-1	65	+905	+29	−1010	−33	−4
	3-2	52	+1460	+47	−1130	−36	+11
	45-8	54	+4290	+138	−1120	−36	+102
	45-7	52	+1535	+50	−955	−31	+20
	45-6	49	+1020	+33	−1190	−38	−5

45-5	54	+1330	+43	-1350	-44	-1
3-1	56	+970	+31	-1670	-54	-23
12	57	+868	+28	-955	-31	-3
5	54	+1220	+39	-356	-11	+28
53	51	+766	+25	-917	-30	-5
6	52	+1350	+43	-1680	-54	-11
45-11	53	+1410	+45	-120	-40	+5
3-5	54	+1540	+50	-1820	-59	-9
Twin Butte Vista						
48-4		+468	+23	-841	-27	-39
48-2		255	+13	-1040	-52	-5
4-2		1485	+74	-1380	-69	-5
3		965	+48	-1140	-57	-9
0		1670	+83	-1070	-54	-29
4-3		1640	+82	-2280	-117	-35

[a] At day 0, 8.4-mm-diameter cores were taken and a new layer of silicon carbide was placed onto the surface of mats which had had another layer applied 31 days before. After another 31 days, cores were taken and protein measured on the mat that had developed on top of the second silicon carbide layer, to provide a measure of growth (protein gain column). Protein was also measured on the first layer that was present between the first and second silicon carbide horizons. The decrease in protein in this layer from day 0 gave a measure of decomposition rate (protein loss column). (From Doemel and Brock, 1977.)

diffusion is probably minimal, so that to a first approximation we can consider that decomposition occurs completely *in situ*.

It is likely that decomposition is completely an anaerobic process. Our data on the microdistribution of sulfide within these mats (Doemel and Brock, 1976) show that sulfide is present below about 3 mm from the surface of the mat. Since sulfide is not stable in the presence of oxygen, we can conclude that conditions are anaerobic below this point, and since 3 mm is just about the end of the photic zone, it can be concluded that virtually all of the decomposition which occurs is anaerobic. Because of the high temperature of these mats, all eucaryotic organisms are absent, including grazing animals, so that decomposition must occur as a result of an anaerobic food chain. Since methane is produced in these mats (David Ward, personal communication), and thermophilic methogenic bacteria have been isolated (J. G. Zeikus, personal communication), complete mineralization of the organic carbon to methane and carbon dioxide probably occurs.

Model of Mat Growth and Formation

Figure 11.21 summarizes our current understanding of the manner of formation of these alkaline hot spring algal-bacterial mats. Beginning with a new surface, there is an initial colonization by both the *Chloroflexus* and *Synechococcus,* with the formation of a thin, microscopically visible layer. It is possible that the initial colonization of a new substratum by *Synechococcus* requires the presence of *Chloroflexus* to form a filamentous matrix, as suggested by Brock (1969), but in relatively quiet waters the unicellular alga can probably develop by itself. However, soon after initial colonization, both organisms are present, and remain together throughout the growth of the mat. There is a gradual increase in thickness of the mat over the first several weeks, until the thickness of the mat has built up to the point that self-shading of the *Synechococcus* occurs. From this point on, the thickness of the *Synechococcus* layer remains approximately fixed. In light too dim for net algal photosynthesis, the *Synechococcus* cells die and lyse, so that no microscopically recognizable forms are present. This lysis may be due either to autolysis or to attack by lytic bacteria present in the mat. At the depth where light limits the growth of *Synechococcus,* there is still sufficient light for net photosynthesis by *Chloroflexus,* so that a pure *Chloroflexus* undermat develops. The boundary between the *Synechococcus-Chloroflexus* upper mat and the pure *Chloroflexus* undermat is quite distinct, as shown by the sharp color transition from green to orange. As shown by Bauld and Brock (1973), the *Chloroflexus* undermat is photosynthetically active, and the ability of *Chloroflexus* to develop at lower light intensities than the alga leads to the development of pure, photosynthetically active *Chloroflexus* populations at appropriate depths in the mat. Below about 3 mm even the *Chloroflexus* undermat is virtually inactive

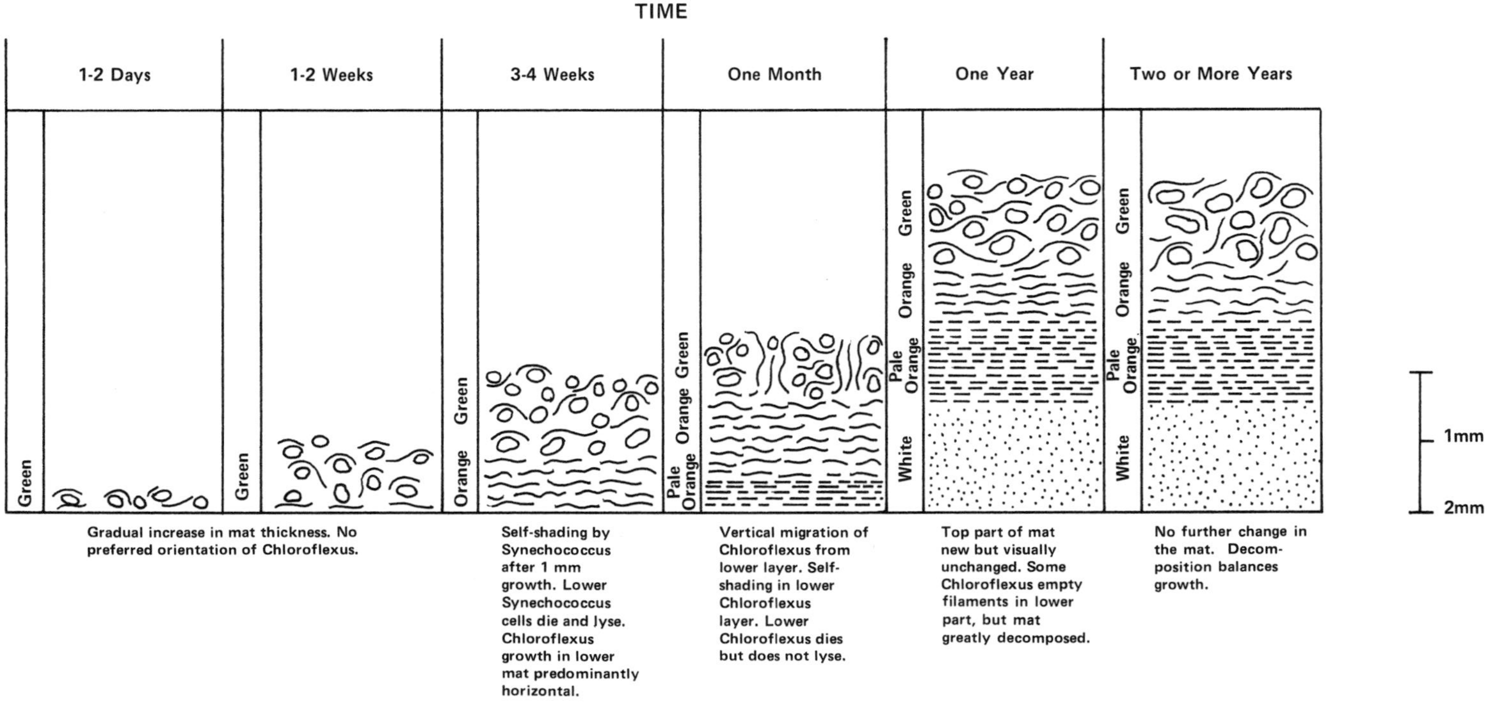

Figure 11.21. Mode of mat growth. (From Doemel and Brock, 1977.)

photosynthetically and is presumably dead. Microscopic examination of the material below 3 mm reveals primarily moribund cells and empty filaments. Thus, the growth of the *Chloroflexus* population also leads eventually to self-shading. [It is possible to obtain pure *Chloroflexus* populations at the surface of mats by experimentally reducing the light intensity to values sufficiently low so that the alga cannot develop (Doemel and Brock, 1974; Madigan and Brock, 1977).]

Eventually a mat is obtained which is several centimeters thick (Figure 11.21), and further change is not seen. At this point, the mat is in steady state, with continued growth being balanced by decomposition. Thus, despite the dynamic nature of the mat, thickness remains essentially constant over many years.

Diurnal changes in the mat are illustrated in Figure 11.22. During the day, *Synechococcus* and *Chloroflexus* develop together, but at night *Chloroflexus* shows positive aerotaxis and glides up on top, so that surface accumulations of pure *Chloroflexus* occur. During the following day, rapid growth by the *Synechococcus* results in repopulation by the alga of the surface of the mat. We know from the culture work of Dyer and Gafford (1961) and Peary and Castenholz (1964) and field studies by Brock and Brock (1968) that the *Synechococcus* of hot springs is able to divide very rapidly, up to 8 doublings a day. The doubling time of *Chloroflexus* has not been measured, but is probably similar, since the two organisms maintain continuously mixed populations. It may be of importance in this regard that *Synechococcus* is an obligate photoautotroph, and hence is able to produce organic carbon only during the day, whereas *Chloroflexus* can grow by

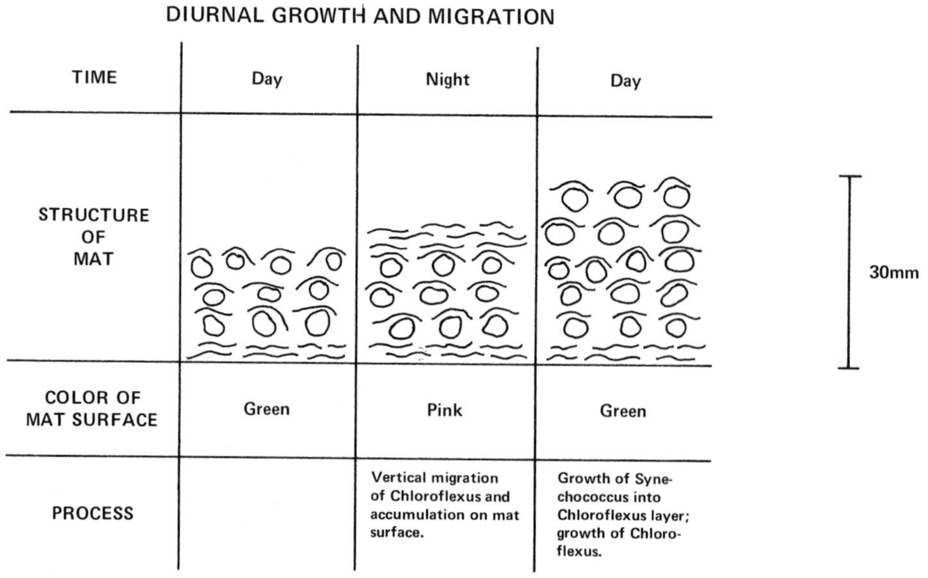

DIURNAL GROWTH AND MIGRATION

TIME	Day	Night	Day	
STRUCTURE OF MAT				30mm
COLOR OF MAT SURFACE	Green	Pink	Green	
PROCESS		Vertical migration of Chloroflexus and accumulation on mat surface.	Growth of Synechococcus into Chloroflexus layer; growth of Chloroflexus.	

Figure 11.22. Diurnal growth and migration in the mat.

GROWTH AND MIGRATION ONTO
DETRITAL MINERALS

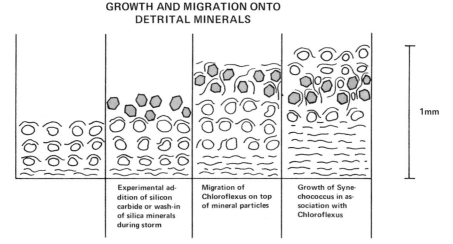

| Experimental addition of silicon carbide or wash-in of silica minerals during storm | Migration of Chloroflexus on top of mineral particles | Growth of Synechococcus in association with Chloroflexus |

Figure 11.23. Growth and migration onto detrital minerals.

three modes of nutrition: photoautotrophic, photoheterotrophic, and heterotrophic (the latter only when O_2 is present). I have presented evidence in this chapter that *Chloroflexus* probably grows in the mats primarily by photoheterotrophic metabolism.

Figure 11.23 describes the model for the behavior of the mat when the surface is covered with a detrital mineral. Small pieces of silica are washed onto the mats periodically as a result of storms, but these detrital minerals are rapidly covered by the mat and become buried. We have studied this phenomenon experimentally by adding thin layers of silicon carbide (carborundum), which is apparently inert biologically, but serves as a substratum for migration of *Chloroflexus*. Within hours after a silicon carbide layer has been added, it has been partially or completely covered with *Chloroflexus*. Soon after, the presence of *Synechococcus* is noted. Since the *Synechococcus* of these hot springs is not motile, it is likely that the inoculum for the *Synechococcus* which develops on top of the silicon carbide layer is derived from the small number of cells that are continually present in the water over the mat. Within a few days or a week, the mat on top of a silicon carbide layer resembles a normal mat, and the presence of the mineral cannot be discerned without coring.

Geological Relevance

Although it is unlikely that precisely similar mats developed in the Precambrian, we can view the Yellowstone mats as models for the kinds of things that might have occurred. Two major conclusions of geological relevance derive from the present work:

    1. *Laminated mats can be formed by photosynthetic bacteria as well as*

by blue-green algae. It has been conventionally assumed that the presence of stromatolitic rocks in Precambrian formations provided evidence for the existence of blue-green algae. In the absence of microfossil evidence, this conclusion seems unwarranted. Clearly, some of the Precambrian formations possess well-preserved fossils that are almost certainly blue-green algae. However, most of the formations with unequivocal blue-green microfossils are 1×10^9 years old or younger. Of the older formations, including the now famous Gunflint chert, the evidence that the microfossils are blue-green algae is minimal. Without knowing of the existence of *Chloroflexus,* Cloud (1965) first suggested that some of the microfossils of the Gunflint could be filamentous bacteria. Indeed, the filamentous organisms from the Gunflint and similar formations (Barghoorn and Tyler, 1965; Cloud, 1965; Cloud and Licari, 1972) are strikingly similar to *Chloroflexus.*

It has been shown in the present work that *Chloroflexus* is responsible for holding the structure of these laminated mats together. Although in most cases this filamentous bacterium lives only together with the blue-green alga, it is possible by experimentally reducing the light intensity to induce the formation of mats consisting of pure *Chloroflexus.* Furthermore, Castenholz (1973) has shown that in high-sulfide springs, blue-green algal development is suppressed and pure *Chloroflexus* populations exist naturally. Thus, mats can form in the absence of blue-green algae, and could have formed in the absence of blue-green algae in the Precambrian. One of the reasons for postulating that stromatolitic rocks had a photosynthetic origin was that the vertical orientation, nodular and laminar appearance, and other features of these structures suggested a phototrophic growth process. When it was thought that the only photosynthetic filamentous phototrophs were blue-green algae, it was natural to attribute formation of stromatolites to them. The existence of *Chloroflexus* and the demonstration that it can form stromatolitic structures alters the need for this interpretation.

These results emphasize the critical importance of developing criteria for evaluating the microbiology of Precambrian fossiliferous rocks. Simply relating observations on such rocks to presumed evolutionary sequences, without objective evidence, is of little value, and may introduce many sources of confusion into the interpretation of the fossil record. I do not necessarily feel that *Chloroflexus* itself was responsible for the formation of Precambrian stromatolitic formations. Its existence, however, shows that *some* sort of photosynthetic bacterium *could* have been the prime agent in the formation of stromatolites. This itself should provide sufficient reason for reexamining present and previous interpretations. Since photosynthetic bacteria are anoxygenic, our data suggest caution in concluding from biological evidence alone the time at which photosynthetic oxygen evolution first occurred.

2. *Complete decomposition of organic matter can occur under anaerobic conditions and in the absence of grazing animals.* Many of the interpre-

tations of how organic matter accumulates in the geosphere assume that decomposition is inhibited under anaerobic conditions. This is standard reasoning when discussing the origin of oil or coal (Krauskopf, 1967). Grazing animals have been assumed to play an important role in destruction of stromatolitic structures and mats (Awramik, 1971; Stanley, 1973). Our data provide the first clear evidence of *complete* decomposition of cellular organic matter under anaerobic conditions. Thus, grazing animals are not required for decomposition, although they greatly speed up the rate (Brock et al., 1969).

It is true that anaerobic food chains are well known in sewage digestion and rumen fermentation (Hobson et al., 1974), but in neither of these habitats does the process go to complete mineralization. In sludge digestors, there is a vast residue of undigested material that must be disposed of, and in the rumen the fermentation acids formed are removed into the blood stream and transported to aerobic tissues for oxidation. Input-output balances have not been carried out in marine or freshwater sediments, where it may be presumed that anaerobic food chains also exist.

A number of organic materials formed by living organisms are not completely mineralized anaerobically. Thus, lignin, polyaromatic rings, porphyrins, aliphatic hydrocarbons, and other constituents almost certainly do not decompose completely under anaerobic conditions (Breger, 1963). Any of these substances formed by the organisms in the mats we have studied must be in such small amounts that they do not contribute to the mass of the mat. Unfortunately, we were not able to carry out specific assays for individual compounds of these types, so that we cannot state that they are not present. But from the viewpoint of the formation of an organic structure, which might become preserved and fossilized, our work shows that anaerobic conditions alone are not sufficient. Further work on the biochemistry and biogeochemistry of decomposition in these hot spring mats would be very desirable.

References

Awramik, S. M. 1971. Precambrian columnar stromatolite diversity: reflection of metazoan appearance. *Science* **174**, 825–827.

Barghoorn, E. S. and S. A. Tyler. 1965. Microorganisms from the Gunflint chert. *Science* **147**, 563–577.

Bauld, J. and T. D. Brock. 1973. Ecological studies on *Chloroflexus,* a gliding photosynthetic bacterium. *Arch. Microbiol.* **92**, 167–284.

Bauld, J. and T. D. Brock. 1974. Algal excretion and bacterial assimilation in hot spring algal mats. *J. Phycol.* **10**, 101–106.

Bott, T. L. and T. D. Brock. 1970. Growth and metabolism of periphytic bacteria: methodology. *Limnol. Oceanogr.* **15**, 333–342.

Breger, I. A. (editor). 1963. *Organic Geochemistry.* Macmillan, New York, 658 pp.

Brock, T. D. 1967a. Microorganisms adapted to high temperatures. *Nature* **214**, 882–885.

Brock, T. D. 1967b. Relationship between standing crop and primary productivity along a hot spring thermal gradient. *Ecology* **48**, 566–571.

Brock, T. D. 1969. Vertical zonation in hot spring algal mats. *Phycologia* **8**, 201–205.

Brock, T. D. 1970. High temperature systems. *Ann. Rev. Ecol. System.* **1**, 191–220.

Brock, T. D. and M. L. Brock. 1968. Measurement of steady-state growth rates of a thermophilic alga directly in nature. *J. Bacteriol.* **95**, 811–815.

Brock, M. L., R. G. Wiegert, and T. D. Brock. 1969. Feeding by *Paracoenia* and *Ephydra* (Diptera: Ephydridae) on the microorganisms of hot springs. *Ecology* **50**, 192–200.

Castenholz, R. W. 1969. Thermophilic blue-green algae and the thermal environment. *Bacteriol. Rev.* **33**, 476–504.

Castenholz, R. W. 1973. The possible photosynthetic use of sulfide by the filamentous phototrophic bacteria of hot springs. *Limnol. Oceanogr.* **18**, 863–876.

Cloud, P. E. 1965. Significance of the Gunflint (Precambrian) microflora. *Science* **148**, 27–35.

Cloud, P. E. and G. R. Licari. 1972. Ultrastructure and geologic relations of some two-aeon old nostocacean algae from northeastern Minnesota. *Am. J. Sci.* **272**, 138–149.

Copeland, J. J. 1936. Yellowstone thermal Myxophyceae. *Ann. N.Y. Acad. Sci.* **10**, 101–106.

Doemel, W. N. and T. D. Brock. 1974. Bacterial stromatolites: origin of laminations. *Science* **184**, 1083–1085.

Doemel, W. N. and T. D. Brock. 1976. Vertical distribution of sulfur species in benthic algal mats. *Limnol. Oceanogr.* **21**, 237–244.

Doemel, W. N. and T. D. Brock. 1977. Structure, growth and decomposition of laminated algal-bacterial mats in alkaline hot springs. *Appl. Environ. Microbiol.* **34**, 433–452.

Dyer, D. L. and T. D. Gafford. 1961. Some characteristics of a thermophilic blue-green alga. *Science* **134**, 616–617.

Gebelein, C. D. 1969. Distribution, morphology, and accretion rate of recent subtidal algal stromatolites, Bermuda. *J. Sediment. Petrol.* **39**, 49–69.

Golubic, S. 1973. The relationship between blue-green algae and carbonate deposits. In *The Biology of Blue-Green Algae,* N. G. Carr and B. A. Whitton, eds. Blackwell Scientific Publ., Oxford, pp. 434–472.

Haupt, W. 1965. Perception of environmental stimuli orienting growth and movement in lower plants. *Ann. Rev. Plant Physiol.* **16**, 267–290.

Hobson, P. N., S. Bousfield, and R. Summers. 1974. Anaerobic digestion of organic matter. *Crit. Rev. Environm. Cont.*, June 1974 (Chemical Rubber Co., Cleveland, Ohio), 131–191.

Krauskopf, K. B. 1967. *Introduction to Geochemistry*. McGraw-Hill, New York, 721 pp.

Madigan, M. T. and T. D. Brock. 1977. Adaptation of hot spring phototrophs to reduced light intensity. *Arch. Microbiol.* **113**, 111–120.

Monty, C. L. V. 1967. Distribution and structure of recent stromatolitic algal mats, Eastern Andros Island, Bahamas. *Ann. Soc. Geol. Belg.* **90**, 55–100.

Peary, J. A. and R. W. Castenholz. 1964. Temperature strains of a thermophilic blue-green alga. *Nature* **220**, 720–721.

Pierson, B. K. and R. W. Castenholz. 1971. Bacteriochlorophylls in gliding filamentous procaryotes from hot springs. *Nature* **233**, 25–27.

Pierson, B. K. and R. W. Castenholz. 1974. A phototrophic gliding filamentous bacterium of hot springs, *Chloroflexus aurantiacus,* gen. and sp. nov. *Arch. Microbiol.* **100**, 5–24.

Stanley, S. M. 1973. An ecological theory for the sudden origin of multicellular life in the late Precambrian. *Proc. Nat. Acad. Sci.* **70**, 1486–1489.

Walter, M. R. 1972. A hot spring analog for the depositional environment of Precambrian iron formations of the Lake Superior region. *Econ. Geol.* **67**, 965–980.

Walter, M. R. (editor). 1976. *Stromatolites.* Elsevier, Amsterdam.

Walter, M. R., J. Bauld, and T. D. Brock. 1972. Siliceous algal and bacterial stromatolites in hot spring and geyser effluents of Yellowstone National Park. *Science* **178**, 402–405.

Walter, M. R., J. Bauld, and T. D. Brock. 1976. Microbiology and morphogenesis of columnar stromatolites *(Conophyton, Vacerrilla)* from hot springs in Yellowstone National Park (Chapter 6.2). In *Developments in Sedimentology, Vol. 20, Stromatolites,* M. R. Walter, ed. Elsevier Scientific Publishing Co., Amsterdam, pp. 275–310.

Weed, W. H. 1889. Formation of travertine and siliceous sinter by the vegetation of hot springs. *U.S. Geol. Survey Rept.* **9**, 613–676 (1887–1891).

Weller, D., W. Doemel, and T. D. Brock. 1975. Requirement of low oxidation-reduction potential for photosynthesis in a blue-green alga (*Phormidium* sp.). *Arch. Microbiol.* **104**, 7–13.

Wickstrom, C. E. and R. W. Castenholz. 1973. Thermophilic ostracod: aquatic metazoans with the highest known temperature tolerance. *Science* **181**, 1063–1064.

Chapter 12
A Sour World: Life and Death at Low pH

I discussed the origin of acidic environments in Chapter 1. As noted, a solfatara is a location where elemental sulfur is precipitating out, due to surficial oxidation of H_2S rising with steam from within the earth. Solfataras are generally areas with minimum amounts of ground water, and the circulation of water through the habitat is very shallow. Thus, except for the acidity and high-sulfate content, the chemistry of acidic geothermal waters is often closer to normal ground water than to deep geothermal water. A typical solfatara, Roaring Mountain, is shown in Figure 12.1a and an acid lake associated with another solfatara is shown in Figure 12.1b. Many solfataras are characterized by crumbling rock, hollow ground, and bleached color, due to acid attack on the rock-forming minerals and to leaching of color-forming components such as iron. The rate at which rocks disintegrate in solfatara areas is amazingly high. During the 10 years that I was in Yellowstone, I actually watched several large rocks disintegrate, and the boulder shown in Figure 12.2a, about the size of an automobile, split into two and collapsed. The power of acid is even more dramatically illustrated by the Grand Canyon of the Yellowstone River (Figure 12.2b), which is riddled with solfataric ground in the area where it is the deepest.

In considering the pH of acid soils, it is important to note that usual methods of measuring soil pH seriously underestimate the pH of very acid soils, such as those of coal mine spoils, cat clays, and solfataras, in which acidity is due to the presence of free sulfuric acid (Doemel and Brock, 1971). Standard methods for measuring soil pH involve making a slurry of soil with (usually) distilled water at a 1:2, 1:5, or 1:10 dilution, and reading the pH of this slurry. Implicit in this procedure is the assumption that the soil is buffered and hence that pH does not change with dilution. This assumption is erroneous in the case of the sulfuric acid soils.

In our work on the microbiology of very acid soils, we wanted to know the pH values to which the microorganisms were actually subjected, that is, the pH of the soil water. To measure this, a slurry of equal parts of soil and water was made and from this slurry a series of dilutions in water was made. The pH of each dilution was measured using a Corning combination glass electrode, and a graph was prepared relating hydrogen ion concentration to dilution (Figure 12.3). In the case of the sulfuric acid soils, a straight line was generally obtained. This line was then extrapolated back to zero dilution and the hydrogen ion concentration of the soil obtained. If the pH of the soil water were desired, the moisture content of the soil was determined and the dilutions then corrected to represent dilutions of the soil. (For example, if the moisture content of the soil were 10%, then the initial dilution would be 1:20 rather than 1:2.) As also shown in Figure 12.3, an acidic agricultural soil, being buffered, does not change pH significantly with dilution, and hence its pH can readily be estimated by the standard procedure.

This procedure was compared with standard procedures for several coal mine spoil and solfatara soils. It was found that the pH value as measured is lower than estimated by standard dilution in distilled water. The pH obtained by standard dilution in KCl or CaCl₂ is lower than that obtained by dilution in distilled water, but reflects exchange of hydrogen ions by diluent cations and hence does not necessarily represent the pH of the soil water.

Lower pH Limit for Living Organisms

What kinds of organisms are present in acid environments, and are there any limits at all? The answer to the latter question is yes, at least for certain kinds of organisms (Table 12.1). Fish, for instance, do not live in very acid waters. The lowest pH at which fish are able to carry out their whole life cycle is somewhere around 4, and only a few kinds of fish can tolerate such values. The Monogahela River, a large river in West Virginia strongly affected by acid mine drainage, has in some parts a pH value around 3.5 and has no fish, but in the upper reaches where acid mine drainage has not had its full effect, carp can be found at pH values around 4. In the southern Swedish lakes, which have become acidic over the past two or three decades as a result of acid rain, most fish have been wiped out, although species vary in sensitivity (Almer et al., 1974). The roach (*Leuciscus rutilus*) has been the most sensitive, showing disturbances in reproduction when the pH drops below 5.5. Other sensitive fish are the brown trout (*Salmo trutta*), the arctic char (*Salvelinus alpinus*), the minnow (*Phoxinus phoxinus*), and the pike (*Esox lucius*). The perch (*Perca fluviatilis*) is also threatened by acidification, partly because the small fish upon which it feeds are being wiped out (Almer et al., 1974). Interestingly, when acidification results in the elimination of fish, invertebrate animals often increase

Figure 12.1. (a) Roaring Mountain, a typical solfatara in Yellowstone National Park. The acid (pH 2) lake in the foreground derives its water from seepage and surface flow from Roaring Mountain; (b) Sour Lake, in the Mud Volcano area. pH 1.8 and rich in CO_2 (note bubbles). It receives its water primarily by seepage, probably from Moose Pool (see Figure 2.5).

greatly in numbers, apparently because they are less sensitive to acidification and are able to develop better in the absence of competition from (or predation by) fish.

It is likely that frogs and other vertebrates are also unable to live in such acid waters, although we do not have any solid data on this point. Surprisingly, water birds do not find acid waters particularly unattractive. Many of the acid lakes in Yellowstone (pH 2–3) have summer populations of Canada geese, and I have similarly seen ducks on New Zealand volcanic lakes with pH values as low as 2. Turbid Lake in Yellowstone (pH around 3) has a large nesting population of Canada geese, although the birds likely do not drink the acid water since there are freshwater streams nearby. A detailed study of birds on acid lakes would be quite desirable and might lead to some new knowledge of adaptability of living organisms to extreme environments.

Although many invertebrates are quite sensitive to acid conditions, and become wiped out when acidification occurs, others are surprisingly tolerant and seem to thrive in acid waters. In the Swedish lakes which have become acidified, all the species of *Daphnia* seem to be sensitive to low

Figure 12.1b. Legend see opposite page.

pH, but *Bythotrephes longimanus* is mostly found in lakes with low pH values (5.4 and lower) (Almer et al., 1974). Acid lakes and streams which develop in coal mine spoil banks are also frequently heavily populated with particular invertebrate species. For instance, in central Missouri, Parsons (1968) found the rotifer, *Keratella quadrata,* only in acidic waters and Harp and Campbell (1967) found midge larvae of *Tendipes plumosus* at pH values as low as 2.3, although emergence was inhibited by pH values of 2.8–3.0.

Figure 12.2. (a) A large boulder along the Gibbon River that crumbled within a few years (arrows) due to attack by acid and steam. The boulder was about 2 meters across. Note also, the crack developing in the smaller boulder on the left (small white arrow). (b) Grand Canyon of the Yellowstone River, illustrating the high erosion that can occur in areas with solfataric ground.

The most extensive observations of invertebrates in acid waters have been made in acid streams in Yellowstone Park. In the warm waters draining geothermal habitats, algae usually grow profusely and constitute a food source for herbivorous insects. No matter what the pH of the water in which the algae are growing, insects seem to be present, although the ones in acid waters are usually different from the ones in neutral waters. One of these insects is a fly, *Ephydra*. When we first observed this fly we were surprised to find that the fly on neutral and acid algae seemed to be the same. Even the experts on the taxonomy of this group of flies could not tell the flies from the two habitats apart. However, when we moved flies from

Figure 12.2b. Legend see opposite page.

the neutral to the acid algae, they quickly died, although the flies from the acid algae could grow well at neutral pH. Thus the two fly populations were different, even though the adult flies looked identical. When we examined the larvae from these two fly populations, distinctive differences could easily be seen, and subsequently it was possible for the taxonomists to detect very small differences in the adults. There are, in reality, two species, one living on neutral algae, which is called *Ephydra bruesii*, and another living on acid algae, called *E. thermophila* (Wirth, 1971). Interestingly, *E. thermophila* can live on neutral algae in the laboratory, although it does not do so in nature, whereas *E. bruesii* is unable to tolerate acid

Table 12.1. Lower pH Limits for Different Groups of Organisms

Group	Approximate lower pH limit[a]	Examples of species found at lower limit
Animals		
Fish	4	Carp
Insects	2	Ephydrid flies
Protozoa	2	Amoebae, Heliozoans
Plants		
Blue-green algae	4	*Mastigocladus, Synechococcus*
Vascular plants	2.5–3	*Eleocharis Sellowiana*
		Eleocharis acicularis
		Carex sp.
		Ericacean plants
		(Heather, blueberries, cranberries, etc.)
Mosses	3	*Sphagnum*
Eucaryotic algae	1–2	*Euglena mutabilis*
		Chlamydomonas acidophila
		Chlorella sp.
	0	*Cyanidium caldarium*
Fungi	0	*Acontium velatum*
Bacteria	0.8	*Thiobacillus thiooxidans*
		Sulfolobus acidocaldarius
	2–3	*Bacillus, Streptomyces*

[a] Lower pH limits are only approximate, and may vary depending on other environmental factors.

conditions. Evolution has apparently been at work, permitting a fly that probably arose on neutral algae to extend its range into acid conditions, where competition from other flies would be absent.

The fly *Ephydra thermophila* of acid habitats consitutes the base of an interesting and surprisingly complex food chain which has been studied in some detail by Collins (1972, 1975). The fly carries out its complete life cycle on the acid algal mats. The female flies lay large numbers of creamy orange, ellipsoidal eggs on the surface of the mat. After 2–3 days, the eggs hatch and the larvae, which are actively motile, move down into the depths of the mat. The larvae develop through three instars within the mat, and then after 11–21 days move to the surface of the mat and form pupae. The pupa undergoes metamorphosis to the adult fly, and within 6–9 days the pupae hatch and the adult flies emerge. After 2–3 days, the female adults are fertile and begin laying eggs, and mature adults remain alive for 11–23 days. In the Yellowstone thermal areas, the flies continue active throughout the year, since, of course, the deep cold winter does not affect the thermal activity which is the base of the whole system. In fact, during the winter the flies are often more conspicuous than in summer, since they are unable to leave the mats without freezing, and hence become concentrated in large numbers on the mats. The eggs, pupae, and adult flies are all subject to predation by various other animals which live on or near the acid algal mats. The eggs are preyed upon by larvae and adult females of the biting midge *Bezzia setulosa,* which is a voracious feeder throughout the year.

A small wasp, genus *Kleidotoma,* lays its eggs within the fly pupae and is probably responsible for some pupal mortality. Adult flies are preyed upon by tiger beetles, which range out over the algal mats capturing flies. The tiger beetles live in small burrows which they construct in the gravelly substrate (also acidic) that surrounds the springs, and during the summer days when the weather is warm enough they become active and capture adult flies. These tiger beetles are not seen in winter, and their manner of overwintering is not known. Another predator of adult flies is the wolf spider, *Trochosa avara,* which is more commonly seen at night than during the day. In the summer, these spiders range over the entire thermal basin, feeding on adults, but in the winter, when the area adjacent to the thermal areas is cold and unfavorable for life, these spiders are concentrated into the narrow strips along the edges of the hot streams. A large predator of the adult flies is a shore bird, the killdeer *(Charadrius vociferous),* which lives

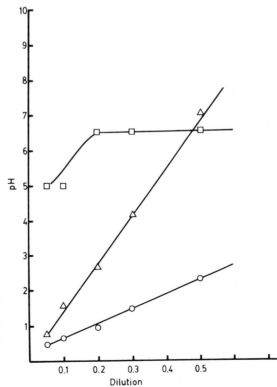

Figure 12.3. The dilution curve of three soils. Squares, clay soil, pH 5.3; triangles, solfatara soil, pH 1.90; circles, mine waste soil, pH 1.30. (From Doemel and Brock, 1971.

in the barren areas surrounding the springs and ranges out on the larger algal mats in search of flies. These birds, like the insects, do not seem to mind the high acidity of the algal mats, most of which range between pH 2 and 3. We thus see the surprising diversity of animal life that is able to live in a sour world at low pH. Unfortunately, we know very little about how these various animals have adapted to life in dilute sulfuric acid. Some of them may actually be acidophiles, living nowhere else, but others are surely adaptable to a wide variety of environments, and are opportunistic invaders of acid habitats. This is almost certainly true of some of the larger animals, such as the killdeer, who may not even drink the acid water. But the ephydrid flies certainly do spend their whole lives in acid water, and hence must be able to drink as well as eat in an acid environment. How they tolerate low pH would make an interesting study in biochemical evolution.

One thing seems well established about all organisms living in acid environments: the intracellular pH is not the same as the extracellular pH (see Chapter 5). Although the environmental pH may be as low as pH 1, the intracellular pH is always close to neutrality. To maintain the viability of the cell, it is essential that this be so, since there are several key biochemical components of cells that are very acid labile, and would be rapidly destroyed if the cellular pH should become acidic. Examples of such acid-sensitive components are chlorophyll, adenosine triphosphate (ATP), DNA, RNA, and nicotinamide adenine dinucleotide (NAD), all vital materials for cell function. Thus organisms living at low pH have developed mechanisms for keeping hydrogen ions (H^+) out of the cell. This could either be through the development of a very efficient hydrogen ion pump, so that hydrogen ions leaking into the cell are extruded as fast as they enter, or it could be through the development of a cell membrane impermeable to hydrogen ions. In fact, a combination of both methods might be used.

Lower pH Limit for the Existence of Blue-green Algae

It is widely assumed that the blue-green algae are highly adaptable and are very tolerant of environmental extremes. However, the environmental tolerance of blue-green algae is, in many respects, not well substantiated. Only with respect to high temperatures is it clear that these algae are considerably more tolerant and adaptable than eucaryotic algae (see Chapter 3). I present here evidence that in very acidic environments blue-green algae are completely absent, whereas eucaryotic algae often proliferate exceedingly well (Brock, 1973).

The data presented here are based on extensive observations of acidic habitats, both natural and man-made, throughout the world; detailed studies of acidic habitats in Yellowstone National Park; observations of algal distributions in natural pH gradients; and enrichment culture studies with media of various pH values and inocula from habitats of various pH values. The pH values were determined either directly at the site or soon after collection, by using an Orion battery-operated pH meter and combination glass electrode. The pH meter was always standardized with a buffer of pH near that of the sample.

Both thermal and nonthermal habitats were studied. Thermal habitats are particularly favorable for determining the lowest pH limit for blue-green algae because at temperatures above about 56–60°C eucaryotic organisms (both photosynthetic and nonphotosynthetic) are always absent and only procaryotic organisms are present (see Chapter 3). Since some species of blue-green algae are able to grow at temperatures up to 70–73°C, the only oxygen-producing photosynthetic organisms at temperatures above 60°C would be blue-green algae. Observations in over 200 habitats of pH less than 4 throughout the world revealed that at temperatures above 56°C no photosynthetic organisms at all are found, and at temperatures from 40°C to 56°C only the eucaryotic alga *Cyanidium caldarium* was found (see Chapter 9). At temperatures below 40°C in these thermal effluents a wide variety of other eucaryotic algae were found, but blue-green algae were absent.

Observations were also made on natural pH gradients in thermal habitats. One pH gradient studied in some detail is in Waimangu Cauldron, New Zealand, where the alkaline waters of Trinity Terrace enter the acidic waters of a hot lake (Brock and Brock, 1970). Since the lake is too hot for *Cyanidium caldarium,* it is devoid of algae where the pH is too low for blue-green algae, and the distribution of blue-green algae in relation to the pH gradient can be observed by directly measuring the pH where algal mats are seen. The lowest pH at which blue-green algae were found was about 4.8 to 5.0. Another pH gradient was examined in the Semi-Centennial Geyser-Obsidian Creek area of Yellowstone National Park. In the mixing zone where alkaline springs enter Obsidian Creek a small pH gradient develops; at pH values over 5 only blue-green algae were seen, and at pH values below 4 only eucaryotic algae were seen. In the pH range between 4 and 5 both blue-green algae and eucaryotic algae were present.

To extend observations to nonthermal environments, studies were made on a large number of lakes of various pH values in Yellowstone National Park. Acid lakes are common throughout the regions of the park where geothermal activity exists, and many of these lakes have normal temperature regimes (Figure 12.4). The acidity of these nonthermal habitats is due to the oxidation of sulfide, which enters these lakes through underwater springs and gas vents, or to the drainage from acidic springs, for which the lakes serve as catchment basins. The acidities, temperatures, and hydrography of the Yellowstone lakes appear to be stable, in contrast to the properties of volcanic crater lakes in many parts of the world, which often fluctuate because of eruptions. Thus, the Yellowstone lakes provide favorable environments for observing the colonization of algae as a function of

Figure 12.4. Clear Lake in the Canyon Area of the Park. pH 2, clear water, and with good populations of eucaryotes, but no blue-green algae.

pH. Twenty-two lakes with pH values ranging from 1.9 to 8.6 were studied. Sampling of benthic algal populations was done from the shoreline. If visible algal mats were present these were sampled; but if no mat was visible, samples of mud, sand, or gravel were taken. Detailed samplings were also taken of benthic algae from Obsidian Creek, since this creek is very acidic (pH 2.35) at its source near Roaring Mountain but becomes progressively alkaline as it flows toward the Gardner River due to inflow of waters from alkaline geysers and hot springs or from nonthermal creeks.

Every sample except one had microscopically visible algae, usually in relatively high numbers. The data presented in Table 12.2 show that although all samples containing algae had eucaryotic algae, blue-green algae were not found in samples from any waters with pH 4.8 or less. To increase the sensitivity of the observations, enrichment cultures were set up with an inoculum from each sample with the medium adjusted to the pH of the water from which each sample was taken. Cultures were incubated in the light at room temperature and at 35–37°C, the latter being used because it is selective for blue-green algae. All cultures yielded good to excellent algal growth. After 2 weeks incubation, all cultures were examined microscopically for the presence of eucaryotic and procaryotic algae. (The pH of each culture was measured at the end of the 2-week incubation and it was

Table 12.2. Presence of Eucaryotic and Blue-Green (Procaryotic) Algae in Lakes and Streams of Various pH Values in Yellowstone National Park[a]

Location	pH	Temperature (°C)	Natural samples Eucaryotic	Natural samples Procaryotic	Cultures Eucaryotic	Cultures Procaryotic
Sour Lake	1.9	27	+	−	+	−
Unnamed Lake[b]	2.1	22	+	−	+	−
Unnamed pond[c]	2.3	29	+	−	+	−
Obsidian Creek[d]	2.35	23	+	−	+	−
Clear Lake	2.5	17	+	−	+	−
Sieve Lake	2.5	40	+	−	+	−
Obsidian Creek	2.7	18	+	−	+	−
Beaver Lake	2.7	16	+	−	+	−
Beaver Lake	2.8	14	+	−	+	−
Obsidian Creek	2.85	24	+	−	+	−
Nymph Lake[d]	2.9	22	+	−	+	−
Obsidian Creek	3.2	17	+	−	+	−
Turbid Lake	3.25	25	+	−	+	−
Sedge Creek	3.4	17	+	−	+	−
Nuphar Lake	3.8	12	+	−	+	−.
North Twin Lake	4.1	17	+	−	+	+
Obsidian Creek	4.5	14	+	−	+	−
Nymph Lake	4.8	22	+	−	+	+
Lake 7847[e]	5.0	30.5	+	+	+	+
Obsidian Creek	5.5	10	+	−	+	+
South Twin Lake	5.5	12	+	+	+	+
Nymph Lake	5.7	22	+	+	+	+
Nymph Lake	5.8	32	+	+	+	+
Obsidian Creek	5.9	16	+	+	+	+
Obsidian Creek	5.95	16	+	+	+	+
Obsidian Lake	6.1	20	+	+	+	−
Obsidian Creek	6.3	12	−	−	+	+
Scaup Lake	6.4	17	+	+	+	+
Beach Lake	6.6	20	+	+	+	+
Harlequin Lake	6.7	17	+	+	+	+
Yellowstone Lake	6.7	16	+	−	+	+
Obsidian Creek	6.9	15	+	+	+	+
Squaw Lake	7.7	17	+	−	+	+
Rush Lake	7.7	17	+	+	+	+
Goose Lake	7.85	17	+	+	+	+
Feather Lake	8.35	17	+	+	+	+
Swan Lake	8.65	20	+	+	ng	ng

[a]Cultures were set up in duplicate and incubated at both room temperature and approximately 37°C. A plus sign indicates that at least one culture was positive at either temperature; ng indicates no growth. [b]Small lake with no outlet south of the Mud Volcano area, near Lake 7847 (U.S. Geological Survey Quadrangle, Canyon Village, Wyoming). [c]Small turbid pond at Mary Bay, Yellowstone Lake, north of East Entrance Road. [d]At Nymph Lake and Obsidian Creek various pH values were found at different sites. [e]Lake 7847 is south of the Mud Volcano area and is at an elevation of 7847 feet. (From Brock, 1973.)

usually within a few tenths of the original pH.) As can be seen in Table 12.2, growth of blue-green algae occurred in culture at pH values down to 4.1, but at no pH below 4 were any blue-green algae seen.

In a further attempt to enrich for blue-green algae at low pH, samples from a number of waters were pooled and the mixtures were used to inoculate culture media of various pH values. Again, incubations were at both room temperature and approximately 37°C. In these enrichment cultures, no blue-green algae were obtained at pH 5 or below, although excellent growth of blue-green algae occurred at pH 6 or higher.

Some studies have also been done on soil algae, since soils of a wide variety of pH values occur in Yellowstone Park. Although blue-green algae are widespread in soils of neutral and alkaline pH, in soils with a pH of 5 or less only eucaryotic algae have been found.

In addition to observations of naturally acidic habitats, a number of observations were made in streams in southern Indiana subject to acid pollution due to drainage from coal refuse piles or spoil banks. In many of these streams, extensive algal growth occurs at pH values of 4 or less, but only eucaryotic algae were present. Observations were also carried out on the acid waters issuing from the leaching dumps used for the beneficiation of low-grade copper ores in Montana (Anaconda Co.) and Arizona (Duval Corp.). Extensive benthic algal populations live in the effluent channels of such dumps, where the water has a pH of 2–2.5 and copper concentrations of 750–1000 μg/ml. Blue-green algae were never observed in these channels, although a variety of eucaryotic algae were present.

The observations reported here are consistent with other reports of the rarity of blue-green algae in environments of acid pH. For instance, Fogg (1956), in his review on the physiology and biochemistry of the blue-green algae, concluded that the blue-green algae show a preference for alkaline conditions. Lund (1962, 1967), in his two reviews on soil algae, noted that blue-green algae prefer neutral or alkaline soils and are absent from acid soils. Prescott (1962) made an extensive survey of lakes of various pH values in Michigan and Wisconsin and concluded that blue-green algae were rare or absent in lakes with pH as low as 5, and green algae were almost the exclusive components. Gessner (1959), in his review of the algal flora of acidic habitats, concluded that the Cyanophyceae were the rarest algal class in acid bogs. Rosa and Lhotsky (1971) did an extensive microscopic and cultural study of algae from acid soils (pH 3.2 to 4.4) of the Iser Mountains, Czechoslovakia, and found no blue-green algae, although eucaryotic algae were widespread. Negoro (1944) reported a few blue-green algae from acid waters in Japan, although these were in the minority and it is not clear that Negoro distinguished these reputed blue-green algae from eucaryotes. The main phycocyanin-containing alga found by Negoro was the eucaryote *Cyanidium caldarium,* which he erroneously classified as a blue-green alga. In Ueno's study of the acid lakes of Japan (1958) blue-green algae were never found. In his study of the algae of acid mine waters,

Bennett (1969) found blue-green algae to occur only rarely and never abundantly.

From the results described above, I conclude that blue-green algae do not exist at pH values below 4, although eucaryotic algae not only exist but grow profusely. Thus, there seems to be an acid barrier that eucaryotic algae, but not blue-green algae, have been able to overcome. It should be emphasized that the eucaryotic algae present in acid habitats are both profuse in numbers and taxonomically diverse. Algae from four major divisions are found: Rhodophyta *(Cyanidium caldarium),* Bacillariophyta (diatoms, *Pinnularia, Eunotia),* Euglenophyta *(Euglena mutabilis),* and Chlorophyta (many species). Among the Chlorophyta, algae from four orders are found—Volvocales *(Chlamydomonas),* Chlorococcales *(Chlorella),* Ulotrichales *(Ulothrix),* and Zygnematales *(Zygogonium).* Thus, it seems that eucaryotic algae have been evolutionarily successful in acidic environments, whereas procaryotic algae have been excluded.

The evolutionary implications of this conclusion seem vast. If we assume that blue-green algae have never been able to live in acidic environments, it seems evident that in the portion of the Precambrian when blue-green algae existed and eucaryotic algae did not, acidic habitats would have represented uncolonized niches within which eucaryotic algae, if they evolved, could have reproduced without competition from blue-green algae. It might have been in such acidic environments that the first eucaryotic algae arose. Although it is not known with certainty why acid is detrimental to blue-green algae, one possibility is that it affects the photosynthetic apparatus, which in these algae is usually at the periphery of the cell adjacent to the plasma membrane. Chlorophyll especially is very acid labile and decomposes to pheophytin under mildly acidic conditions. In eucaryotic algae, chlorophyll and the photosynthetic apparatus are segregated in chloroplasts, which are membrane-bounded structures surrounded by cytoplasm. Since the cytoplasm is probably of neutral pH even in eucaryotic algae living at acid pH, the cytoplasm provides an environment of neutral pH within which the chloroplast can exist. I suggest that if eucaryotic algae evolved they could have invaded acidic environments, where they would be free from competition, so that the evolution of the chloroplast would have immediately provided a selective advantage. It is then conceivable that eucaryotic algae could have radiated from acidic environments into the wide variety of other environments in which they exist today. It is irrelevant for this hypothesis whether eucaryotic algae first arose from a symbiotic association of a blue-green alga with a nonphotosynthetic organism or whether a *de novo* origin of the eucaryotic algae occurred.

The ecological implications of the above conclusions are also clear. Blue-green algal blooms should never occur in acid lakes, and the pollution of lakes and streams with acid mine drainage should eliminate blue-green algae from these waters. Since even in mildly acidic waters (pH 5 to 6) blue-green algae are uncommon, mild acidification of lakes may control or eliminate blue-green algal blooms. Since in rice cultivation nitrogen fixation by blue-green algae may be beneficial, rice culture in acidic soils may be less favorable than in neutral and alkaline soils, and liming of such soils may considerably improve rice yields.

The Eucaryotic Alga *Zygogonium*

In many of the acidic thermal areas of Yellowstone National Park, a species of *Zygogonium* forms extensive dark purple mats over the surface of the substrate (Lynn and Brock, 1969). Although *Zygogonium* does not live in warm waters, it is found adjacent to thermal features, and always in habitats that are quite acidic. Although at one time there was some doubt about the validity of the genus *Zygogonium,* it is now well established and 14 species are recognized. Our material at Yellowstone agrees with the general description of the genus in the size and shape of the cell wall and in the structure of the chloroplast. The most prominent features of mature cells are the purple vacuolar pigment and the single, large "dumbbell-shaped" chloroplast. Mature cells are up to 40 μm in length and up to 18 μm in width. A distinct mucilaginous region up to 5 μm in thickness is often present exterior to the cell wall. The cell wall is up to 5 μm in width and is thickened at the ends in a manner typical of the replicate end walls of *Spirogyra*. The chloroplast contains the same purple pigment present in the large central vacuole of the cell and contains at both ends prominent starch-ensheathed pyrenoids. The single spherical nucleus is centrally located adjacent to the thinnest portion of the "dumbbell-shaped" chloroplast with one or more radiating cytoplasmic strands in apparent contact with it. Branches are rare, and most filaments appear untapered, although occasionally narrower filaments are seen. A phase photomicrograph of typical filaments is shown in Figure 12.5.

Zygogonium mats have been observed in many thermal areas of Yellowstone Park with acidic conditions such as Norris Geyser Basin (many locations), Amphitheater Springs, Mud Volcano, Sylvan Springs, Shoshone Geyser Basin, the small thermal basin south of Artist's Point in the Canyon area, and several areas along the Norris-Mammoth road. A photograph of a typical *Zygogonium* mat is shown in Figure 12.6 and this mat is diagrammed in Figure 12.7. Such mats always develop in areas where there is a moist surface created by the presence of small acidic springs or seeps. In none of these mats is there a marked flow of water. Even if the springs themselves are hot, the small volume of flow leads to a rapid cooling of the water and the temperature of the surfaces of the *Zygogonium* mats usually approaches air temperature. The uphill boundaries of these mats are probably controlled by the availability of ground water, whereas the lower edges are often delineated by the warm water in effluent channels of adjacent hot

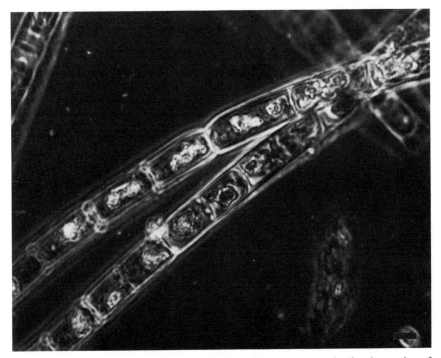

Figure 12.5. Phase contrast photomicrograph of *Zygogonium,* the dominant alga of many low-temperature acid habitats. Marker bar 10 μm.

springs. Occasionally *Zygogonium* tufts have been found in hotter water, but the cells are usually devoid of pigment and look moribund, suggesting that they have washed down into the hotter water and been killed. Values for pH within a number of *Zygogonium* mats were measured using a glass electrode and values ranged from 2.0 to 3.1. All species of *Zygogonium* are found in acid habitats such as bogs and acid mineral soil. However, pH values below 3 are ordinarily found only in thermal areas or in areas receiving drainage from acid mine wastes.

We studied two *Zygogonium* mats in some detail in order to obtain an idea of areal extent, temperature, and pH throughout the mat. The mat diagrammed in Figure 12.7 occurs in a small thermal basin just northwest of Norris Junction, and can be easily seen from a car. A transect made across this mat showed that there was very little variation in temperature and pH. It was this mat that was used in the experimental studies reported below.

The second mat studied in detail is at the lower end of the Sylvan Springs thermal basin (Figure 12.8) (see also Figure 2.4). This was the largest single mat observed and its areal extent has been estimated at about 3000 m². The temperature in different regions of this mat varied from 20°C to 31°C and the pH varied from 2.4 to 3.1.

Winter observations of several *Zygogonium* mats were made by W. N.

Figure 12.6. *Zygogonium* mat (central dark area) northwest of Norris Junction, Yellowstone National Park. This mat remained essentially unchanged for at least 8 years. (From Lynn and Brock, 1969.)

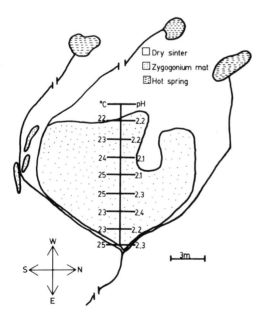

Figure 12.7. Diagram approximately to scale of the *Zygogonium* mat northwest of Norris Junction as it appeared in the summer of 1968. (From Lynn and Brock, 1969.)

Figure 12.8. Diagram approximately to scale of the *Zygogonium* mat at the lower end of the Sylvan Springs thermal basin. See also Figure 2.4. (From Lynn and Brock, 1969.)

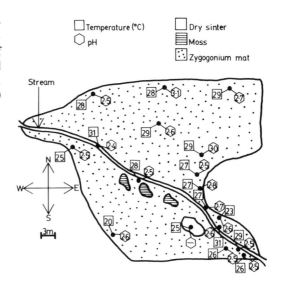

Doemel and H. Freeze on 4 February 1969. The mats were snow-free and visually unchanged from the summer. (At the time of the observations the snow depth away from thermal areas was 6–8 feet.) Temperatures in the mats were somewhat lower than in the summer, varying between 10°C and 20°C, and pH values were 2.5–3.0. Microscopically *Zygogonium* filaments resembled those of material collected in the summer.

The relationships of this species of *Zygogonium* to the key environmental factors are of interest. In habitats where *Zygogonium* has been found, trees are rare or absent, so that the mats are exposed to intense sunlight. The thickness of the mats varies from a few millimeters up to 6 cm. In the thicker mats, only the surface layer of the mat has purple pigment; however, the filaments that live a centimeter or so underneath the surface usually appear bright green or light yellow, suggesting that light plays some role in pigment formation, and that the purple pigment may play some photoprotective role. Both Lagerheim and Fritsch have also reported that the purple pigment of *Zygogonium* was only found in the surface layers (Fritsch, 1916). The effect of light on photosynthesis was studied by using neutral density filters to reduce the light intensity of some of the vials during incubation with $^{14}CO_2$. Incubations were in filtered mat water (pH 2.4) and as seen in Figure 12.9, maximal photosynthesis was obtained at full sunlight, which was about 1.3 g·cal/cm²/min total radiation. These results show that photosynthesis may not be light saturated even at full sunlight.

The temperature optimum for $^{14}CO_2$ fixation was measured using standard techniques. The results of this experiment, presented in Figure 12.10, indicate an optimum temperature for photosynthesis of 25°C, a temperature near that of the mat itself.

To study the effect of pH on photosynthesis, aliquots of filtered mat

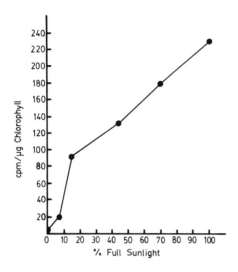

Figure 12.9. Effect of light intensity on photosynthesis by *Zygogonium*. (From Lynn and Brock, 1969.)

water were adjusted to pH values between 0.5 and 10.0, using 1 N H_2SO_4 for pH values between 0.5 and 2.0 and 1 N NaOH for values of 3.0–10.0. Incubation was at 25°C. The results (Figure 12.11) show a broad pH optimum between 1.0 and 5.0. Since the pH of the habitat was about 2.5, the alga is living at a pH at which it can photosynthesize optimally, but it could probably survive and grow at values considerably higher and lower than those at which it is found.

The high water-holding capacity of *Zygogonium* mats has been noted by previous workers. West (1916) noted that *Zygogonium* mats living on bare sand or peat of British heaths "have great absorptive capacity, greedily taking up water, and in this way they regulate the moisture of the surface soil, the thriving of some of the smaller phanerogams depending to a great

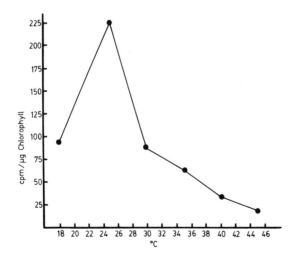

Figure 12.10. Effect of temperature on photosynthesis by *Zygogonium*. (From Lynn and Brock, 1969.)

Figure 12.11. Effect of pH on photosynthesis by *Zygogonium*. (From Lynn and Brock, 1969.)

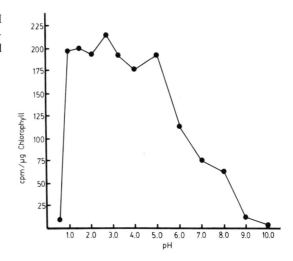

extent on [its] presence." Fritsch (1916) noted that *Zygogonium* cells retain much water when air dried, perhaps due to the thick sheath, but when completely dry they absorb water rapidly and increase in size, even after many months of desiccation. In most of the habitats studied by us, ground water seemed adequate so that the mats did not dry out underneath, but the surface of the mat which was exposed to the direct sunlight often became dry and hard. When such dried mats were moistened they quickly took up water, and one quantitative measurement of the water-holding capacity of a mat showed that 1 g of algal material took up 1.2 g of water.

Zygogonium appears to be a primary colonizer of the moist acid soil and continued growth of the mat results in the establishment of an aquatic environment beneath the mat. The entrapped water within the mat serves as a habitat for the development of other algal forms, especially species of *Euglena* and *Chlamydomonas,* which are often found at high population densities in the lower layers of the thinner mats. The mat also serves as a repository for the eggs of the brine fly *Ephydra bruesii*, although the larvae and adults of this fly probably do not consume *Zygogonium* filaments as their main source, as noted earlier in this chapter. The *Zygogonium* mat is probably also an ideal habitat for the establishment of terrestrial plant forms, since many moss and fern gametophytes have been observed on moist portions of the mat, and older, thicker portions of the mat often have an inseparable matrix of moss protonema interwoven with the *Zygogonium* filaments.

It is surprising that although these extensive *Zygogonium* mats are quite common in Yellowstone Park, they have not been previously reported. One explanation may be that phycological investigations have previously concentrated on the Cyanophyta. According to a personal communication received from Dr. Francis Drouet of the Academy of Natural Sciences, Philadelphia, Walter H. Weed collected *Zygogonium* from Yellowstone

Park. Two of Weed's specimens are now in the herbarium of Dr. Drouet: No. 60, Tepid Spring, Turbid Lake, Norris Geyser Basin, Summer 1887, and No. 71, at 130°F, Deep Creek, Summer 1888. Although the temperature of the latter is too high for *Zygogonium,* it could represent material washed into a higher temperature environment from a mat, as we have described above. Tilden (1898) presented observations on eucaryotic algae from Yellowstone without mentioning *Zygogonium.* A careful reading of her paper suggests that she may have seen *Zygogonium* but classified it as *Rhizoclonium hieroglyphicum* (Ag.) Kg. var. *atrobrunnem.* She reported this dark brown form from two locations in the Norris Geyser Basin, where *Zygogonium* is very common. In the absence of conjugation, it is possible that these two forms can be confused. Indeed, West and West (1894), in their report of *Zygogonium pachydermum* from a warm stream in the crater of Grande Soufriere, Dominica, note that "This species has a thick membrane which at first sight reminds one of a *Rhizoclonium. . . ."* I visited this same area of Dominica (the so-called Valley of Desolation) in November 1971 and found small *Zygogonium* mats to be widespread. Microscopically the material resembled that seen in Yellowstone, and pH values ranged from 3.2 to 3.9 in different locations (all nonthermal).

Bacteria

The bacteria also have some important members restricted to acidic habitats and these bacteria provide some of the most dramatic examples of biochemical evolution. As noted in Chapter 10, boiling springs at neutral pH all usually have bacteria growing in them, but acid boiling springs are apparently sterile. It seems as if organisms are unable to overcome two extreme factors at once, although they can overcome each separately. However, once an acid boiling spring has cooled only a few degrees, it can be colonized by bacteria. In Yellowstone Park, where water boils at about 92.5°C, boiling acid springs with temperatures just below 90°C frequently contain high population densities of *Sulfolobus acidocaldarius,* which is the sole occupant of the environmental niche defined by low pH and high temperature (see Chapter 6). This interesting organism is nutritionally versatile, able to use both sulfur and ferrous iron as energy sources, as well as a wide variety of organic compounds. When growing on sulfur or iron it grows as an autotroph, obtaining all of its cell carbon from CO_2, and it is the most thermophilic autotroph known. Once temperatures have dropped to about 55°C, *Sulfolobus* is no longer able to grow, and another group of sulfur-oxidizing bacteria appear, members of the genus *Thiobacillus.* There are quite a few species of thiobacilli, but only two are restricted to low pH environments, *Thiobacillus thiooxidans* and *Thiobacillus ferrooxidans.* Both of these species oxidize sulfides and sulfur to sulfuric acid. In addition, *T. ferrooxidans* oxidizes ferrous iron, hence the derivation of its species name. Both *T. thiooxidans* and *T. ferrooxidans* are present in geothermal habitats once the temperature has dropped below 55°C, and are found in solfatara soils and acid springs; solfataras are probably natural reservoirs of these bacteria. Thiobacilli are not the only bacteria present in low pH environments of normal temperatures. Several bacteria are present that use organic materials as their sources of carbon and energy (Millar, 1973). In fact, the diversity of bacteria in acid environments is surprisingly high, suggesting that it has not been especially difficult for members of this group of organisms to evolve the ability to grow in these extremely acidic environments, once the temperature has dropped to the normal levels. It is only in hot, acid environments that the diversity of bacteria is so low, and *Sulfolobus* develops exclusively.

This does not complete the list of bacteria adapted to acid pH values. Canners have known for a long time that certain spore-forming bacteria of the genus *Bacillus* are able to live in acidic foods. These bacteria are usually classified as the species *Bacillus coagulans,* and are not acidophilic, but merely acid tolerant, since they can also grow at neutral pH. They are probably normal soil organisms that have extended their range to lower pH values. However, Robert Belly found in acid hot springs an organism closely resembling *Bacillus coagulans* which was a true acidophile, unable to grow at neutral pH (Belly and Brock, 1974). This organism was also mildly thermophilic, being able to grow at temperatures up to 55°C. This type of *Bacillus coagulans* could probably also occur in canned foods, but would be missed in any bacteriological testing if culture media were used that were not adjusted to acid pH values. Another *Bacillus,* previously unknown and thus classified as a new species, *Bacillus acidocaldarius,* is also found in acid hot springs, but at even higher temperatures, being able to grow up to 65°C (Darland and Brock, 1971).

Does this complete the list of bacteria able to live in acidic environments? Undoubtedly not. A systematic attempt to uncover all of the acidophilic bacteria has never been made. The new organisms which have been discovered have appeared incidentally during ecological and biogeochemical studies of acidic environments. A concerted effort to probe the species diversity of acidic environments would undoubtedly lead to the discovery of many more new species, and probably even new genera. Why have these organisms escaped discovery for so long? Partly it is because scientists have not studied acidic environments extensively, but more importantly it is because improper cultural techniques have been used. Since all of these acidophilic bacteria are unable to grow at neutral pH, they will of course not be cultured on the usual bacteriological media, which are of neutral or slightly alkaline pH. Indeed, since for many years bacteriological textbooks stated that bacteria do not grow at acid pH values, whereas fungi do, it was natural that bacteriologists would ignore the possibility of acidophilic bacteria. Even the discovery of an undoubted acidophilic bacterium, *Thiobacillus thiooxidans,* did not change the thinking of most bacteriologists, since it was assumed that this sulfur-oxidizing bacterium, an autotroph, was unique. But now that we know that a variety of hetero-trophic bacteria are acidophilic, there is no reason not to search further, using culture media adjusted to acid pH values. The diversity of bacteria we already know indicates to us that low pH is not an evolutionary barrier for the group as a whole, although certain kinds of bacteria are apparently unable to evolve acidophilic species.

Rate of Sulfuric Acid Production in Yellowstone Solfataras

Acid production in solfataras occurs both in the soil and in the springs. Most of the springs are nonflowing, but exchange studies (described below) showed that water flows through the springs quite rapidly by seepage. Knowledge of the existence and activity of thermophilic thiobacilli and *Sulfolobus* in acid springs and acid soils (Fliermans and Brock, 1972; Mosser et al., 1973) suggested that the bulk of the acid produced in solfataras was due to bacterial activity. Studies on the fractionation of stable sulfur isotopes by Schoen and Rye (1970) had indicated bacterial involvement in sulfuric acid production in solfataras.

Schoen (1969) measured rate of sulfuric acid production in several Yellowstone solfataras by measuring the rate of runoff of water into nearby streams and by determining the sulfuric acid content of the runoff water. This method provides a good way of obtaining an overview of rate of production, but could be in error if not all the acid water were present in the runoff. We developed a technique for measuring growth rates of *Sulfolobus* in acid springs by adding chloride ion and measuring the dilution rate of chloride with time. The use of this method to measure the growth rate of *Sulfolobus* in a number of Yellowstone springs is described in Chapter 6. Since the method provided a measure of the water turnover time, and since sulfuric acid should behave in a manner similar to chloride, the method provided a means of measuring the rate of sulfuric acid production in nonflowing springs. The data obtained were then compared to those of Schoen, obtained by runoff measurements.

The procedure was to obtain samples of water for chemical and micro-biological determinations several times before chloride enrichment, to assess the steady-state nature of the springs. All of the springs studied were very low in chloride, so that enrichment could be easily accomplished. Initially, the dimensions of the springs were measured with a weighted line, in order to calculate approximate volumes. Then, sufficient sodium chlo-ride was added to increase the chloride concentration by about 10-fold, and samples were taken periodically to determine the rate at which the chloride was diluted. To obtain rapid mixing, the salt was dissolved in spring water to create a brine that was poured onto the surface of the pool at several locations; the brine quickly mixed with the spring water. Water samples were removed every 15 minutes during the first hour after the salt was added, in order to ascertain that mixing was complete. Thereafter, samples were removed at intervals ranging from 1 day to several weeks, depending on the rate of chloride dilution. Even in large springs, rapid mixing of the salt could be carried out satisfactorily (Figure 12.12). For instance, in Moose Pool (volume 1.1×10^6 liters) and Sulfur Caldron (volume 7.6×10^5 liters), the large amounts of salt needed (around 80 kg) were added within a 15-minute period and when the first samples were taken, about 30 minutes after the salt addition had begun, complete mixing was found.

Figures 6.18 and 6.19 show the dilution rates determined for some of the smaller springs. Loss of chloride is exponential with time, as would be expected for a dilution process occurring at a constant rate. The half-times for chloride dilution calculated from these data are given in Table 12.3. Most of the rates were quite rapid, except for the two large springs, Moose Pool and Sulfur Caldron. Chemical analyses were performed on water taken at the time chloride was added and again at the time that the chloride concentration had diluted to the natural level again. During this time

Figure 12.12. Procedure for addition of salt brine to Moose Pool. Within 15 minutes, 90 kg of salt were mixed homogeneously through the water volume (1.1×10^6 liters).

Table 12.3. Relation of Flow Rate to Volume and Area of Solfataric Springs

Location	Volume (liters)	Water surface area (m²)	Turnover time	Flow rate (liters/hr)	Flow rate (liters/m²/hr)
Norris Geyser Basin					
Locomotive	2422	24	16hr	76	3.2
White Bubbler	52	0.78	10hr	2.6	3.3
Vermillion	273	2.4	17hr	8.0	3.3
26-5	257	1.35	13hr	10	7.4
21-1	97	1.77	14hr	3.5	2.0
Sylvan Springs					
59-1	453	3.33	24hr	9.4	2.8
59-2	22	1.91	24hr	0.5	0.26
54-5	595	1.96	24hr	12.4	6.3
56-3	95	1.13	34hr	1.4	1.2
58-3	80	1.76	13hr	3.1	1.8
Mud Volcano					
Moose Pool	1.1×10^6	300	28days	818	2.7
Sulfur Caldron	7.6×10^5	100	35days	452	4.5

(From Brock and Mosser, 1975.)

period, the natural concentrations of the various chemical species remained essentially constant. These results show that the springs in question are steady states, with water and the various chemical species flowing through the springs at essentially constant rates. The dilution rates in some springs were determined twice, and similar dilution rates were obtained, which suggests that flow rate does not change significantly with time. Because the experiments were done during the maximum stream-flow period of mid-June to mid-July, the flow rates may not be representative of other periods of the year.

Figure 12.13 shows the dilution rates determined for Moose Pool and Sulfur Caldron. The rates are much slower than those for the small pools, and half-times of 28 and 35 days were calculated (Table 12.3).

Although the turnover rates of the large springs are much less than those of the smaller springs, because of the large volumes of the large springs, more water is flowing through them per unit of time. It can be seen in Table 12.3 that Sulfur Caldron and Moose Pool had volumes of water flowing through them an order of magnitude larger than that of the smaller springs. When this calculation is considered, the low turnover times of the large springs do not seem to be unreasonable. Locomotive Spring, of intermediate volume, had a volume flow rate intermediate between the large and small springs.

Because sulfuric acid production is an aerobic process, it may be more meaningful to relate volume flow rate to the area of the spring surface. Thus flow rates per unit area of pool surface are set out in Table 12.3. The flow rates of all springs vary with the area of the pool. Except for Spring 59-2, the range of flow rates per unit area of pool surface varies only about threefold, even though the volumes vary about 100,000-fold.

These data can be compared with those of Schoen (1969), derived by a different method. Schoen estimated flow rates of several acid areas in Yellowstone National Park by measuring the volume of water flowing out of the streams that drain each area and by determining the area of the acid-producing land by measuring on aerial photographs the approximate area of acid-altering terrain. Although none of the springs we studied was in the specific areas studied by Schoen, the different acid areas in Yellowstone may be generally comparable. My calculations of Schoen's data indicate the following flow rates: Amphitheater Springs, 2.0 l/m²; Norris Ranger Cabin, 2.7 l/m²; Norris Junction, 0.72 l/m²; and Obsidian Creek, 9.8–11.2 l/m². Schoen's estimates are close to those presented in Table 12.3.

Because the springs are steady-state systems, it is possible to calculate the rate at which sulfuric acid moves through the pools from the chemical

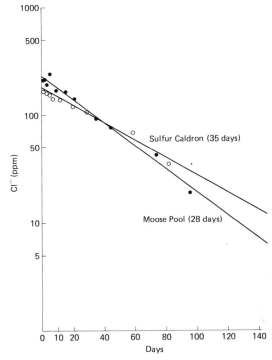

Figure 12.13. Dilution rates of chloride after addition to two large springs, Moose Pool and Sulfur Caldron. Calculated half-times for chloride dilution are given in parentheses. (From Brock and Mosser, 1975.)

Table 12.4. Rate of Sulfuric Acid Flow through the Various Springs

Location	$SO_4^{=a}$ (mg/liter)	Flow[b] (liter/hr)	$SO_4^{=c}$ (g/day)	Spring area (m^2)	$SO_4^{=}$ (g/m^2/day)
Norris Geyser Basin					
Locomotive	231	76	420	24	17.5
White Bubbler	324	2.6	20	0.78	25.6
Vermillion	400	8.0	77	2.4	32
26-5	396	10	94	1.4	67
21-1	151	3.5	12.7	1.8	7.1
Sylvan Springs					
59-1	344	9.4	78	3.3	23.6
59-2	1,516	0.5	18	1.9	9.5
54-5	472	12.4	140	2.0	70
56-3	291	1.4	9.8	1.1	8.9
58-3	506	3.1	38	1.8	21.1
Mud Volcano					
Moose Pool	2,260	818	44,352	300	148
Sulfur Caldron	2,260	452	24,504	100	245
Data of Schoen (1969)					
Roaring Mountain	379	125	1,180,000	1.28×10^5	9.2
Amphitheater Springs	285	144,000	950,000	7.2×10^4	13.2
Norris Ranger Cabin	160	18,360	71,000	6.7×10^3	10.6
Norris Junction	440	7,920	82,000	1.1×10^4	7.5

[a]Chemical assay, from Table 6.3.
[b]From turnover time and volume, Table 12.3.
[c]Calculated from the data in the two columns immediately to the left.
(From Brock and Mosser, 1975.)

assays for sulfate and the flow rate. These calculations are presented in Table 12.4 and are compared to data of Schoen. The values obtained are a composite of sulfuric acid production by bacterial action within the pools and of underground seepage of sulfuric acid into and out of the pools through the surrounding soil. Despite wide variations in concentration of sulfate, flow rate, and area, the rate of sulfuric acid movement through the pools varies less than 50-fold. In addition, the data of Schoen fall within the range of the data we calculate for many individual springs. Only the two large springs seem to be different. In these springs, greater amounts of sulfuric acid move through per unit surface area, but because these springs are much greater in volume (and probably considerably deeper per unit area), it is not surprising that their rates per unit area would be greater. As noted below, most of the sulfuric acid in these two springs is produced *in situ*, whereas in the small springs it is produced in the surrounding soil.

As another approach to measuring the hydrological conditions of acid thermal springs, and to confirm the data from the chloride-dilution experiments, we drained three pools by siphoning and allowed them to refill.

The three pools drained were 54-5, 56-3, and 59-2, with volumes of 595, 95, and 22 liters, respectively. Steam vents lined with silica were seen in the bottom of 54-5 and 56-3 and in Spring 59-2, steam rose from several small sources in the gravelly bottom. The rate of return of water in Spring 59-2 was considerably slower than that calculated from the chloride rate, but the other two springs recovered at the predicted rate.

The history of drainage and refilling of Springs 54-5 and 56-3 is given in Table 12.5. The first water reentering the pools was hotter than the water in the fully recovered springs. Because steam vents were seen in the dry bottom, which had temperatures almost at the boiling point, cool or cold spring water entering the pools is probably heated by steam. The temperature of each spring is probably a function of temperature and rate of inflowing water, temperature and rate of upflow of vapor, surface area of the spring, rate of discharge from surface and subsurface, and evaporation rate.

Table 12.5. Changes in Temperature and pH during Draining and Refilling of Two Springs

Time	Temperature (°C)	pH
Spring 54-5		
Before draining	83	1.85
Completely drained	91 (steam)	—
1 minute (first water)	85	2.0
1 hr (½ recovered)	81.5	—
2 hr (¾ recovered)	80.0	1.95
1 day (fully recovered)	81.5	1.90
2 days	82.0	1.85
3 days	81.5	1.85
5 days	81.0	1.80
15 days	82.0	1.80
Spring 56-3		
Before draining	61.5	1.90
Completely drained	93 (steam)	—
4 minutes (first water)	82.5	2.00
15 minutes (1/10 recovered)	77.2	—
1 hr (¼ recovered)	74.5	—
2.5 hr	72.0	2.00
4 hr (½ recovered)	69.5	2.00
1 day (fully recovered)	62.5	2.10
2 days	61.5	2.00
6 days	61.0	1.85
8 days	62.0	1.90
16 days	63.0	1.95

(From Brock and Mosser, 1975.)

As seen in Table 12.5, the water entering drained pools is virtually of the same pH as was the water standing in the springs. Because pH and sulfate concentration are directly related at the low pH values involved, it seems reasonable to conclude that most of the sulfate was present in the water that entered the small springs and was not produced *in situ.* A small amount of sulfate was probably produced *in situ,* as shown by the slightly higher pH values of the water entering the drained springs than the water after steady state had been restored. The conclusion that most of the sulfuric acid in the smaller springs is not produced *in situ* seems reasonable when the assays for the various sulfur species are considered in Table 6.3: the levels of elemental sulfur and sulfide in the two springs are more than 10-fold lower than the level of sulfate sulfur. Spring 54-5 has 157 ppm sulfate sulfur and 2.1 ppm of reduced sulfur compounds, while Spring 56-3 has 96 ppm of sulfate sulfur and 8.2 ppm of reduced sulfur compounds. Because both springs are in steady state, the only way that sulfate sulfur can be higher than reduced sulfur is if the excess is entering the springs from outside. Similar reasoning can be used for most of the other small springs listed in Table 6.3.

Unfortunately, it would not be possible to drain a large spring and measure the concentration of sulfate in the newly inflowing water. From the chemical analyses in Table 6.3, however, it seems more likely that most of the sulfate sulfur in these large springs arises as a result of bacterial oxidation of elemental sulfur. Both Moose Pool and Sulfur Caldron have around 700 ppm of sulfate sulfur and over 1000 ppm of reduced sulfur species. Also, studies using $^{35}S^0$ have shown that the bacteria in these two pools are very active in oxidizing elemental sulfur to sulfate. The rates calculated from these radioisotope studies are 60,000 g sulfate/day from Moose Pool, and 57,000 g/day from Sulfur Caldron [calculations derived from data of Mosser et al. (1973), and unpublished data]; both rates are more than sufficient to account for the rate of sulfuric acid production in these pools. Also, the volume flow rates of these two springs are very high (Table 12.3) and it seems likely that rates of sulfuric acid production in the nearby soil would be insufficient to maintain the rates of sulfuric acid production needed.

It can be concluded that the nonflowing springs of acid-altered geothermal areas are steady-state systems, because during the time that added chloride was being exponentially diluted, the other chemical parameters measured (sulfur compounds, iron, pH) remained essentially constant. The water and its dissolved constituents were being replaced at relatively rapid rates, which implies that all of the measured constituents were being replaced from underground sources or were being produced within the springs as rapidly as they were being removed. It is likely that the sulfuric acid production rates measured in the small springs relate mainly to the rate of seepage of acid water into the springs, and that very little of the acid in the springs is formed by bacterial action *in situ.* On the other hand, the

Figure 12.14. Plant-free area (nonthermal) in the seepage zone of Lemonade Creek, illustrating the lower pH limit for higher plants. Measurements of pH in the roots of sedges, grasses, and pines growing at the border of the plant-free zone showed values around 3.0–3.5, which is presumably the lowest pH value at which higher plants grow.

much larger production rates in the large springs are probably the result of bacterial production *in situ*. As seen in Table 12.4, the rate of sulfuric acid production in these large springs is many times that in the smaller springs, and even on an areal basis they are producing at a considerably faster rate. Thus, the presence of a few large springs in a basin will tend to dominate the production rate.

It seems possible that the reason that little sulfuric acid is produced *in situ* in the smaller springs is because turnover rates are rapid, so that there is insufficient reduced sulfur species for bacterial oxidation. In the large springs, on the other hand, turnover rates are low, and elemental sulfur can build up to high amounts as a result of oxidation of hydrogen sulfide. Another consequence of low turnover rate is that the bacteria are not washed out very rapidly, thus permitting them to grow to high population densities.

It is also useful to contrast the spring habitat as a site for bacterial development with the soil habitat. It is well established that both *Thiobacillus* and *Sulfolobus* grow in acid geothermal soils (Schoen and Ehrlich, 1968; Fliermans and Brock, 1972), and because much of the area in these habitats is terrestrial rather than aquatic, considerably greater amounts of bacteria should be present. However, soil is generally not as favorable for bacterial development as water because of moisture stress. Thus even though greater numbers of bacteria may be present in soil, sulfuric acid production per single bacterial cell is almost certainly higher in the aquatic environment. This conclusion agrees with the data that show the rate of sulfuric acid production in the large springs (Table 12.4) and also agrees with the demonstration by radioisotope techniques of the high degree of bacterial activity in the large springs (Mosser et al., 1973). It also agrees with the high values of sulfuric acid production measured in the Japanese Noboribetsu crater lakes, which are also predominantly aquatic systems (Schoen, 1969).

Note

I have not discussed the lowest pH limit for higher plants. Although we made no systematic studies on this question, casual observations (Figure 12.14) suggest a lower pH limit around 3.0–3.5. The Yellowstone acid areas provide excellent habitats for a detailed study of this question.

References

Almer, B., W. Dickson, C. Ekstrom, E. Hornstrom, and V. Miller. 1974. Effects of acidification on Swedish lakes. *Ambio* **3**, 29–36.
Belly, R. T. and T. D. Brock. 1974. Widespread occurrence of acidophilic strains of *Bacillus coagulans* in hot springs. *J. Appl. Bacteriol.* **37**, 175–177.
Bennett, H. D. 1969. Algae in relation to mine water. *Castanea* **34**, 306–328.

Brock, T. D. 1973. Lower pH limit for the existence of blue-green algae: evolutionary and ecological implications. *Science* **179**, 480–483.

Brock, T. D. and M. L. Brock. 1970. The algae of Waimangu Cauldron (New Zealand): distribution in relation to pH. *J. Phycol.* **6**, 371–375.

Brock, T. D. and J. L. Mosser. 1975. Rate of sulfuric-acid production in Yellowstone National Park. *Geol. Soc. Am. Bull.* **86**, 194–198.

Collins, N. C. 1972. Population biology of the brine fly *Ephydra thermophila* (Diptera: Ephydridae) associated with acid seepages in Yellowstone National Park, Wyoming. Ph.D. dissertation, University of Georgia, Athens.

Collins, N. C. 1975. Tactics of host exploitations by a thermophilic water mite. *Miscell. Publ.* **9**, 250–254.

Darland, G. and T. D. Brock. 1971. *Bacillus acidocaldarius* sp. nov., an acidophilic thermophilic spore-forming bacterium. *J. Gen. Microbiol.* **67**, 9–15.

Doemel, W. N. and T. D. Brock. 1971. pH of very acid soils. *Nature* **229**, 574.

Fliermans, C. B. and T. D. Brock. 1972. Ecology of sulfur-oxidizing bacteria in hot acid soils. *J. Bacteriol.* **111**, 343–350.

Fogg, G. E. 1956. The comparative physiology and biochemistry of the blue-green algae. *Bacteriol. Rev.* **20**, 148–165.

Fritsch, F. E. 1916. The morphology and ecology of an extreme terrestrial form of *Zygnema* (*Zygogonium ericetorum* Kütz.) Hansg. *Ann. Bot.* **30**, 135–149.

Gessner, F. 1959. *Hydrobotanik, Vol. II.* VEB Deutscher Verlag der Wissenschaften, Berlin, 701 p.

Harp, G. L. and R. S. Campbell. 1967. The distribution of *Tendipes plumosus* (Linné) in mineral acid water. *Limnol. Oceanogr.* **12**, 260–263.

Lund, J. W. G. 1962. Soil algae. *In Physiology and Biochemistry of Algae,* R. A. Lewin, ed. Academic Press, New York, pp. 759–770.

Lund, J. W. G. 1967. Soil algae. *In Soil Biology,* A. Burges and F. Raw, eds. Academic Press, New York, pp. 129–148.

Lynn, R. and T. D. Brock. 1969. Notes on the ecology of a species of *Zygogonium* (Kütz.) in Yellowstone National Park. *J. Phycol.* **5**, 181–185.

Millar, W. N. 1973. Heterotrophic bacterial population in acid coal mine water; *Flavobacterium acidurans. Int. J. Syst. Bacteriol.* **23**, 142–150.

Mosser, J. L., A. G. Mosser, and T. D. Brock. 1973. Bacterial origin of sulfuric acid in geothermal habitats. *Science* **179**, 1323–1324.

Negoro, K. 1944. Untersuchungen über die Vegetation der mineralogen-azidotrophen Gewässer Japans. Science Report of the Tokyo Bunrika Daigaku, Sect. B6, 232–373.

Parsons, J. D. 1968. The effects of acid strip-mine effluents on the ecology of a stream. *Arch. Hydrobiol.* **65**, 25–50.

Prescott, G. W. 1962. *Algae of the Western Great Lakes* (rev. ed.). W. C. Brown Co., Dubuque, Iowa, 977 p.

Rosa, K. and O. Lhotsky. 1971. Edaphische Algen und Protozoen im Isergebirge, Tschechoslowakei. *Oikos* **22**, 21–29.

Schoen, R. 1969. Rate of sulfuric acid formation in Yellowstone National Park. *Geol. Soc. Am. Bull.* **80**, 643–650.

Schoen, R. and G. G. Ehrlich. 1968. Bacterial origin of sulfuric acid in sulfurous hot springs. *XXIII Int. Geol. Congr.* **17**, 171–178.

Schoen, R. and R. O. Rye. 1970. Sulfur isotope distribution in solfataras, Yellowstone National Park. *Science* **170**, 1082–1084.

Tilden, J. E. 1898. Observations on some West American thermal algae. *Bot. Gaz.* **25**, 89–105.

Ueno, M. 1958. The disharmonious lakes of Japan. *Verh. Int. Ver. Limnol.* **13**, 217–226.

West, G. S. 1916. *Algae, Vol. 1*. Cambridge University Press.

West, W. and G. S. West. 1894. On some freshwater algae from the West Indies. *J. Linnean Soc.* **30**, 264–280.

Wirth, W. W. 1971. The brine flies of the genus *Ephydra* in North America (Diptera:Ephydridae). *Ann. Entomol. Soc. Am.* **64**, 357–377.

Chapter 13
The Firehole River

The Firehole River, which flows through the main geyser basins of Yellowstone National Park (Figure 13.1), has been the main avenue of tourist travel through Yellowstone for over 100 years. It would be a remarkable river in any setting, but here in the valley of the geysers, its beauty and excitement are unsurpassable. Immediately as one glimpses the Firehole through the trees during the descent into the Upper Geyser Basin, one knows that this is not an ordinary mountain river, but something special. However, the Firehole River is more than a tourist attraction: it is of great scientific interest because it is the best example known of a naturally thermally polluted river. In these days when the long-range consequences of environmental pollution are constantly debated, it is not difficult to see the value for study of a river such as the Firehole, which has been receiving thermal inputs for hundreds of years and may, consequently, reveal to us something about how ecosystems can adapt to pollutants, given enough time. And the Firehole is especially interesting when it is understood that despite the fact that its temperature is raised almost 15°C as a result of thermal inputs (about what the largest power plant could do to its cooling water), it is still considered an excellent trout stream, one that attracts fly fishermen from all over the world.

The description of the Firehole by C. W. Cook of the 1870 Folsom-Cook-Peterson expedition sets the tone: "We followed up the river five miles, and there found the most gigantic hot springs we had seen. They were situated along the river bank, and discharged so much hot water that the river was blood warm a quarter of a mile below. . . . The waters from the hot springs in this valley, if united, would form a large stream; and they increase the size of the river nearly one-half. Although we experienced no bad effects from passing through the 'Valley of Death,' yet we were not

Figure 13.1. The Firehole River in the middle of the Upper Geyser Basin. Old Faithful Geyser is at the upper right of the picture.

disposed to dispute the propriety of giving it that name. It seemed to be shunned by all animated nature. There were no fish in the river, no birds in the trees, no animals—not even a track—anywhere to be seen; although in one spring we saw the entire skeleton of a buffalo that had probably fallen in accidentally and been boiled down to soup" (Bonney and Bonney, 1970). In actuality, the Geyser Basins teem with life; the reason Cook and party found no fish is that the Firehole Falls constituted a migration barrier. When trout were introduced into the river in 1889, they thrived in the ice-free waters.

The Firehole was given its name by early trappers, but there are conflicting accounts of the origin of the word. By one account, the name was given to directly express the steamy and firey nature of the Firehole basin, whereas by another account, the word is a variant of "Burnt Hole," the name given to the region around the present-day West Yellowstone,

Montana, which suffered a serious forest fire in the early 1800s. The Madison River which flows through the West Yellowstone basin is really the lower part of the Firehole River, so that the name Burnt Hole might have been transferred upstream and altered into "Firehole" (Bonney and Bonney, 1970). (To the early trappers, "hole" was any broad, generally flat valley between the mountains, viz., Jackson's Hole, Gardner's Hole.) The trapper Jim Bridger, a legend in this part of the West, advanced the interesting theory that the Firehole River was hot because it flowed so rapidly over its bottom that it was heated by friction. When Nathaniel Langford visited Yellowstone with the Washburn expedition of 1870, he recalled Jim Bridger's tale: "Mr. Hedges and I forded the Firehole River a short distance below our camp . . . taking off our boots and stockings, we selected for our place of crossing what seemed to be a smooth rock surface in the bottom of the stream . . . When I reached the middle of the stream I paused a moment and turned around to speak to Mr. Hedges . . . when I discovered from the sensation of warmth under my feet that I was standing upon . . . a hot spring that had its vent in the bed of the stream. I exclaimed to Hedges: 'Here is the river which Bridger said was *hot at the bottom*'." (Langford, 1870).

General Features of the River

The Firehole River arises as a typical mountain stream on the slopes of the Continental Divide (see Figure 13.2). It flows about 12 km before reaching the first real thermal feature, Lone Star Geyser, but this geyser has little water output and scarcely influences the temperature of the river. It is below Kepler Cascades that the Firehole first enters real geyser country, the upper Geyser Basin which contains such large and famous geysers as Old Faithful, Castle, Grand, Giantess, Rocket, and Grotto, plus probably a hundred more of a size that would make them noteworthy in other parts of the world. Just above Lone Star Geyser, the midsummer temperature of the Firehole is around 10°C in July, typical of streams in this high Rocky Mountain area, but by the time the river has passed through the Upper Geyser Basin its temperature has increased to around 14°C. After flowing through the Biscuit Basin its temperature rises to around 17°C, and it receives another big boost in temperature to around 22°C at Midway Geyser Basin, where the huge Excelsior Geyser Crater contributes so much water that the flow of the river is visibly increased. The impressive Excelsior runoff is visible to any Yellowstone tourist, as the Grand Loop Road passes just on the opposite side of the river. As Lt. Doane, a leader of the expedition of 1870, noted, the thermal stream entering the river here is big enough to run a grist mill. But as yet the Firehole has not reached the Lower Geyser Basin, the largest in the Park. Here it receives input from such famous thermal features as Great Fountain Geyser, Hot Lake, and

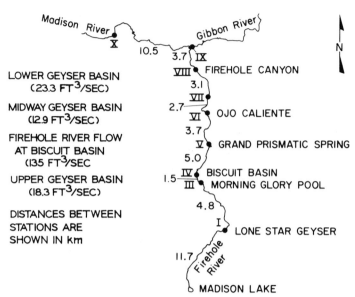

Figure 13.2. Sketch map of the Firehole River showing the locations of the main thermal inputs.

Ojo Caliente. The geysers of the River Group are here, a large number of superheated springs that lie along both sides of the river. At this point, the summer temperature of the river reaches 25–26°C, just perfect for swimming, a fact not missed by a large number of summer visitors. In contrast, it is impossible to enter the river above Kepler Cascades without waders or a wet suit. The Lower Geyser Basin is essentially the end of the thermal area. A total of about 55 cubic feet per second of thermal water has entered the river below Lone Star Geyser, and since the river itself only has a flow of about 135 cubic feet per second, about 30% of the volume of water in the river is from geysers and hot springs (Allen and Day, 1935). From here on, the river gradually cools, but it never drops back to its original temperature, as soon below it is joined by the Gibbon River at Madison Junction, forming the Madison River. Eventually the Madison River joins with the Jefferson and Gallatin at Three Forks, Montana, to make the Missouri River, and on flows the water to the sea.

    In all, during its 15-km excursion through the Yellowstone Geyser Basins, the temperature of the river has been raised about 15°C, about what a large electric power plant would do to the cooling water that it takes through its plant. However, no power plant has the impact on its river that the Geyser basins have on the Firehole, since power plants never use *all* of the water in a river, but only a fraction, so that the increase in the average temperature of the river is only a few degrees, whereas in the case of the Firehole, the *whole river* is increased in temperature by 15°C, a condition that would be achieved by the electric power industry only if it used virtually all of the water in a river for cooling purposes. In the Firehole River, as opposed to the thermally polluted river, there are no cool deep holes in which organisms can hide; if they live in the river, they are heated by it.

For a variety of reasons, the Firehole is an excellent river for studies on thermal pollution. Because there are no cool places for organisms to hide, research studies are easy to make on the Firehole, as the scientist is quite certain of the temperatures that the organisms he is studying are experiencing. The Firehole has long-range stability also; the thermal input of the Yellowstone geyser basins has remained essentially constant over the past 100 years, so that the temperature is predictable, within limits, from year to year. Because of this predictability, we do not have to spend an inordinate amount of time checking temperatures. Finally, it is quite certain that no environmental protection authority is going to turn off the flow of heat, so that long-range studies can be initiated without the worry that the habitat is going to change. Only a major geological catastrophe could alter the essential features of the Yellowstone area, although, of course, continual small changes in the activity of single thermal features do occur.

Thermal Regime of the River

The temperature at any point along the Firehole River is not constant, but can vary both seasonally and diurnally. The diurnal fluctuations are rather minor; the river is warmed during the day by the heat of the sun, and cools at night, the range over a 24-hour period being about 5°C (Figure 13.3). The annual range is somewhat wider, although not perhaps as wide as might be thought. The river is not at its coldest in the middle of the winter, despite

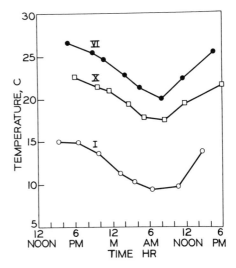

Figure 13.3. Diurnal temperature variation at three stations in the Firehole River on 16–17 August 1971. The Roman numerals refer to station locations given on Figure 13.2. (From Boylen and Brock, 1973.)

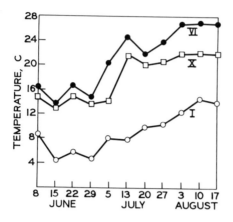

Figure 13.4. Temperature during the summer of 1971 at three stations on Firehole River. The Roman numerals refer to station locations shown on Figure 13.2. (From Boylen and Brock, 1973.)

the fact that the Yellowstone region has long, hard, snowy winters. The coldest season for the river is in early to mid-June, when the runoff from melting snow is at its peak (Figure 13.4). In the winter, all of the precipitation that falls is held on the ground as snow, and the river is at its lowest water level, and temperatures remain moderately high. The warmest period is in early to middle August, when all snow melt water has gone and the summer sun has had a month or so to add its contribution to the heat budget of the river.

Chemical Alteration of the River

The geysers and hot springs affect more than just the temperature of the river. These thermal waters are fairly mineralized, and contribute large amounts of arsenate, sodium, chloride, and bicarbonate to the river water (Table 13.1). When snow melt is high, the effect is the least, but by mid-August the chemistry of the river has been profoundly altered. However, the nitrogen budget of the river is only slightly affected since the thermal waters are lower in nitrate and ammonium than normal mountain river water. Thus, the concentrations of nitrate and ammonium are actually lowered somewhat in the stretch of river where the maximum input of hot spring water occurs. Because of these chemical alterations, it is important to distinguish thermal from chemical effects on the organisms of the river.

Biological Effects

That the organisms living in the river are profoundly affected by the inputs of geyser and hot spring water is well established. This was first surmised by David Starr Jordan, the famous nineteenth century ichthyologist and first president of Stanford University, who made a survey of the Yellow-

Table 13.1. Temperature and Water Analysis of Firehole River at Different Sampling Stations[a]

Station	Temperature (°C)	pH	Alkalinity (mg/l CaCO₃)	Conductivity (μmhos/cm)	Arsenate + Phosphate (μg/l)[b]	Nitrate (μg/l)	Chloride (mg/l)	Ammonium (μg/l)
I	12	7.4	15	65	50	240	7	70
III	17	7.6	32	160	125	250	22	60
IV	20	7.7	46	220	200	270	25	50
V	23	8.1	64	285	295	210	34	55
VI	26	8.3	90	420	485	190	57	45
VII	25	8.3	88	410	485	210	53	50
VIII	24	8.3	84	385	425	140	50	45
X	21	7.8	88	360	345	190	41	60

[a]Data are arithmetic averages for the period 20 July–17 August 1971, when water depth was at a minimum. (From Boylen and Brock, 1973.)
[b]The phosphate assay used by Boylen and Brock also measures arsenate, a significant component in Yellowstone thermal waters (Robert Stauffer, personal communication), and much of the increase seen in this column from Station I through Station VI is probably due to arsenate.

stone fish species for the U.S. Fish Commission (Jordan, 1889). However, the first quantitative data on biological effects was that of K. B. Armitage, who determined standing crops of riffle insects at different locations along the river (Armitage, 1958, 1961). Armitage found a much higher standing crop in the stretch of river that flowed through the geyser basins than he found above the geyser basins at Lone Star Geyser. He suggested that increased bicarbonate in the river from the geysers and hot springs resulted in increased plant growth which indirectly resulted in increased growth of riffle insects. He also showed (Armitage, 1961) that certain genera were found primarily in the colder water whereas others seemed to prefer warmer water. The data of Armitage interested John Wright of Montana State University, and in the middle and late 1960s a group of his students and associates carried out studies on the ecology of the Madison River system, although their main work was in the Madison River rather than the Firehole (Roeder, 1966; Todd, 1967; Wright and Horrall, 1967; Wright and Mills, 1967). As a result of Wright's work, I became aware of the potential of the Firehole River for examining some of the long-range consequences of thermal pollution. My own work had been concentrated on the biology of the hot springs and geysers themselves, and for over 10 years we have carried out a variety of studies in the geyser basins. From our laboratory in West Yellowstone, Montana, we drove along the Firehole River every day to reach our study areas. It was not long before we were asking ourselves questions about thermal effects on Firehole ecosystems, and with the financial aid of the Atomic Energy Commission we were able to carry out such studies. Our work on the Firehole was initiated in collaboration with J. G. Zeikus, now a faculty member at the University of Wisconsin, and was pushed considerably further by Charles Boylen, now at Rensselaer Polytechnic Institute. Our work has concentrated on effects on the microbiota, especially algae and bacteria, and since the algae are at the base of the Firehole River food chain, first attention should be given to them. Since in a river, organisms suspended in the water are usually drifting rather than actively growing, focus is placed on bottom-living (benthic) organisms. Benthic organisms not only make the major contribution to the function of river ecosystems, but their environments can be more precisely specified, since the places where they live and grow, being fixed to the bottom, are exactly known.

Algal Studies

Boylen measured standing crop of benthic algae at a series of stations along the river, above, within, and below the geyser basin (Boylen and Brock, 1973). Standing crop was quantified by measurement of chlorophyll-a, the predominant chlorophyll in most algae. Stones of about 200 cm^2 area were removed from the rubble at the bottom of the river and placed in heavy plastic bags, top surface down, and acetone added to completely submerge the algal mat. After overnight extraction in the dark, the chlorophyll in the acetone extract was quantified by measurement of its absorption at 663 nm, the major absorption peak of chlorophyll in the red light region. After extraction, the stones were allowed to dry and the area previously occupied by the algal mat could be seen as a white residue; this area was measured and chlorophyll content per unit area calculated. At the same time that these measurements were made, samples were taken to determine the algal species present.

Boylen found that algal standing crop correlated directly with temperature of the habitat, showing a minimum in the cold upper reaches of the river, rising sharply as the river flowed through the geyser basins, and falling off again as the river gradually cooled below the geyser basins (Figure 13.5). In fact, it was not really necessary to do these quantitative studies to realize this fact, since simple observation of the river bottom easily confirms that algae are less abundant in the cool than in the warmer waters. Boylen also found marked differences in algal species at different locations. In the colder waters, diatoms predominated, whereas in the warmest waters, filamentous green algae of the genera *Spirogyra, Oedogonium, Cladophora,* and *Stigeoclonium* dominated. Blue-green algae, often thought to be favored by thermal pollution, were present at some stations, but never were dominant except at Biscuit Basin (an area receiv-

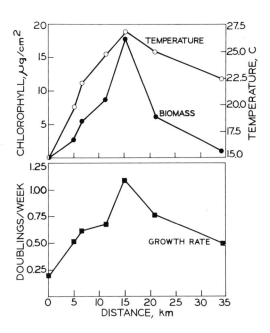

Figure 13.5. Comparison of temperature with biomass (measured as chlorophyll-a) and growth rate (measured as shown in Figure 13.6) of algal populations growing on rocks in the Firehole River, as a function of distance down the river. The points correspond to the station numbers given in Figure 13.2 and the temperature data represent average values for the midsummer period. (From Boylen and Brock, 1973.)

ing only moderate temperature increases), where a mat of *Nostoc* nodules was present. Other work in our laboratory has shown that it is only when water temperatures are pushed up to 40°C that blue-green algae become dominant (Brock, 1975).

Algal standing crop is a function of three major variables: algal growth rate, washout rate, and grazing by herbivores. In early summer when the river is swollen by snow melt water, washout is probably the predominant process, and the existing mats are torn loose and the rocks swept virtually clean. As the river level drops, increases in algal biomass can occur, and by early August the standing crop is back to normal levels. Since grazing does not seem to be a major factor, standing crop in August is determined primarily by growth rate. Thus, the fact that standing crop is highest in the warmest parts of the river is probably due to the fact that growth rate is most rapid in these areas.

Boylen confirmed that algal growth rates were fastest in the warmer parts of the river by direct measurements. Growth rates were measured in two ways, both involving quantification of algal biomass by chlorophyll extraction. In the first, stones were removed at intervals through the summer after erosion of the algal mats by the high currents of June had swept the stones clean of algae. In the second, marked clean stones (obtained from a nearby glacier deposit) were placed in the river to provide sites for algal colonization and were removed at intervals for chlorophyll extraction (Figure 13.6). Both methods showed the same result: growth rate was markedly faster in the warmer than the cooler sections of the river (Figure 13.5).

Although the data relating river temperature to algal growth rates are quite clear-cut, the interpretation is not direct. As we noted earlier, the

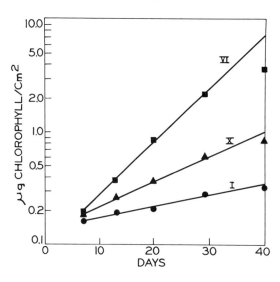

Figure 13.6. Growth rates of algal populations on clean stones placed 22 June 1972 at Stations I, VI, and X. For a summary of all the data, see the lower part of Figure 13.5 (From Boylen and Brock, 1973.)

geyser basins not only cause an increase in the temperature of the river, but an increase in certain nutrients (Table 13.1). At one time we thought that the geyser basins caused a major increase in the phosphate concentration of the river, but Robert Stauffer of the U.S. Geological Survey has now shown that the phosphate assay used is not specific for phosphate, but assays arsenate as well, and it is arsenate rather than phosphate that the hot spring waters are rich in. Thus, it is likely that little increase in phosphate occurs in the river as it flows through the geyser basins. The increase in bicarbonate is real, however, and significant, and since bicarbonate and its equilibrium products are the main carbon source for algae, the increases in algal standing crop could be due to bicarbonate enrichment of the water. Although it is not easy to test this point experimentally, Boylen attempted to do so by observing the effects of bicarbonate enrichment on photosynthesis by algal populations from different parts of the river. Limnologists have used such a nutrient-enrichment approach to determine factors potentially limiting algal growth in lakes. Boylen did find the bicarbonate additions stimulated algal photosynthesis, but the stimulation by bicarbonate was even greater in the algae from the warmest section of the river than in those from the unheated portion, so that bicarbonate limitation can probably not explain the higher algal growth rates in the warmest parts of the river. Although nutrients may play some part in the increased growth rate, it seems reasonable to conclude that temperature itself is a factor in the marked increase in algal growth rate and standing crop in the warm sections of the river.

That increases in temperature of the kind we are dealing with should increase algal growth rate is perhaps not surprising. There are very few algae that are optimally adapted to cold waters, and even those that are must cope with the fact that low temperatures slow down such physical processes as diffusion and turbulent mixing, which aid in bringing nutrients to the algal cell. Even the few algae that have been shown to be optimally adapted to low temperatures (mainly found in the Antarctic and in high alpine lakes) do not grow exceptionally fast. If water warms, these algae will be quickly replaced by faster growing algae adapted to the higher temperatures. To look at the phenomenon of temperature adaptation in the Firehole algae, Boylen determined the effect of temperature on photosynthesis of populations taken from various sections of the river. Results obtained at the coldest and warmest stations are shown in Figure 13.7 and a summary of the results for all the stations is given in Figure 13.8. The results were quite clear-cut. The algal population at each location in the river had a temperature optimum close to that of its habitat. Additionally, the temperature optima of the algal populations did not change throughout the summer as the temperature of the water changed, but remained close to that of the temperature of their habitats at the warmest time of the year, in early August. This was true even when studies were done with populations

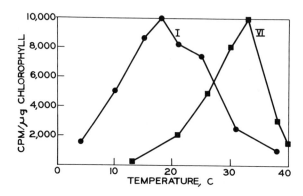

Figure 13.7. Effect of temperature on photosynthesis by algal populations taken from Stations I and VI. (From Boylen and Brock, 1973.)

collected at the time of maximum snow melt, when the temperature of the river was low. Thus it appears as if the algae retain their midsummer temperature optima throughout the year, and do not adapt to the low temperatures that they experience during the spring. In effect, the temperature optimum of the algae at any location in the river is determined by the maximum temperature that this location reaches in the warmest time of the year. This is a far-reaching conclusion, since it shows that algae are not able to keep up with changes in the temperatures of their habitats, so that much of the year they are growing at suboptimal temperatures. This conclusion is of considerable importance for predicting the consequences of thermal pollution, since it means that if thermal additions (as, from a power plant) were to lead to warming of a body of water during the colder parts of the year, then the algae present, already preadapted to these warmer temperatures, would immediately grow at faster rates, thus leading to increases in algal standing crop. James Hoffman and I have already obtained evidence that exactly fits this prediction from studies at a power plant outfall on Lake Monona in the city of Madison, Wisconsin (Brock and Hoffman, 1974). Thus the Yellowstone observations seem to be quite applicable to man-made thermal habitats.

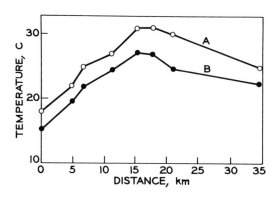

Figure 13.8. Comparison of temperature optima for photosynthesis by algae at different stations (line A) compared with the average midsummer temperatures (line B). (From Boylen and Brock, 1973.)

Bacterial Studies

The hot springs and geysers are habitats for a unique group of extremely thermophilic bacteria, capable of living and growing rapidly at temperatures of 70–90°C (Brock and Freeze, 1969; Bott and Brock, 1969; Brock et al., 1971). As discussed earlier, these bacteria are of considerable evolutionary interest because they provide living evidence of the adaptability of organisms to extreme environments. However, none of these bacteria are capable of reproducing at normal temperatures, and would not be expected to become established in the Firehole River. The bacteria in the Firehole River, like the algae, are the normal sorts of organisms found in mountain streams and lakes, although in the heated portions of the river their activities are modified by the increased temperature. In 1970, J. G. Zeikus carried out a series of studies on bacterial growth rates in the unheated and heated portions of the Firehole River (Zeikus and Brock, 1972). In addition, Zeikus set up some studies to determine whether the thermophilic bacteria living in geysers and hot springs might be useful as indicators of thermal additions. The rationale here was that these organisms were being carried by spring and geyser waters from their high-temperature habitats into the Firehole River. Although unable to reproduce in the river because of the temperature, these organisms remain alive and drift down the river. They are easy to detect, because they can be specifically counted if the agar medium on which they are plated is incubated at high temperature.

In Zeikus's work, two bacteria were studied, *Bacillus stearothermophilus,* a spore-forming gram-positive rod which has a temperature optimum around 55–60°C, and *Thermus aquaticus,* a gram-negative nonsporing rod which has a temperature optimum of 70–75°C. Both organisms live at appropriate temperatures in the run-off channels of geysers and hot springs. Plate counting done at 55°C permits quantification of *B. stearothermophilus,* and incubation at 70°C can be used to count *T. aquaticus.* As a measure of the normal river bacteria, plate counts can also be done at 25°C. These counting studies were done at a series of stations along the Firehole River, from above the geyser basins down through the recovery zone in the Madison River.

Results of a typical series of counts of the thermophilic bacteria drifting down the river are seen in Figure 13.9. Note that at Station I, above the geyser basins, no thermophilic bacteria were seen, whether platings were done at 55°C or 70°C. Low populations of *B. stearothermophilus* were seen at Station II, the first station with thermal influence, and then the numbers of this organism increased dramatically, peaking at Station V, just above the warmest section of the river. In the recovery zone, counts of this organism dropped to almost undetectable levels. *Thermus aquaticus* also was absent in the cold section of the river, and numbers rose in the heated sections, reaching a peak at Station VI and then dropping to again undetectable levels at Station X in the Madison River. The increases of thermophilic

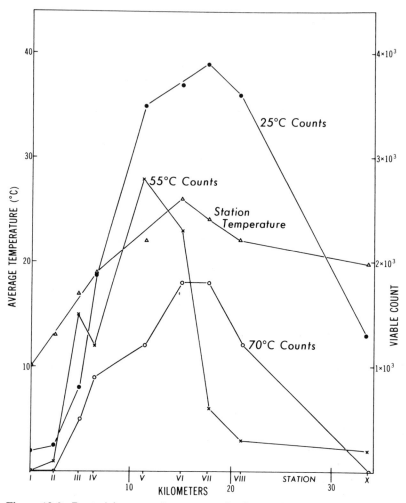

Figure 13.9. Bacterial counts of water samples from different stations performed by incubating parallel series of plates at three different temperatures, 25°C, 55°C, and 70°C. Numbers per milliliter of water. (From Zeikus and Brock, 1972.)

bacteria in the heated sections of the river are thus a result of the direct input of thermal water to the river, and these organisms are excellent indicators of thermal additions. The causes of the decrease in numbers of thermophilic bacteria below the geyser basins are unknown, but could result from death, predation, or sedimentation, or (most probably) a combination of all three.

Figure 13.9 also shows a marked increase in numbers of bacteria detectable by plating at 25°C. This increase undoubtedly reflects an increase in bacterial production in the river itself, since organisms capable of growing at 25°C are not capable of growing at high temperatures. This increased bacterial production in the warmer sections of the river is almost certainly a reflection of the increased algal productivity and the consequent increase in organic nutrients needed by these heterotrophic bacteria.

These results show that counts of thermophilic bacteria can be used as a simple way of measuring thermal additions. What advantage might this procedure have over simply measuring the temperature of a river? The use of bacterial indicators permits a distinction between two kinds of thermal pollution, that due to addition of large amounts of moderately warm water and that due to additions of small amounts of quite hot water. Addition of large amounts of moderately warm water, as is done by power plants, would not result in increases in thermophilic bacteria, since the temperature of the warmed water is at most only 10–15°C higher than ambient, so that even in midsummer, temperatures of the added water would not rise above 40°C. Thus, thermophilic bacteria would not be expected to be present in power plant effluents. However, many manufacturing operations produce rather small amounts of quite hot water, temperatures over 55°C, and sometimes up to boiling. It is in this kind of water that the thermophilic bacteria thrive, and counts in the receiving water could provide a simple means of detecting such kinds of industrial pollution. The net effect on the temperature of the river may be no different with additions of small amounts of very hot water and large amounts of moderately warm water, but we can distinguish these two situations by counting thermophilic bacteria. It is now well established (Brock and Boylen, 1973; Brock and Yoder, 1971) that thermophilic bacteria such as *Thermus aquaticus* are widespread in hot water heaters and steam condensate lines. These bacteria find the temperatures in these systems close to optimal, and presumably obtain their nutrients for growth from the small amounts of organic matter present in the waters. When such hot waters are discharged, the bacteria are carried along and can be detected by plating and incubation at 70°C. In one study in Indiana, Brock and Yoder (1971) found *Thermus aquaticus* in a small, spring-fed creek that was being polluted by a leaking steam condensate line from the heating system of Indiana University. The condensate water itself had ranged in temperature from 48°C to 56°C and had very high counts of *Thermus aquaticus*. Despite the small size of the effluent, its effect on the creek could be detected through its input of *T. aquaticus* to the creek water. In another branch of the same creek, thermal pollution from an unknown source under the Jordan Hall of Biology resulted in input of *T. aquaticus* which could also be detected by plating. Thus, this organism is a simple and sensitive indicator of thermal additions from high-temperature sources.

Fish in the Firehole River

In the Rocky Mountains, fish means trout. This parochial view of what constitutes fish life has resulted in rather narrowly circumscribed research studies on the Firehole River, with primary emphasis on sport fisheries.

David Starr Jordan made pioneering observations on the fish of Yellow-stone in 1889 for the U.S. Fish Commission. The motivation for this work, as he says, was the thought "that much could be done towards enhancing the attractions of the great national 'pleasuring ground' by the stocking of those of its various streams and lakes which are now destitute of fishes." Stocking of fish has been a traditional activity of the wildlife manager, and has been practiced throughout the world. Fish obviously cannot jump from lake to lake, and the only means of dispersal is by swimming. Lakes connected by rivers can be colonized if there are no barriers such as waterfalls, large rapids, or other obstructions. Yellowstone Park is a prime example of an area where migration barriers prevented fish colonization of waters that were otherwise suited for fish life. Yellowstone is a high plateau, separated from the surrounding lowlands by sheer cliffs and steep descents, and every river leaving the well-watered uplands drops through waterfalls or large rapids. In the Yellowstone River there are the Upper and Lower Falls (109 and 308 feet, respectively) at the entrance to the Grand Canyon of the Yellowstone. On the south side of the Park, Lewis Falls of the Lewis River is 80 feet high, and there is a series of waterfalls on the Bechler and Falls Rivers in the southwest corner of the Park. In the North, Osprey Falls in the Gardner River is 150 feet high, and Tower Falls (135 feet) is a famous tourist attraction of Tower Creek. In the Madison River drainage, Virginia Cascades (60 feet) and Gibbon Falls (80 feet) break the flow of the Gibbon River, and the Firehole Falls (60 feet) sits at the lower end of the Firehole River.

It seems to be generally agreed by nineteenth century observers that fish (i.e., trout) were entirely absent in the regions of Yellowstone above the major waterfalls except in the Yellowstone River-Yellowstone Lake drain-age where native cutthroat trout were present. Thus, fish were present in the Madison River, but were absent in the Firehole, which is separated from the Madison by the 60-foot-high Firehole Falls. David Starr Jordan confirmed this fact during his rather hurried reconnaissance in early Octo-ber 1889, and provided a list of all of the rivers and lakes in Yellowstone that were devoid of fish.

Even while Jordan's reconnaissance was underway, stocking of fish was in progress. In September 1889, 1000 brown trout *(Salmo trutta)* were planted in the cold, clean section of the Firehole River above Kepler Cascades, and rainbow trout *(Salmo gairdneri)* were planted in the Gibbon River above Virginia Cascades (Benson et al., 1959). In later years, fish stocking became a way of life for the Park manager, and thousands of fingerlings were planted yearly during the early post World War II period. Then it was learned that stocking was not really necessary to maintain the Yellowstone fisheries, and in 1953 stocking was terminated (except for a token stocking of grayling in Ice Lake in 1961) and the waters allowed to return to their original self-sustained basis (Sharpe, 1970). Today, manage-ment of the Yellowstone fisheries is accomplished through size and bag

limits, restrictions on kinds of lures and baits, and catch-and-release programs. Because of stocking and because fish census concerns only catcha-
ble-size fish, it is difficult to draw any firm conclusions about the thermal
impact on fish in the Firehole River. Thus Jordan's observations, made
before stocking became an established practice, provide the only insight on
this point. Unfortunately, his observations of the Firehole were only
cursory and did not involve temperature measurements. Jordan did note
the presence of several fish in Witch Creek, a warm creek that drains the
Heart Lake Geyser Basin. At a temperature of 31°C, he recorded the Utah
chub (*Leuciscus atrarius,* now named *Gila atraria*), the fish he considered
most tolerant of hot water. This fish is a member of the family Cyprinidae, a
large family that contains such environmentally tolerant fishes as the carp,
goldfish, and minnow. Another fish he found in Witch Creek was the sucker
Agosia nubila and concluded: "The chubs ascend until they reach water
fairly to be called hot, and the sucker is not far behind." However, 31°C is
not especially hot as far as fish life is concerned, as a variety of desert
pupfish (family Cyprinodontiformes) are able to carry out their whole life
history in waters with temperatures close to 40°C (Deacon and Minckley,
1974). Pupfish are common in warm springs and desert pools throughout
the Great Basin states of Utah, Arizona, and Nevada, but have not yet
spread into Yellowstone, although they would surely grow well in the many
warm streams. Indeed, there is almost certainly an uncolonized niche in the
warm Yellowstone streams and in the cooled waters of hot springs and
geysers where pupfish would grow rapidly if they were introduced (I do not
advocate such an introduction; I like Yellowstone the way it is).

However, the Firehole never reaches temperatures much above 26–27°C
except in the areas immediately around hot spring outfalls (Argyle, 1966),
so that what we are concerned with is the ability of fish such as trout to
multiply in waters that are only moderately warm. It is often thought that
trout will only live and reproduce in cold water. Santaniello (1971) gives the
maximum recommended temperature for spawning, egg development, and
growth of trout as 20°C and Snyder (1969) gives a preferred temperature
range of 12–18°C and an upper lethal temperature of 25°C for juveniles and
28°C for adults. Yet is is well established that trout live and reproduce in the
Firehole River at temperatures considerably higher than the recommended
values and close to the upper lethal temperatures. Benson et al. (1959)
found that adult brown and rainbow trout, as well as whitefish *(Prosopium
williamsoni),* were growing rapidly in both the Madison and Firehole
Rivers. The catch of brown trout was primarily 3- and 4-year-olds, while
the catch of rainbow trout was 2- and 3-year-olds. Whitefish catch represented even older age classes. Benson et al. (1959) concluded: "Hot springs
and geysers entering the Firehole River do not appear to restrict the
number of brown and rainbow trout." David Starr Jordan had drawn a
similar conclusion based on his brief observations in 1889: "The water of
the geysers and other calcareous and silicious springs does not appear to be

objectionable to fishes. In Yellowstone Lake trout are especially abundant about the overflow from the Lake Geyser Basin [now called West Thumb Geyser Basin]. The hot water flows for a time on the surface, and trout may be taken immediately under these currents. Trout have also been known to rise to a fly through a scalding hot surface current. They also linger in the neighborhood of hot springs in the bottom of the lake. This is probably owing to the abundance of food in these warm waters, but the fact is evident that geyser water does not kill trout." These words echo the tale attributed to Jim Bridger that in Yellowstone Lake he could catch a trout in the underlying cold waters and, having hooked him, cook him on the way out! (Chittenden, 1964). In considering the Firehole itself, Jordan could also find no detrimental reason why trout should not grow: "Even at the midway geyser basin the stream is probably not too warm for trout." Thousands of fishermen who have cast flies in this section of the river would probably agree (Figure 13.10). Indeed, as Benson et al. (1959) note: "The Madison River drainage . . . includes the Madison, Firehole, and Gibbon Rivers . . . and has had a reputation for providing some of the best trout fishing in the western United States." (Considering the high arsenate values in the Firehole River, I am not certain that it is advisable to eat fish caught in the river.)

Conclusion

We have seen in this chapter that the Firehole River is an excellent model for analysis of the long-range consequences of thermal pollution. In its upper reaches it is a typical cold mountain river, and becomes markedly heated at it flows through the Yellowstone geyser basins. The thermal loading to the river from the geysers and hot springs is quite stable, and temperatures at defined locations in the river are predictable. During its 15-km excursion through the Yellowstone geyser basins, the temperature of the river is raised about 15°C, about what a large electric power plant would do to the cooling water that it takes through its plant.

Addition of hot spring and geyser water to the Firehole River results in marked biological effects. Algae, bacteria, and invertebrates all grow markedly faster and reach higher standing crops in the heated portions of the river, although it is not certain that all of the increase in biological activity is due to heating alone, since the geyser and hot spring waters are moderately mineralized. Only bicarbonate has been identified as a possible algal nutrient added by the thermal waters, but it is concluded that temperature itself is a factor, perhaps the major factor, in the marked increase in algal growth rate in the heated sections of the river. The increases in bacterial and invertebrate growth can be attributed primarily to the increased nutrients made available to them from algal photosynthesis.

The algae do not adapt to the changing temperatures that occur during

Figure 13.10. The Firehole River just above the Midway Geyser Basin. The cloud of steam in the upper left arises from Grand Prismatic Spring, the largest spring in the Park. The trout fisherman on the right provides the scale.

the annual cycle, but remain preadapted throughout the year to the warmest temperatures, experienced in early to middle August. Thus, the maximum temperature which the river reaches is seen as the key factor controlling the temperature response of the algal populations. A similar conclusion has been reached from studies in lakes and streams receiving power plant effluents.

Thermophilic bacteria living in the hot springs and geysers are carried into the Firehole River and can be quantified by selective plating techniques. These organisms can serve as indicators of location and extent of thermal additions to the river. These bacteria may also be useful as indicators of addition of high-temperature waters in man-influenced systems, such as occur from industrial and manufacturing operations, since the same bacteria found in the Yellowstone hot springs also live in hot water heaters and steam condensate lines.

Although it is commonly assumed that trout cannot live and reproduce in warm waters, it is well established that trout are successful in the Firehole River. In fact, the Firehole River is world renowned as a trout stream. Before 1889 trout did not exist above Firehole Falls, but were planted at that time and a number of subsequent plantings have also been made, although no stocking is occurring at the present time. If the trout in the warmest sections of the river represent a new race adapted to higher temperatures, an opportunity exists for the study of the genetic and physiologic processes involved in temperature adaptation. Such studies could be of use in predicting the long-range consequences of thermal pollution. Unfortunately, despite the great interest in this topic, very little work has been done since David Starr Jordan made the first scientific survey of Yellowstone fish in 1889.

Note added in proof. At long last a thorough study of trout in the Firehole River has been done: Kaya, Calvin M. 1977. Reproductive biology of rainbow and brown trout in a geothermally heated stream: The Firehole River of Yellowstone National Park. *Trans. Amer. Fish. Soc.* **106**, 354–361.

References

Allen, E. T. and A. L. Day. 1935. *Hot Springs of the Yellowstone National Park.* Carnegie Institution of Washington Publication No. 466, Washington, D.C. 525 pp.

Argyle, R. 1966. *Thermal Dispersion in Lotic Waters of Yellowstone National Park.* Masters thesis. Colorado State University, Fort Collins. 122 pp.

Armitage, K. B. 1958. Ecology of the riffle insects of the Firehole River, Wyoming. *Ecology* **39**, 571–580.

Armitage, K. B. 1961. Distribution of riffle insects of the Firehole River, Wyoming. *Hydrobiologia* **17**, 152–174.

Benson, N. G., Cope. O. B., and Bulkley, R. V. 1959. *Fishery Management Studies on the Madison River System in Yellowstone National Park.* Special Scientific Report-Fisheries No. 307, United States Dept. of Interior Fish and Wildlife Service, 29 pp.

Bonney, O. H. and L. Bonney. 1970. *Battle Drums and Geysers.* The Swallow Press, Inc., Chicago, 622 pp.

Bott, T. L. and T. D. Brock. 1969. Bacterial growth rates above 90°C in Yellowstone hot springs. *Science* **164**, 1411–1412.

Boylen, C. W. and T. D. Brock. 1973. Effects of thermal additions from the Yellowstone geyser basins on the benthic algae of the Firehole River. *Ecology* **54**, 1282–1291.

Brock, T. D. 1975. *Predicting the Long-Range Consequences of Thermal Pollution from Studies on Natural Geothermal Systems.* International Atomic Energy Agency Symposium, Oslo, Norway, published by I.A.E.A., Vienna, Austria, pp. 599–622.

Brock, T. D. and K. L. Boylen. 1973. Presence of thermophilic bacteria in laundry and domestic hot-water heaters. *Appl. Microbiol.* **25**, 72–76.

Brock, T. D., M. L. Brock, T. L. Bott, and M. R. Edwards. 1971. Microbial life at 90 C: the sulfur bacteria of Boulder Spring. *J. Bacteriol.* **107**, 303–314.

Brock, T. D. and H. Freeze. 1969. *Thermus aquaticus* gen.n. and sp.n., a non-sporulating extreme thermophile. *J. Bacteriol.* **98**, 289–297.

Brock, T. D. and I. Yoder. 1971. Thermal pollution of a small river by a large university: bacteriological studies. *Proc. Indiana Acad. Sci.* **80**, 183–188.

Brock, T. D. and J. Hoffman. 1974. Temperature optimum of algae living in the outfall of a power plant on Lake Monona. *Trans. Wis. Acad. Sciences, Arts and Letters* **62**, 195–203.

Chittenden, H. M. 1964. *The Yellowstone National Park.* Edited and with an introduction by R. A. Bartlett, from the 1895 edition. University of Oklahoma Press, Norman, 208 pp.

Deacon, J. E. and W. L. Minckley. 1974. Desert fishes. In *Desert Biology, Vol. II,* G. W. Brown, Jr., ed. Academic Press, New York, pp. 385–488.

Jordan, D. S. 1889. A reconnaissance of the streams and lakes of the Yellowstone National Park, Wyoming, in the interest of the United States Fish Commission. *Bull. U.S. Fish Comm.* **9**, 41–63.

Langford, N. P. 1870. *The Discovery of Yellowstone Park.* Reprinted by University of Nebraska Press, Lincoln, 1872, Foreword by A. L. Haines, 125 pp.

Roeder, T. S. 1966. Ecology of the diatom communities of the upper Madison River system, Yellowstone National Park. Ph.D. dissertation, Montana State University, Bozeman.

Santaniello, R. M. 1971. Water quality criteria and standards for industrial effluents. In *Industrial Pollution Control Handbook,* H. F. Lund, ed. McGraw-Hill, New York, pp. 4-23–4-40.

Sharpe, F. P. 1970. *Yellowstone Fish and Fishing.* Yellowstone Library and Museum Association, 49 pp. (Available from the Yellowstone National Park, Wyoming.)

Snyder, G. R. 1969. Heat and anadromous fishes. Discussion of paper by R. E.

Nakatani. In *Biological Aspects of Thermal Pollution,* P. A. Krenkel and F. L. Parker, eds. Vanderbilt University Press, Nashville, Tenn., pp. 318–337.

Todd, E. H. 1967. Primary productivity of the Madison River in Yellowstone National Park, Wyoming. Ph.D. dissertation, Montana State University, Bozeman.

Wright, J. C. and R. M. Horrall. 1967. Heat budget studies on the Madison River, Yellowstone National Park. *Limnol. Oceanogr.* **12**, 578–583.

Wright, J. C. and I. K. Mills, 1967. Productivity studies on the Madison River, Yellowstone National Park. *Limnol. Oceanogr.* **12**, 568–577.

Zeikus, J. G. and T. D. Brock. 1972. Effects of thermal additions from the Yellowstone geyser basins on the bacteriology of the Firehole River. *Ecology* **53**, 183–290.

Chapter 14
Some Personal History

It would sound reasonable if I were to say that the research work I have discussed in this book began as a result of a grand design, with a vision of the goals in mind. Unfortunately, this would not be true. This work began the day I took a detour through Yellowstone National Park on my way to Seattle. As many, I had avoided Yellowstone for years, having visions of vast crowds of tourists gaping at Old Faithful. This time, in the early summer of 1964, I decided to see for myself, and after a bit of climbing in the Grand Tetons, a detour was made up onto the Yellowstone plateau. The first stop was the West Thumb Geyser Basin, a small but attractive thermal area on the west shores of Yellowstone Lake. The weather was marvelous, as it often is, and the waters of Yellowstone Lake were a deep blue, but what really hit me were the algal mats, bright orange, red, and green, spread out along the silica channels under sheets of hot, steaming water. I knew that algae lived in thermal environments, but I wasn't prepared for this, these thick gobs of living protoplasm sticking tenaciously together in the rippling water. As I arrived, a naturalist was giving a talk, and afterwards I stopped him briefly to inquire about the algae. Aside from the fact that they were blue-green algae, he could tell me little. I had little time to inquire further, and went on my way to the west coast, where I spent the summer doing marine microbiology at the Friday Harbor Laboratories. The work there went well, but I couldn't keep Yellowstone out of my mind, and on the return trip to Indiana I stopped again, this time a little longer, and took some samples. Not knowing really what I was doing, or why, I didn't get anything significant, but my interest in hot springs was aroused.

The next step on the road to Yellowstone went via Iceland. That year, for some reason, I signed on for a visit to Surtsey, the new volcanic island that had just appeared off the south coast of Iceland. My visit to Iceland

and Surtsey was planned for 2 weeks in late July in 1965, and I didn't really know what I was going to do, as I had no experience in these types of environments. I decided that before I went to Iceland, I should return to Yellowstone and spend a couple of weeks looking at the environments there, which were at least somewhat comparable. I wrote to the Chief Naturalist, got some encouragement and a form to fill out, upon which I was asked to give the title of my research project. Without too much thought, but with enough to know that the title sounded good, I wrote: "Thermal control of primary productivity." This title seemed to relate to Yellowstone and its algae, and could probably be converted into a reasonable research study. I planned 2 weeks for late June 1965, and began to think what I really wanted to do. From the beginning, my research in Yellowstone involved the expert assistance of my former wife, Louise Brock. She did much of the early methodological work, while I was still busy working on the molecular aspects of mating in yeast, genetics of streptococci, and the mode of action of antibiotics. She was working without pay (such were the nepotism laws in those days), and I was careful never to use grant funds from other projects for the Yellowstone work. Our transportation, housing, and subsistence costs all came out of our own pocket, and at the time I wasn't even smart enough to deduct it from my income tax. I viewed the trip as a working vacation, which in some ways it was, although we worked as hard those 2 weeks as ever.

The work went well, and at the end of 2 weeks we had obtained enough data to write a paper, which I eventually combined with some of the Iceland work and published in *Nature* (Brock and Brock, 1966, see references to Chapter 9). Looking back on that paper, and especially on the first draft which got rejected by *Science,* I find it hard to believe how naive I was, but if nothing else, the paper crystallized my thinking on the kind of research work that might be done. From the beginning, I viewed the hot springs as model systems in which studies in microbial ecology could be most easily done. I quote from the first two sentences of that 1966 paper: "In analyzing the influence of specific environmental factors on growth, it is desirable to study habitats where only a single factor varies. Hot springs provide suitable habitats where temperature is the only variable, so that the relation of temperature to biological development can be measured directly." However, I was also aware of the evolutionary implications of the work, as the final discussion of this paper suggests. Strangely, I never thought about our Yellowstone work as of relevance to the U.S. space program, and never obtained research funds from the National Aeronautics and Space Administration, although later, when we had set up our Yellowstone field laboratory, I hosted a number of space-oriented scientists.

After that summer of 1965, my research work was at a crossroads. The yeast work was going very well, but I knew that I wanted to concentrate on microbial ecology, an important field that clearly had a great future. I had done a fair bit of marine microbiology, and considered concentrating in this area, yet the Yellowstone habitats beckoned strongly. I was about to write up a research grant for one of these two areas but I could not decide which. I remember clearly asking Louise which she would prefer, Yellowstone or Friday Harbor. Without hesitation she said Yellowstone. I sat down and wrote the grant proposal for the National Science Foundation, entitled "Biochemical Ecology of Yellowstone Hot Springs." It was funded for June 1, 1966, although someone at NSF had the presence of mind to change the title to "Biochemical Ecology of Microorganisms in Thermal Environments," a title which made the work sound less like a junket. I never tried to find out who had made the change, but I kept it through several renewals until I finally became embarrassed with the pretentiousness of "Biochemical Ecology" and changed it to "Microbial Ecology." In one form or another, this NSF grant was renewed for 10 years, and was a major factor permitting me to do good research in Yellowstone.

That initial proposal, written in the fall of 1965, set the theme that I followed for many years: "The objectives of the present proposal are to understand at the biochemical, physiological and ecological levels the relationships of microorganisms to the thermal environment, using the Yellowstone hot springs as model ecosystems."

> Studies on primary productivity are proceeding in a variety of environments, and the importance of such studies is now well established. In terms of the exploding populations on earth, fundamental knowledge of the controls of primary productivity are urgently needed, but to date such controlling factors are known only in a general way. The present work will provide detailed knowledge under nearly ideal conditions of thermal controls on productivity. The importance of model systems in physiology, genetics and molecular biology is well established, and the Yellowstone hot springs provide a unique model system for studying primary productivity. . . . From an evolutionary point of view, an understanding of the mechanisms of adaptation of living organisms to high temperature is significant. . . . From a practical point of view, the present work may be of importance to an understanding of the problems of water pollution. Thermal pollution will become a more serious problem with the development of atomic power plants. In addition, these studies relate to water pollution in a more general sense, since hot spring waters with a high mineral content resemble polluted streams in chemical characteristics. . . . Many of the studies proposed here under nearly ideal natural conditions can be expected to apply more or less directly to the more variable polluted stream.

I am typing this chapter on my birthday in 1976, about 11 years after the above words were written. I have not read those words since the time they were finally mimeographed to send to NSF in late 1965. I am amazed at how closely the theme of this work held to those sentiments. Only one point of relevance which I did not foresee at the time was the variety of new organisms that we would discover. Without even trying, four new genera have been discovered in thermal environments, representing probably four whole new orders of bacteria (three of these were discovered in my laboratory). The industrial applications of these and other thermophilic organisms is clearly of importance, but did not occur to me at the time.

My research proposal went to the Environmental Biology Program at NSF (later changed to General Ecology). Fortunately, it was submitted at a time when interest in the environment was awakening, and increasing funds were being made available for this area, and I was able to ride the wave to its crest. (I hope it is also true that the proposal was a good one.) I had no real idea that the proposal would be funded, but I knew that if I were to do what I proposed the following summer, I must begin to make plans in the winter. I took a chance, and began to line up housing in West Yellowstone, Montana, the closest town and the only place where a research project on Yellowstone thermal areas could be reasonably based. Then, I went off to Europe for 3 months. When I returned to Bloomington in mid-May, I got news that my NSF grant was funded, and that the starting date could be June 1. With hardly enough time to unpack my bags, I got ready to go to Yellowstone. We were off and running; it was 10 years before we could stop. Later, additional research funds became available from Indiana University, the Atomic Energy Commission (now part of the Department of Energy), and after I moved to Wisconsin, from the Wisconsin Alumni Research Foundation.

However, research is done by individuals, and a lot of people have been involved in this work. I owe all of them something, and some of them an awful lot. It seems appropriate to provide at least some recognition of the personal accomplishments of these individuals, many of whom have gone on to be independent and well-established scientists. Following is a list of those who have been involved directly in the Yellowstone work.

Personnel Involved in Yellowstone Research Project

1965 T. D. Brock. M. L. Brock.
1966 T. D. Brock. M. L. Brock. Robert Lenn. Hudson Freeze. Sally Murphy. W. D. P. Stewart (visitor).
1967 T. D. Brock. M. L. Brock. W. N. Doemel. Nancy Doemel. John Murphy. Sally Murphy.
1968 T. D. Brock. M. L. Brock. W. N. Doemel. Nancy Doemel. Thomas Bott. Pat Holleman. Raymond Lynn. Mercedes Edwards (visitor).
1969 T. D. Brock. M. L. Brock. W. N. Doemel. Maxine Cohen. Thomas Bott. Gary Darland. Mercedes Edwards (visitor).
1970 T. D. Brock. M. L. Brock. W. N. Doemel. Gary Darland. Carl Fliermans. Frances Foy. John Bauld. Jeffrey Davidson. J. L. Mosser. Michael Tansey. Katherine Middleton. J. G. Zeikus. William Samsonoff (visitor). Deborah Dellmer (visitor).

1971   T. D. Brock. K. M. Brock. Robert Belly. Michael Tansey. David
       Smith. Richard Weiss. John Bauld. Bonnie Bauld. Charles Boylen.
       Carl Fliermans. Katheryn Boylen. Malcolm Walter (visitor). Kimio
       Noguchi (visitor). Kasho Aikawa (visitor). Tatsuo Goto (visitor).
1972   T. D. Brock. K. M. Brock. J. L. Mosser. Ann Mosser. Jack Meeks.
       John Bauld. David Smith. Bonnie Bauld. Charles Boylen. Katheryn
       Boylen. Robert Belly. Malcolm Walter (visitor).
1973   T. D. Brock. J. L. Mosser. Ann Mosser. B. B. Bohlool. W. N.
       Doemel. Robert Belly. James Zabrecky. John Conner. Donald
       Weller.
1974   T. D. Brock. W. N. Doemel. J. L. Mosser. Ann Mosser. Donald
       Weller. Patrick Remington. James Cook. Sandra Petersen. Susan
       Entemann Cook. Charlene Knaack Singh. Stjepko Golubic (visitor).
       John Haury (visitor). (See Figure 14.1.)
1975   T. D. Brock. W. N. Doemel. David Ward. Stephen Zinder. Michael
       Madigan. Charlene Knaack Singh. Timothy Parkin.

Bibliographic Note

During the ten years that we worked in Yellowstone, our research generated a total of 90 research papers and review articles. Most of these have
been cited in this book in one or another chapter, as they were appropriate,
but for one reason or another, a few papers have not been cited. For
bibliographic completeness, I list all uncited references here.

Brock, T. D. and M. L. Brock. 1967. The hot springs of the Furnas Valley, Azores.
    Intern. Revue ges. Hydrobiol. **52**, 545–558.
Brock, T. D. and M. L. Brock. 1968. Relationship between environmental tempera-
    ture and optimum temperature of bacteria along a hot spring thermal gradient. *J.
    Appl. Bacteriol.* **31**, 54–58.
Brock, T. D. and M. L. Brock. 1968. Life in a hot-water basin. *Natural History* **77**,
    47–53.
Brock, T. D. and M. L. Brock. 1969. The fate in nature of photosynthetically
    assimilated ^{14}C in a blue-green alga. *Limnol. Oceanogr.* **14**, 604–607.
Brock, T. D. 1971. Microbial adaptation to extremes of temperature and pH. In:
    Biochemical Responses to Environmental Stress (Bernstein, I. A. ed.) Plenum
    Press, N.Y., pp. 32–37.
Brock, T. D. 1972. One hundred years of algal research in Yellowstone National
    Park. In: *Taxonomy and Biology of Blue-green Algae.* (T. V. Desikachary, ed.)
    Center for Advanced Study in Botany, Madras, India, pp. 393–405.
Brock, T. D. 1976. Environmental microbiology of living stromatolites. In: *Develop-
    ments in Sedimentology, 20, Stromatolites.* (Walter, M. R. ed.) Elsevier,
    Amsterdam, pp. 141–148.
Brock, T. D. 1976. Biological techniques for the study of microbial mats and living
    stromatolites. In: *Developments in Sedimentology, 20, Stromatolites.* (Walter,
    M. R. ed.), Elsevier, Amsterdam, pp. 21–29.

Public Service

As our work in Yellowstone Park progressed, it was natural that we would have contact with the naturalist and research staff of the Park. Frankly, our research was not the kind that got most of the Park personnel excited, because it did not relate to day-to-day management problems. However, the naturalist staff showed keen interest, and were quite curious to learn more about life in the geyser basins. I gave several lectures to groups of summer naturalists, to provide them with background for their interpretive programs. Then Bill Dunmire, Chief Naturalist in 1969–1971, asked me to write a booklet on Life in the Geyser Basins, to be sold at the museums and other interpretive facilities distributed throughout the Park. This booklet, well illustrated with color photos, was attractive and proved to be quite successful.

In 1968, I wrote an article for *Natural History* magazine on life in the geyser basins, and this article was seen by a number of people who were interested in the overall idea of life in extreme environments. As a result of this article, a diorama display on life in hot springs was prepared at the American Museum of Natural History, in New York City, and I advised the museum people on the preparation of this display.

Movies and Television

My article for Natural History magazine was read by Bert von Bork, a top photographer for Encyclopaedia Britannica, and I was contacted by him regarding the preparation of a movie on our Yellowstone work. I was attracted to this project, partly because it would enlarge the sphere of knowledge about microbial ecology, and partly because it would instruct another group on the significance of the Yellowstone thermal habitats. The Encyclopaedia Britannica film operation was well-financed, and Bert von Bork and his assistant Ulf Bergstrom spent quite a bit of time in Yellowstone, both in summer and winter. They spent a whole month in the summer of 1971 with us, and most of the footage was shot then, and we worked with them on editing and completing this film. Although I am a little embarrassed about how closely they oriented this film around me personally, it does a fair job of indicating something about life in geothermal habitats, and is visually quite attractive. This movie is available from Encyclopaedia Britannica under the title: Ecology of a Hot Spring: Life at High Temperatures (EB No. 3143). Another film that Bert and Ulf did at the same time, more because of the visual appeal of Yellowstone than for scientific purposes, is called Geyser Valley (EB No. 3142).

By 1973, our work had become well-enough known at the National Science Foundation that the public relations people there took an interest in it. I was asked to cooperate in making a short television program, giving

Figure 14.1. The Yellowstone research group in the summer of 1974, a golden year for research productivity. Left to right: W. N. Doemel, Charlene Knaack Singh, Katherine M. Brock, T. D. Brock (holding Emily Brock), Patrick Remington, Donald Weller, Stjepko Golubic (visitor), James Cook, John Haury (visitor), Sandra Petersen, Susan Entemann (later Cook), Jerry L. Mosser, Anne G. Mosser. The laboratory trailers are in the background. Despite the wooded appearance, this was no natural area. The trailers were situated in the dusty town of W. Yellowstone, Montana, which received about one million visitors during the 3-month summer season. On each side of the laboratory were motels.

some of the rationale for studies such as ours. A group called Audio Productions made a whirlwind trip to Yellowstone in early June 1974, were blessed with good weather, and managed to shoot some footage of me talking at various thermal locations. They then pieced this together with footage they bought from the Encyclopaedia Britannica and made a short (5.5 minute) movie for television called Life's Outer Limits (available as No. 1120 from Filmedia, Inc., 48 W. 48th St., New York, N.Y. 10036).

The West Yellowstone Laboratory

It became obvious to me during my first summer, 1965, that any significant research in Yellowstone would require some sort of laboratory facility nearby. During 1966, some preliminary inquiries were made with the Park Service staff about a laboratory inside the Park, but it was clear that this did not meet any Park Service program. Indeed, the impression was given that even if all of the money came from other sources, the Park Service would not be willing to provide land for a laboratory. At the time, the Park was actively engaged in eliminating private installations, and it was not about to permit a new one. It thus became clear to me that a facility outside the Park was essential. The only town within reasonable distance of the main study areas is W. Yellowstone, Montana, and we were already using rented cabin facilities in this town for our research work. W. Yellowstone had the additional advantage that a variety of cabin facilities were available for living, and there was a post office, commercial airport, stores, etc. We were renting cabins from Ed Daley, and he volunteered to rent us a space where we could install a laboratory trailer. In the fall of 1967, we purchased a very inexpensive mobile home (dimensions 10×50 feet) for $3700, which was installed on Ed Daley's lot. We used this during the winter of 1967–1968, and the early part of the summer of 1968. I was looking for a lot to buy in W. Yellowstone, and at this time a lot became available next to Daley's. Somehow, I managed to get Indiana University to buy this lot, and I bought another mobile home, this one 12×55 feet, for $4300. As I recall, the University paid $11,000 for the lot, and we put another $4000 into it to get water and sewer. By mid-summer 1969, we had both of our trailers installed on this lot, and had fixed them up quite well, using primarily old furniture and some simple carpentry. Most of the work of fixing up these trailers for laboratory work was carried out by Louise Brock and Bill Doemel, but lots of other people helped. These two mobile homes are shown in the background of Figure 14.1.

It would serve no purpose to go into detail on how these laboratories were set up and used, but the whole operation was an eye-opener to me, as it showed how inexpensively laboratory facilities could be constructed. For the price of about $25,000 we had a nice private facility, completely under our control, in which we could do virtually any kind of ecological or microbiological study. Indeed, we even did a little biochemistry in this facility, and a lot of water chemistry. We moved in expensive equipment as we needed it for a summer's work, then took it back to campus for the winter. And since we had little investment in permanent facilities, we could junk the whole operation when we were finished, and sell the lot with (presumably) little loss in money.

The Decision to Quit

Every year we solved a number of problems but opened up as many new ones. This is, of course, the way of research, and it was clear that we could keep going in Yellowstone forever. However, I have always felt that one stayed fresh by doing new things, and after 10 years, there was no doubt that I was getting a little jaded. More importantly, there was a vast array of interesting and important microbiological problems awaiting to be studied in more normal environments, and it seemed reasonable that some of the techniques and approaches we developed in Yellowstone could be applied to other situations. Since my laboratory at Madison was within a half block of one of the best-studied lakes in the world, Lake Mendota, and since this lake was famous for its massive blue-green algal blooms, it was a logical decision to switch from geothermal habitats to Wisconsin lakes. Thus, the summer of 1975 was my last summer studying the thermal areas. I didn't leave Yellowstone without a fair bit of regret, but I have always been more interested in looking ahead than behind. Now that I have had two field seasons studying Wisconsin lakes, the decision to leave seems even better than it did. Among other things, I have been able to convert a 3-month field season into a 12-month one, as our close proximity to the lakes has made it possible for us to continue work throughout the year. And the logistics have been much easier, especially for the students, who do not have to worry about pulling up stakes and trying to find new places to live for the summer season. And it does seem reasonable that one who is paid (at least in part) by the taxpayers of the State of Wisconsin owes them something in return, and the least I can do is to try to understand something about the causes and consequences of blue-green algal blooms.

I hope very much that you will go to Yellowstone and work. If you do, I hope you will find something in our own work to build upon. Perhaps this book will ease your way, at least a little. Good luck!

Subject Index

Springer Series in Microbiology

Editor: **Mortimer P. Starr,** Department of Bacteriology, University of California, Davis, USA.

The *Springer Series in Microbiology* features textbooks and monographs designed for students and research workers in all areas of microbiology, pure and applied. Monographs by leading experts carefully summarize the state-of-the-art in various specialized aspects of microbiology. Advanced and elementary textbooks by experienced teachers treat particular microbiological topics in a manner consistent with realistic utility in classroom, laboratory, or self-instruction situations.

Bacterial Metabolism

By **Gerhard Gottschalk,** University of Göttingen, Federal Republic of Germany
1978. approx. 320p. approx. 160 illus. cloth, in preparation.

This concise, yet comprehensive textbook surveys the domain of bacterial metabolism and describes the various facets of this subject in terms useful to students and research workers. Emphasis is on those metabolic reactions which occur only in bacteria or those of particular importance for bacteria. Thus, the book describes in detail the energy metabolism of various groups of bacteria, including aerobic and anaerobic heterotrophs as well as chemolithotrophs and phototrophs. In addition, the book examines the various pathways used by bacteria for the degradation of certain organic compounds, the fixation of molecular nitrogen, the biosynthesis of cellular constituents, and the regulation of bacterial metabolism.

Microbial Ecology

Editor-in-Chief: **Ralph Mitchell,** Harvard University

Microbial Ecology, an international journal, features papers in those branches of ecology in which microorganisms are involved. Articles describe significant advances in the microbiology of natural ecosystems, as well as new methodology. In addition, the journal presents reports which explore microbiological processes associated with environmental pollution, and papers which treat the ecology of all microorganisms including pro-karyotes, eukaryotes, and viruses.

Current Microbiology

Editor: **Mortimer P. Starr,** University of California, Davis

Current Microbiology is a new journal devoted to the rapid publication of concise yet thorough research reports that deal with significant facts and ideas in all aspects of microbiology. Using an accelerated publishing technique and the active cooperation of the international editorial board, the journal is published monthly, with each issue containing from 15 to 20 brief papers. The Editorial Board and referees review each contribution at the highest professional level.

Current Microbiology presents contemporary advances in the whole field of microbiology: medical and nonmedical, basic and applied, taxonomic and historical, theoretical and practical, methodological and conceptual. The large $8\frac{1}{4}$ inch by 11 inch double-column format permits easier readability and maximum flexibility for the presentation of detailed tables and high-quality reproductions of half-tone illustrations